TODAY'S TECHNICIAN™

CLASSROOM MANUAL

For Automotive Brake Systems

TODAY'S TECHNICIAN™

CLASSROOM MANUAL

For Automotive Brake Systems

SEVENTH EDITION

Ken Pickerill

Australia • Brazil • Mexico • Singapore • United Kingdom • United States

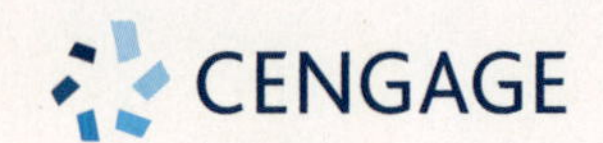

Today's Technician: Automotive Brake Systems, Seventh Edition

Ken Pickerill

SVP, GM Skills & Global Product Management: Jonathan Lau

Product Director: Matthew Seeley

Senior Product Manager: Katie McGuire

Senior Director, Development: Marah Bellegarde

Senior Product Development Manager: Larry Main

Senior Content Developer: Meaghan Tomaso

Product Assistant: Mara Ciacelli

Vice President, Marketing Services: Jennifer Ann Baker

Associate Marketing Manager: Andrew Ouimet

Senior Content Project Manager: Cheri Plasse

Design Director: Jack Pendleton

Cover image(s): Umberto Shtanzman/ Shutterstock.com

Library of Congress Control Number: 2017962929

Book only ISBN: 978-1-3375-6453-3
Package ISBN: 978-1-3375-6452-6

Cengage
20 Channel Center Street
Boston, MA 02210
USA

Cengage is a leading provider of customized learning solutions with employees residing in nearly 40 different countries and sales in more than 125 countries around the world. Find your local representative at **www.cengage.com.**

Cengage products are represented in Canada by Nelson Education, Ltd.

To learn more about Cengage Platforms and Services, visit **www.cengage.com.**

Purchase any of our products at your local college store or at our preferred online store **www.cengagebrain.com**

Notice to the Reader

Printed in the United States of America
Print Number: 01 Print Year: 2018

CONTENTS

PREFACE

The *Today's Technician*™ series features textbooks and digital learning solutions that cover all mechanical and electrical systems of automobiles and light trucks. The content corresponds to the 2017 ASE Education Foundation program accreditation requirements. They are specifically correlated to the Task Lists contained in each level of program accreditation; Maintenance and Light Repair (MLR), Automotive Service Technology (AST), and Master Service Technology (MAST).

Additional titles include remedial skills and theories common to all of the certification areas and advanced or specific subject areas that reflect the latest technological trends. *Today's Technician: Automotive Electricity & Electronics, 7e* is designed to give students a chance to develop the same skills and gain the same knowledge that today's successful technician has. This edition also reflects the most recent changes in the guidelines established by the ASE Education Foundation.

The purpose of the ASE Education Foundation program accreditation is to evaluate technician training programs against standards developed by the automotive industry and recommend qualifying programs for accreditation. Programs can earn accreditation upon the recommendation of ASE Education Foundation. These national standards reflect the skills that students must master. ASE Education Foundation accreditation ensures that certified training programs meet or exceed industry-recognized, uniform standards of excellence.

HIGHLIGHTS OF THIS NEW EDITION—CLASSROOM MANUAL

The text and figures of this edition are updated to show modern brake technology and its applications, including the integration of stability control and active braking systems. The Classroom Manual covers the complete mechanical-hydraulic automotive braking theories. It introduces the reader to basic brake systems as well as advanced electronics utilized in stability control systems. The following chapters cover basic brake physics theories: discussion of newer components and materials, including a section on electric parking brakes, and any braking functions required for passenger cars and light trucks. The reader is introduced to fundamental information on trailer brakes, DOT requirements for trailer brakes, and a brief introduction to air brakes. Chapter 10, Electrical Braking Systems (EBS), simplifies the discussion on traditional antilock brake systems (ABS) while retaining the information for a complete understanding of ABS. Included in this chapter is a detailed discussion of electro-hydraulic brakes including the Teves Mk60/70, Delphi DBC-7, and the newer Bosch 9.0 are introduced in chapter 11, Advanced Braking Systems goes more into depth on stability control and its relationship to traction control and ABS systems. This chapter also explains some of the ancillary systems that make stability control work more effectively, such as electro-hydraulic and fully electric steering and tire pressure monitoring systems. The very latest technologies, such as active braking and intelligent cruise control systems, are introduced. Lastly, the chapter examines regenerative braking systems in use on the latest hybrid vehicles in production today. The Classroom Manual guides the reader from traditional hydraulic brake to the brake system of the future.

HIGHLIGHTS OF THIS NEW EDITION—SHOP MANUAL

Safety information remains in the first chapter of the Shop Manual, placing this critical subject next to the tasks to be accomplished. Chapter 2, Brake Service Tools and Equipment, covers basic tools with more information on brake special tools and equipment. Figures and technical information have been added to cover the use of common shop tools such as on-car brake lathes. Some of the safety information that is pertinent to a particular piece of equipment is still in the chapter, so safety issues are presented just prior to the operation of the equipment. In keeping with typical shop diagnostic procedures and curriculum sequence, Chapter 3 retains the information on related systems that may have a direct impact on the braking system. Updated information on diagnosing electric parking brakes and electric braking systems has been added to this edition. To clarify the diagnosis and repair procedures for electric braking, three major ABS/TCS brands, Delphi DBC-7 and Bosch ABS 9.0 and Teves Mk 60/70, are retained for discussion instead of an individual discussion on all industry ABS offerings. This helps the reader better understand the technical diagnosing and repairing for all ABS/ TCS. This edition of the Shop Manual will guide the student/technician through all the basic tasks in brake system repair and presents a look into the near-term future of electric brakes and vehicle stability systems. The Shop Manual has several additions in the Advanced Braking Systems chapter, Chapter 11. This chapter deals with the diagnosis and repair of stability control systems and the surrounding technologies, such as electric steering, tire pressure monitoring systems, active braking, and intelligent cruise control.

CLASSROOM MANUAL

Features of the Classroom Manual include the following:

Cognitive Objectives

These objectives outline the chapter's contents and identify what students should know and be able to do upon completion of the chapter. Each topic is divided into small units to promote easier understanding and learning.

Terms to Know List

A list of key terms appears in the beginning of the chapter. Students will see these terms discussed in the chapter. Definitions can also be found in the Glossary at the end of the manual.

Margin Notes

The most important terms to know are highlighted and defined in the margin. Common trade jargon also appears in the margin and gives some of the common terms used for components. This helps students understand and speak the language of the trade, especially when conversing with an experienced technician.

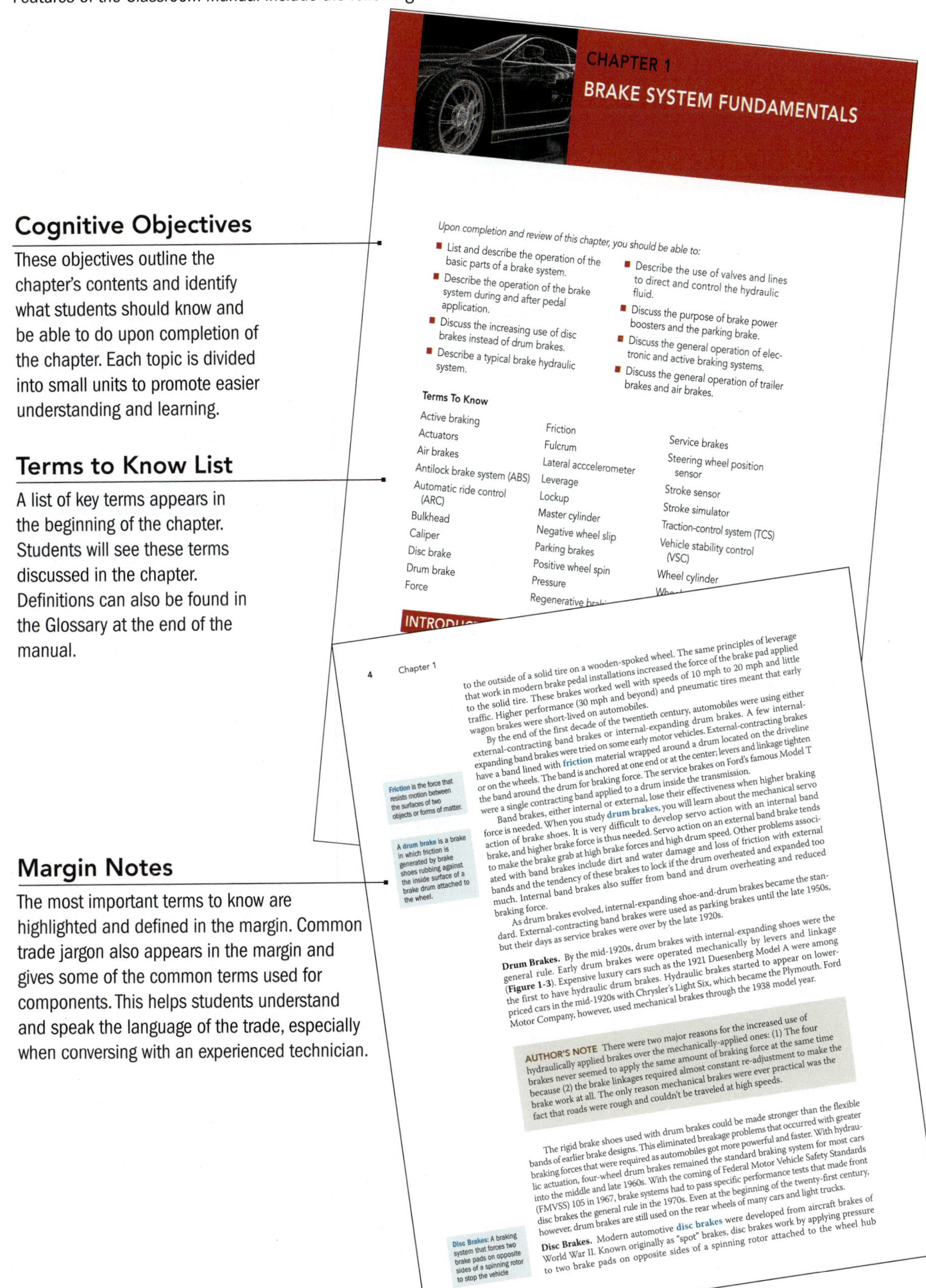
CHAPTER 1

BRAKE SYSTEM FUNDAMENTALS

Upon completion and review of this chapter, you should be able to:

- List and describe the operation of the basic parts of a brake system.
- Describe the operation of the brake system during and after pedal application.
- Discuss the increasing use of disc brakes instead of drum brakes.
- Describe a typical brake hydraulic system.
- Describe the use of valves and lines to direct and control the hydraulic fluid.
- Discuss the purpose of brake power boosters and the parking brake.
- Discuss the general operation of electronic and active braking systems.
- Discuss the general operation of trailer brakes and air brakes.

Terms To Know

Active braking
Actuators
Air brakes
Antilock brake system (ABS)
Automatic ride control (ARC)
Bulkhead
Caliper
Disc brake
Drum brake
Force
Friction
Fulcrum
Lateral accelerometer
Leverage
Lockup
Master cylinder
Negative wheel slip
Parking brakes
Positive wheel spin
Pressure
Regenerative brak...
Service brakes
Steering wheel position sensor
Stroke sensor
Stroke simulator
Traction-control system (TCS)
Vehicle stability control (VSC)
Wheel cylinder
Whe...

INTRODU...

4 Chapter 1

to the outside of a solid tire on a wooden-spoked wheel. The same principles of leverage that work in modern brake pedal installations increased the force of the brake pad applied to the solid tire. These brakes worked well with speeds of 10 mph to 20 mph and little traffic. Higher performance (30 mph and beyond) and pneumatic tires meant that early wagon brakes were short-lived on automobiles.

By the end of the first decade of the twentieth century, automobiles were using either external-contracting band brakes or internal-expanding drum brakes. A few internal-expanding band brakes were tried on some early motor vehicles. External-contracting brakes have a band lined with **friction** material wrapped around a drum located on the driveline or on the wheels. The band is anchored at one end or at the center; levers and linkage tighten the band around the drum for braking force. The service brakes on Ford's famous Model T were a single contracting band applied to a drum inside the transmission.

Band brakes, either internal or external, lose their effectiveness when higher braking force is needed. When you study **drum brakes**, you will learn about the mechanical servo action of brake shoes. It is very difficult to develop servo action with an internal band brake, and higher brake force is thus needed. Servo action on an external band brake tends to make the brake grab at high brake forces and high drum speed. Other problems associated with band brakes include dirt and water damage and loss of friction with external bands and the tendency of these brakes to lock if the drum overheated and expanded too much. Internal band brakes also suffer from band and drum overheating and reduced braking force.

As drum brakes evolved, internal-expanding shoe-and-drum brakes became the standard. External-contracting band brakes were used as parking brakes until the late 1950s, but their days as service brakes were over by the late 1920s.

Friction is the force that resists motion between the surfaces of two objects or forms of matter.

A **drum brake** is a brake in which friction is generated by brake shoes rubbing against the inside surface of a brake drum attached to the wheel.

Drum Brakes. By the mid-1920s, drum brakes with internal-expanding shoes were the general rule. Early drum brakes were operated mechanically by levers and linkage (**Figure 1-3**). Expensive luxury cars such as the 1921 Duesenberg Model A were among the first to have hydraulic drum brakes. Hydraulic brakes started to appear on lower-priced cars in the mid-1920s with Chrysler's Light Six, which became the Plymouth. Ford Motor Company, however, used mechanical brakes through the 1938 model year.

AUTHOR'S NOTE There were two major reasons for the increased use of hydraulically applied brakes over the mechanically-applied ones: (1) The four brakes never seemed to apply the same amount of braking force at the same time because (2) the brake linkages required almost constant re-adjustment to make the brake work at all. The only reason mechanical brakes were ever practical was the fact that roads were rough and couldn't be traveled at high speeds.

The rigid brake shoes used with drum brakes could be made stronger than the flexible bands of earlier brake designs. This eliminated breakage problems that occurred with greater braking forces that were required as automobiles got more powerful and faster. With hydraulic actuation, four-wheel drum brakes remained the standard braking system for most cars into the middle and late 1960s. With the coming of Federal Motor Vehicle Safety Standards (FMVSS) 105 in 1967, brake systems had to pass specific performance tests that made front disc brakes the general rule in the 1970s. Even at the beginning of the twenty-first century, however, drum brakes are still used on the rear wheels of many cars and light trucks.

Disc Brakes. Modern automotive **disc brakes** were developed from aircraft brakes of World War II. Known originally as "spot" brakes, disc brakes work by applying pressure to two brake pads on opposite sides of a spinning rotor attached to the wheel hub

Disc Brakes: A braking system that forces two brake pads on opposite sides of a spinning rotor to stop the vehicle

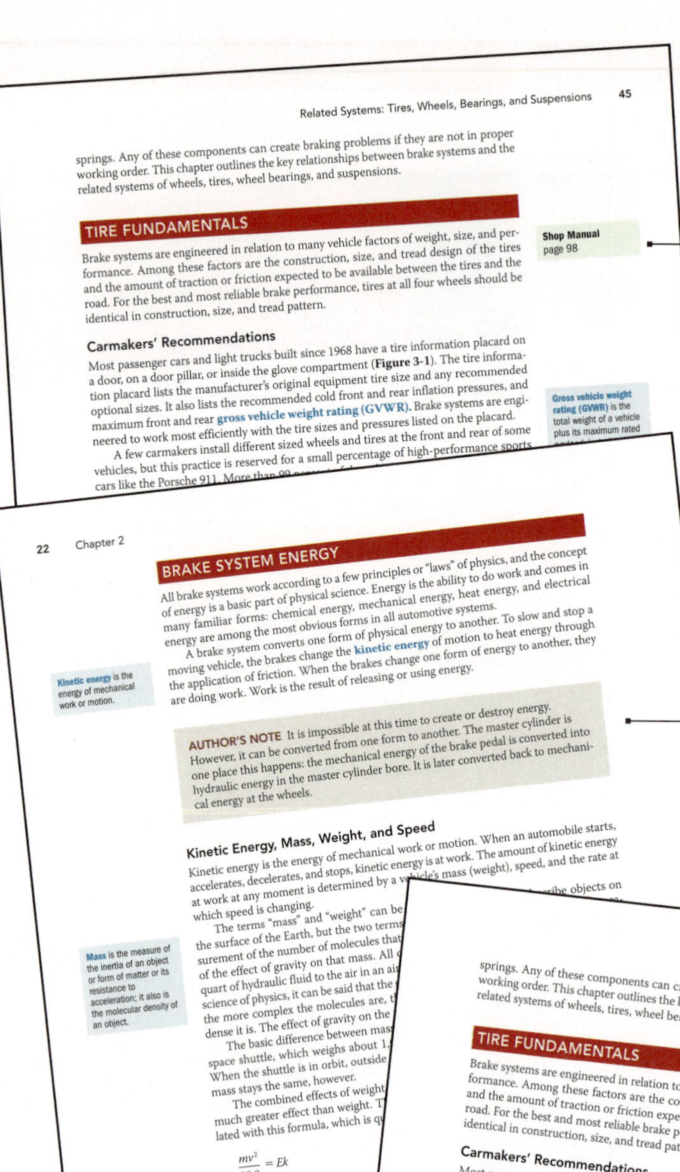

Related Systems: Tires, Wheels, Bearings, and Suspensions 45

springs. Any of these components can create braking problems if they are not in proper working order. This chapter outlines the key relationships between brake systems and the related systems of wheels, tires, wheel bearings, and suspensions.

TIRE FUNDAMENTALS

Brake systems are engineered in relation to many vehicle factors of weight, size, and performance. Among these factors are the construction, size, and tread design of the tires and the amount of traction or friction expected to be available between the tires and the road. For the best and most reliable brake performance, tires at all four wheels should be identical in construction, size, and tread pattern.

Shop Manual page 98

Carmakers' Recommendations

Most passenger cars and light trucks built since 1968 have a tire information placard on a door, on a door pillar, or inside the glove compartment (**Figure 3-1**). The tire information placard lists the manufacturer's original equipment tire size and any recommended optional sizes. It also lists the recommended cold front and rear inflation pressures, and maximum front and rear **gross vehicle weight rating (GVWR).** Brake systems are engineered to work most efficiently with the tire sizes and pressures listed on the placard.

A few carmakers install different sized wheels and tires at the front and rear of some vehicles, but this practice is reserved for a small percentage of high-performance sports cars like the Porsche 911. More than 99

Gross vehicle weight rating (GVWR) is the total weight of a vehicle plus its maximum rated

22 Chapter 2

BRAKE SYSTEM ENERGY

All brake systems work according to a few principles or "laws" of physics, and the concept of energy is a basic part of physical science. Energy is the ability to do work and comes in many familiar forms: chemical energy, mechanical energy, heat energy, and electrical energy are among the most obvious forms in all automotive systems.

A brake system converts one form of physical energy to another. To slow and stop a moving vehicle, the brakes change the **kinetic energy** of motion to heat energy through the application of friction. When the brakes change one form of energy to another, they are doing work. Work is the result of releasing or using energy.

Kinetic energy is the energy of mechanical work or motion.

AUTHOR'S NOTE It is impossible at this time to create or destroy energy. However, it can be converted from one form to another. The master cylinder is one place this happens: the mechanical energy of the brake pedal is converted into hydraulic energy in the master cylinder bore. It is later converted back to mechanical energy at the wheels.

Kinetic Energy, Mass, Weight, and Speed

Kinetic energy is the energy of mechanical work or motion. When an automobile starts, accelerates, decelerates, and stops, kinetic energy is at work. The amount of kinetic energy at work at any moment is determined by a vehicle's mass (weight), speed, and the rate at which speed is changing.

The terms "mass" and "weight" can be ... the surface of the Earth, but the two terms ... surement of the number of molecules that ... of the effect of gravity on that mass. All ... quart of hydraulic fluid to the air in an ai... science of physics, it can be said that the ... the more complex the molecules are, ... dense it is. The effect of gravity on the ...

The basic difference between mass ... space shuttle, which weighs about 1... When the shuttle is in orbit, outside ... mass stays the same, however.

The combined effects of weight ... much greater effect than weight. ... lated with this formula, which is ...

$$\frac{mv^2}{29.9} = Ek$$

where

m = mass (weight) in pou...
v = velocity (speed) in m...
Ek = kinetic energy in fo...

Consider two cars, bo... pounds; the other weighs 4...

Mass is the measure of the inertia of an object or form of matter or its resistance to acceleration; it also is the molecular density of an object.

Cross-References to the Shop Manual

References to the appropriate page in the Shop Manual appear whenever necessary. Although the chapters of the two manuals are synchronized, material covered in other chapters of the Shop Manual may be fundamental to the topic discussed in the Classroom Manual.

Author's Notes

This feature includes simple explanations, stories, or examples of complex topics. These are included to help students understand difficult concepts.

Related Systems: Tires, Wheels, Bearings, and Suspensions 45

springs. Any of these components can create braking problems if they are not in proper working order. This chapter outlines the key relationships between brake systems and the related systems of wheels, tires, wheel bearings, and suspensions.

TIRE FUNDAMENTALS

Brake systems are engineered in relation to many vehicle factors of weight, size, and performance. Among these factors are the construction, size, and tread design of the tires and the amount of traction or friction expected to be available between the tires and the road. For the best and most reliable brake performance, tires at all four wheels should be identical in construction, size, and tread pattern.

Shop Manual page 98

Carmakers' Recommendations

Most passenger cars and light trucks built since 1968 have a tire information placard on a door, on a door pillar, or inside the glove compartment (**Figure 3-1**). The tire information placard lists the manufacturer's original equipment tire size and any recommended optional sizes. It also lists the recommended cold front and rear inflation pressures, and maximum front and rear **gross vehicle weight rating (GVWR).** Brake systems are engineered to work most efficiently with the tire sizes and pressures listed on the placard.

A few carmakers install different sized wheels and tires at the front and rear of some vehicles, but this practice is reserved for a small percentage of high-performance sports cars like the Porsche 911. More than 99 percent of the vehicles on the road are originally fitted with wheels and tires of the same size at each corner. Although manufacturers may recommend one or two optional tire sizes at the rear that are larger than the front original equipment size, a large variation from the carmaker's recommendation can lead to braking problems, as well as problems with other vehicle systems.

For example, an extreme difference in tire diameters from front to rear may produce unequal speed signals from the wheel speed sensors of ABSs. Tires much larger than those recommended by the vehicle maker may produce inaccurate vehicle speed-sensor signals to the PCM or the ABS control module. This same problem exists if all four tires are larger or smaller than the manufacturer's recommendations.

Gross vehicle weight rating (GVWR) is the total weight of a vehicle plus its maximum rated payload, including passengers and full fuel tank.

A BIT OF HISTORY

When radial tires were first introduced in the 70s, there was a lot of resistance by drivers to using the new design. Complaints ranged from "feels funny when driving" to "they don't have enough air in them." Some drivers even went so far as to remove radial tires from a brand-new vehicle and to install bias tires. Two major characteristics of the radial tire overcame this die-hard resistance: a much smoother ride and increased fuel mileage. Lower-profile tires of today have also eliminated most of the comments about the tires "appearing underinflated."

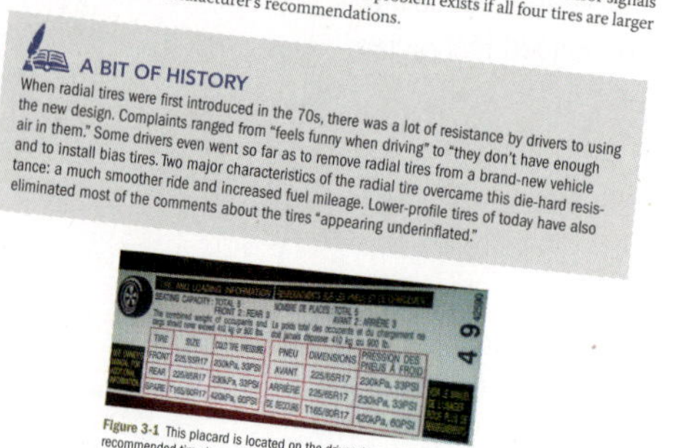

Figure 3-1 This placard is located on the driver door and lists recommended tire size and cold inflation pressure.

A Bit of History

This feature gives the student a sense of the evolution of the automobile. This feature not only contains nice-to-know information, but also should spark some interest in the subject matter.

Summary

Each chapter concludes with summary statements that contain the important topics of the chapter. These are designed to help the reader review the contents.

Review Questions

Short-answer essay, fill in the blank, and multiple-choice questions follow each chapter. These questions are designed to accurately assess the student's competence in the stated objectives at the beginning of the chapter.

92 Chapter 4

In most instances, only one dual-piston cylinder is used with some type of split system. However, some race crews opt for two identical single-piston master cylinders. The two master cylinders act like a split hydraulic system in that one master cylinder serves the front wheels, whereas the other serves the rear wheels. The master cylinders are applied by one brake pedal acting through a balance bar between the pedal lever and the two push-rods. Some race units are equipped with a brake power booster, and others are not. In this case, it is more an issue of weight than of driver endurance.

Of primary importance to race vehicle braking is the type of brake fluid used. On short tracks with a lot of braking, the boiling point of the fluid can be reached quickly and may be sustained for long periods. Brake fluids developed for racing purposes generally have the same chemical properties as conventional fluids, but they have much higher boiling points. Castrol offers a blend of polyglycol ester of dimethyl silane, ethylene polyglycols, and oxidation inhibitors. This blend has a dry boiling point of 450°F(232°C) and helps prevent fluid contamination during operation. Another brand, GS610, offers a fluid with a dry boiling point of 610°F(321°C). There are several manufacturers and suppliers of racing brake components. Brembo is one of the larger manufacturers of racing components, and some of its products are now being installed on some production performance vehicles.

SUMMARY

- Brake fluid specifications are defined by SAE Standard J1703 and FMVSS 116.
- Fluids are assigned DOT numbers: DOT 3, DOT 4, DOT 5, DOT 3/4, and DOT 5.1.
- Always use fluid with the DOT number recommended by the specific carmaker.
- Never use DOT 5 fluid in an ABS or mix with any other brake fluid.
- HSMO fluids are very rare and should never be used in brake systems designed for DOT fluids.
- The brake pedal assembly is a lever that incre... pedal force to the master...
- The h...
- hydraulic systems. Each of the two pistons in the master cylinder has a cup, a return spring, and a seal.
- During application, the piston and cup force fluid ahead of the piston to activate the brakes.
- During release, the return spring returns the piston.
- Fluid from the reservoir flows from the reservoir through the replenishing ... ton cup...

Master Cylinders and Brake Fluid 93

REVIEW QUESTIONS

Essay

1. Explain why DOT 5 brake fluid is not recommended by any manufacturer.
2. Explain why the boiling point of brake fluid is important.
3. Explain why it is not a good idea to mix DOT 5 fluids with DOT 3 and DOT 4.
4. Describe a sure sign of brake fluid contamination with mineral oil.
5. Explain why brake pedal linkage free-play is necessary.
6. Explain the split hydraulic system.
7. Describe a composite master cylinder.
8. Describe a master cylinder cup seal and how it is used.
9. What are the ports in the bottom of the master cylinder reservoir, and what do they do?
10. Explain the advantage of a quick take-up master cylinder.

Fill in the Blanks

1. A fast-fill or quick take-up master cylinder is identified by the dual bore design that creates a ________ or ________ of the casting.
2. DOT 3 and DOT 4 fluids are polyalkylene-glycol-ether mixtures, called ________ for short.
3. Because both DOT 3 and DOT 4 fluids ________ from the air, always keep containers tightly capped.
4. Silicone fluid ________ slightly under pressure, which can cause a slightly spongy brake pedal feel.
5. Polyglycol fluids have a very ________ shelf life.
6. The ________-to-________ hydraulic split system is the oldest split system.
7. Most late-model cars have a ________ split hydraulic system.
8. The master cylinder has two main parts: a ________ and a ________.
9. All master cylinder caps or covers are vented to prevent a ________ ________ as the fluid level drops in the reservoir.
10. The piston assembly at the rear of the cylinder is the ________ piston, and the one at the front of the cylinder is the ________ piston.

Multiple Choice

1. Technician A says the master cylinder changes the driver's mechanical force on the pedal to hydraulic pressure. Technician B says this hydraulic pressure is changed back to mechanical force at the wheel brakes. Who is correct?
   A. A only
   B. B only
   C. Both A and B
   D. Neither A nor B
2. Technician A says choosing the right fluid for a specific vehicle is based on the simple idea that if DOT 3 is good, DOT 4 must be better, and DOT 5 better still. Technician B says most vehicle manufacturers recommend DOT 4. Who is correct?
   A. A only
   B. B only
   C. Both A and B
   D. Neither A nor B
3. Technician A says the dry boiling point of brake fluid is the minimum boiling point of new, uncontaminated fluid. Technician B says polyglycol fluids are hygroscopic, which means that they do not absorb water vapor from the air. Who is correct?
   A. A only
   B. B only
   C. Both A and B
   D. Neither A nor B
4. Technician A says a high-temperature boiling point is the only requirement that brake fluid must meet. Technician B says brake fluid also must resist freezing and evaporation and must pass specific viscosity tests at low temperatures. Who is correct?
   A. A only
   B. B only
   C. Both A and B
   D. Neither A nor B

SHOP MANUAL

To stress the importance of safe work habits, the Shop Manual also dedicates one full chapter to safety. Other important features of this manual include:

Performance-Based Objectives

These objectives define the contents of the chapter and define what the student should have learned on completion of the chapter.

Basic Tools Lists

Each chapter begins with a list of the basic tools needed to perform the tasks included in the chapter.

Special Tools Lists

Whenever a special tool is required to complete a task, it is listed in the margin next to the procedure.

Terms To Know List

Terms in this list are also defined in the Glossary at the end of the manual.

Author's Notes

This feature includes simple explanations, stories, or examples of complex topics. These are included to help students understand difficult concepts.

138 Chapter 4

SERVICE TIP The vehicle's brake light switch must be activated any time the brake pedal is moved downward any amount. There is "no free play" allowed with regard to the brake light switch.

AUTHOR'S NOTE The following procedure is based on a Honda S2000. Other vehicles have similar procedures. Many vehicles do not have an adjustment for pedal height.

Photo Sequences

Many procedures are illustrated in detailed Photo Sequences. These photographs show the students what to expect when they perform particular procedures. They also familiarize students with a system or type of equipment that the school might not have.

252 Chapter 6

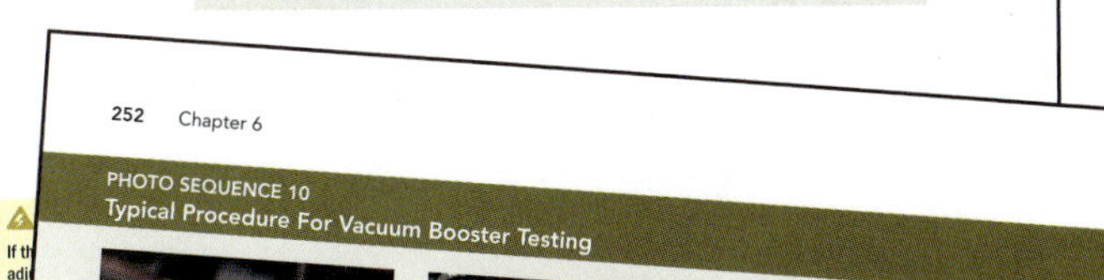

PHOTO SEQUENCE 10
Typical Procedure For Vacuum Booster Testing

P10-1 With the engine idling, attach a vacuum gauge to an intake manifold port. Any reading below 14 in. Hg of vacuum may indicate an engine problem.

P10-2 Disconnect the vacuum hose that runs from the intake manifold to the booster and quickly place your thumb over it before the engine stalls. You should feel strong vacuum.

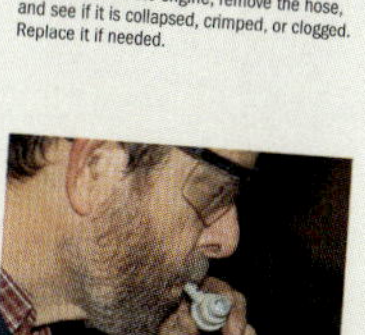

P10-3 If you do not feel a strong vacuum in step 2, shut off the engine, remove the hose, and see if it is collapsed, crimped, or clogged. Replace it if needed.

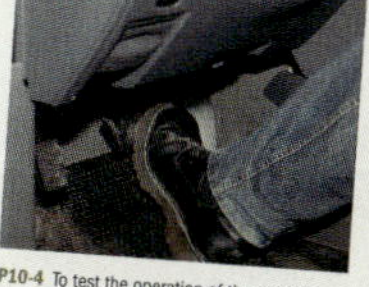

P10-4 To test the operation of the vacuum check valve, shut off the engine and wait for 5 minutes. Apply the brakes. There should be power assist on at least one pedal stroke. If

P10-5 Remove the check valve from the booster.

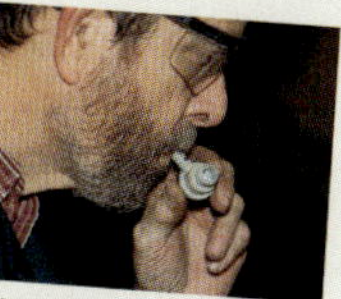

P10-6 Test the check valve by blowing into the intake manifold end of the valve. There should be a complete blockage of airflow.

Margin Notes

The most important terms to know are highlighted and defined in the margin. Common trade jargon also appears in the margins and gives some of the common terms used for components. This feature helps students understand and speak the language of the trade, especially when conversing with an experienced technician.

Hydraulic Line, Valve, and Switch Service 223

replace the parking brake switch. If the lamp is still off, find and repair the open circuit in the wiring harness between the body control computer and the switch.

Brake Fluid Level Switch Test

With the ignition on and the brake fluid level switch closed, the brake warning lamp lights to alert the driver of a low-fluid condition in the master cylinder. Some switches are built into the reservoir body; others are attached to the reservoir cap. Test principles are similar for both types.

Begin by ensuring that the fluid level is at or near the full mark on the reservoir. Turn the ignition on and observe the warning lamp. If it is lit, disconnect the wiring connector at the switch. If the lamp then goes out, replace the switch. If the lamp does not go out, find and repair the short circuit between the switch and the lamp.

To verify that the warning lamp will light when the fluid level is low, manually depress the switch float or remove the cap with an integral switch and let the float drop. If the lamp does not light with the switch closed, check for an open circuit between the switch and the lamp. If circuit continuity is good, replace the switch.

As a final check, disconnect the wiring harness from the switch, and connect a jumper wire between the two terminals in the harness connector. The warning lamp should light. If it does not, find and repair the open circuit between the switch and the body control computer.

Electrical Wiring Repair

Wire size is determined by the amount of current, the length of the circuit, and the voltage drop allowed. Wire size is specified in either the **American Wire Gauge (AWG)** system or in metric cross-sectional area. The higher the number in AWG the smaller the conductor. A 20 gauge is much smaller than a 12 gauge.

When replacing a wire, the correct size wire must be used as shown on applicable wiring diagrams or in parts books. Each harness or wire must be held securely in place to prevent chafing or damage to the insulation due to vibration. Always use **rosin flux solder** to splice a wire, and use insulating tape or **heat-shrink tubing** to cover all splices or bare wires. Rosin flux cleans the connection during soldering without eroding the material as does acid-based flux. Applying heat to shrink tubing causes the tubing to contract and completely seal the wiring and connections. Utility companies used heat-shrink tubing to seal underground electrical supply cables.

Many electrical system repairs require replacing damaged wires. It is important to make these repairs in a way that does not increase the resistance in the circuit or lead to shorts or grounds in the repaired area. Several methods are used to repair damaged wire with many factors influencing the choice. These factors include the type of repair required accessibility of the wiring, the type of conductor and size of wire needed, and the circuit requirements. The three most common repair methods are:

1. Wrapping the damaged insulation with electrical tape (in cases where the insulation is damaged and the wiring is unharmed)
2. Crimping the connections with a solder-less connector
3. Soldering splices

When deciding where to cut a damaged wire, avoid points close to other splices or connections. As a rule, do not have two splices or connections within 1.5 inches (40 mm) of each other. Use a wire of the same size or larger than the wire being replaced.

Crimping. A solderless connection uses a compressed junction to connect two conductors. Some manufacturers require the use of self-sealing solderless connections on all repairs. Crimping selfsealing solder less connections is an acceptable way to splice wire,

An **American wire gauge** (AWG) is a system for specifying wire size (conductor cross-sectional area) by a series of gauge numbers; the lower the number, the larger the wire cross section.

Caution Never replace a wire with one of a smaller size. Using the incorrect size could cause repeated failure and damage to the vehicle electrical system.

Rosin flux solder is solder used for electrical repairs.

Heat-shrink tubing is plastic tubing that shrinks in diameter when exposed to heat.

Service Tips

Whenever a shortcut or special procedure is appropriate, it is described in the text. Generally, these tips describe common procedures used by experienced technicians.

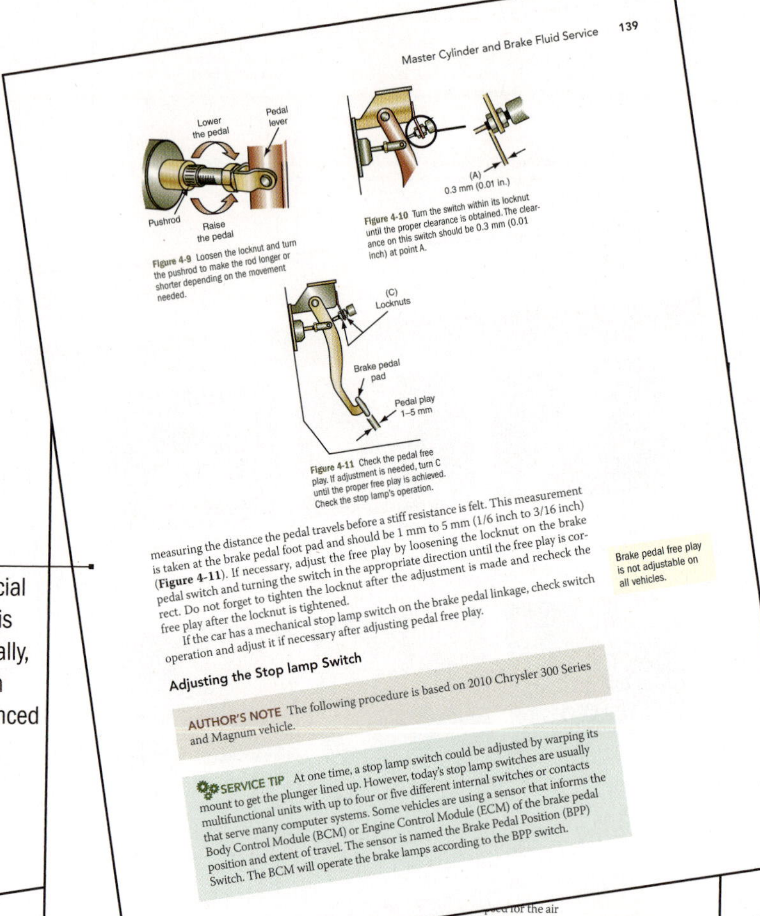

Figure 4-9 Loosen the locknut and turn the pushrod to make the rod longer or shorter depending on the movement needed.

Figure 4-10 Turn the switch within its locknut until the proper clearance is obtained. The clearance on this switch should be 0.3 mm (0.01 inch) at point A.

Figure 4-11 Check the pedal free play. If adjustment is needed, turn C until the proper free play is achieved. Check the stop lamp's operation.

measuring the distance the pedal travels before a stiff resistance is felt. This measurement is taken at the brake pedal foot pad and should be 1 mm to 5 mm (1/6 inch to 3/16 inch) (**Figure 4-11**). If necessary, adjust the free play by loosening the locknut on the brake pedal switch and turning the switch in the appropriate direction until the free play is correct. Do not forget to tighten the locknut after the adjustment is made and recheck the free play after the locknut is tightened.

Brake pedal free play is not adjustable on all vehicles.

If the car has a mechanical stop lamp switch on the brake pedal linkage, check switch operation and adjust it if necessary after adjusting pedal free play.

Adjusting the Stop lamp Switch

AUTHOR'S NOTE The following procedure is based on 2010 Chrysler 300 Series and Magnum vehicle.

SERVICE TIP At one time, a stop lamp switch could be adjusted by warping its mount to get the plunger lined up. However, today's stop lamp switches are usually multifunctional units with up to four or five different internal switches or contacts that serve many computer systems. Some vehicles are using a sensor that informs the Body Control Module (BCM) or Engine Control Module (ECM) of the brake pedal position and extent of travel. The sensor is named the Brake Pedal Position (BPP) Switch. The BCM will operate the brake lamps according to the BPP switch.

Cautions and Warnings

Cautions appear throughout the text to alert the reader to potentially hazardous materials or unsafe conditions. Warnings advise the student of things that can go wrong if instructions are not followed or if an incorrect part or tool is used.

136

... (**Figure 6-29**). The switch ... installation of the booster. Use a screwdriver to ... from the booster pushrod, and slide the pushrod from the pedal ... (refer back to Figure 6-29). Remove the booster's four mounting nuts, and remove the booster from the engine compartment.

Before installing the new booster, ensure that a new booster seal is present on the bulkhead side of the booster (**Figure 6-30**). Slide the booster into place through the bulkhead and tighten the four mounting nuts to specifications. Position the booster pushrod over the pedal pin and install a new retaining clip. Install and adjust the new stop lamp switch. Under the hood, install the master cylinder onto the booster and reconnect all electrical connections. Install the wiper module and other removed components. Connect the battery and road test the vehicle.

Caution: Before even beginning to work on a hybrid or electric vehicle, make certain that you are aware of the procedure to disable the high voltage power supply system according to service information.

SERVICING AN ELECTROHYDRAULIC POWER BOOSTER SYSTEM

Hybrid vehicles, as well as some conventional gasoline vehicles, use an electric brake booster pump (often referred to as a hydraulic power unit **Figure 6-31**) used to pressurize brake fluid for use in a hydraulic booster system, which has the master cylinder

References to the Classroom Manual

References to the appropriate page in the Classroom Manual appear whenever necessary. Although the chapters of the two manuals are synchronized, material covered in other chapters of the Classroom Manual may be fundamental to the topic discussed in the Shop Manual.

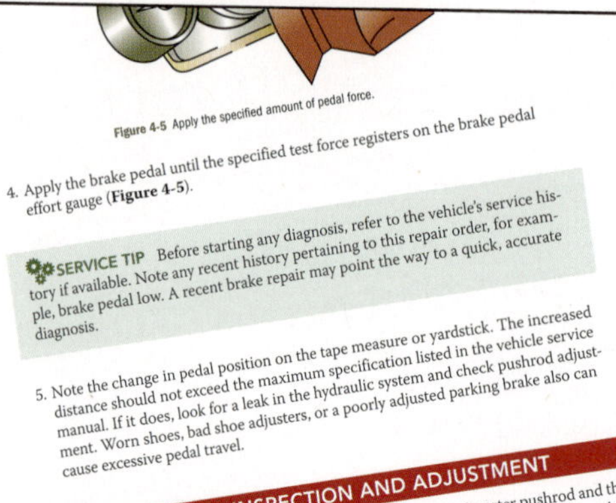

Figure 4-5 Apply the specified amount of pedal force.

4. Apply the brake pedal until the specified test force registers on the brake pedal effort gauge (**Figure 4-5**).

SERVICE TIP Before starting any diagnosis, refer to the vehicle's service history if available. Note any recent history pertaining to this repair order, for example, brake pedal low. A recent brake repair may point the way to a quick, accurate diagnosis.

5. Note the change in pedal position on the tape measure or yardstick. The increased distance should not exceed the maximum specification listed in the vehicle service manual. If it does, look for a leak in the hydraulic system and check pushrod adjustment. Worn shoes, bad shoe adjusters, or a poorly adjusted parking brake also can cause excessive pedal travel.

PEDAL FREE PLAY INSPECTION AND ADJUSTMENT

Classroom Manual page 75

Brake pedal free play is the clearance between the brake pedal or booster pushrod and the primary piston in the master cylinder. A specific amount of free play must exist so that the primary piston is not partially applied when the pedal is released and so that pedal

140 Chapter 4

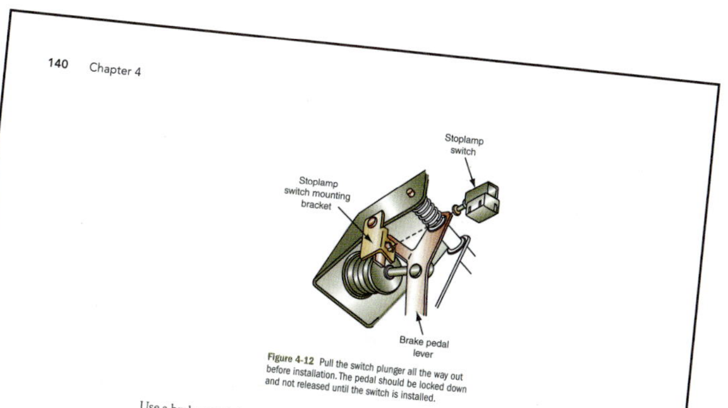

Figure 4-12 Pull the switch plunger all the way out before installation. The pedal should be locked down and not released until the switch is installed.

Use a brake pedal depressor to hold the brake pedal down (check the alignment machine for a depressor). Rotate the stop lamp switch approximately 30 degrees counterclockwise and pull rearward on the switch. It should separate from its mount (**Figure 4-12**). Using hand force only, pull the switch plunger out to its fully extended position. Low clicks should be heard as the plunger ratchets out.

Ensure the brake pedal is down as far as it will go and is firmly held in place. Align the switch's index key to the notch in the bracket and push the switch into place. Rotate the switch about 30 degrees clockwise until it locks.

Apply foot force to the brake pedal and remove the pedal depressor. Allow the pedal to gently rise until it stops. Using gentle hand force, pull up on the brake pedal until it stops moving. This will ratchet the switch plunger to the correct position. The switch adjustment is initially checked by having an assistant observe the brake lights as the brake pedal is depressed and released. However, the final check requires a road test on a road where the cruise control can be safely used. During the road test, engage the cruise control at a safe speed. Once the system is stabilized, depress the brake slightly. The cruise control should turn off. If not, then the switch must be checked and readjusted as needed.

Caution
Do not release the brake pedal by pulling the depressor out and letting the pedal slam up to its stop. The stop lamp switch will not adjust properly and may be damaged.

CUSTOMER CARE A customer's only contact, literally, with the brake system in his or her car is through the brake pedal. Customers tend to judge brake performance by "pedal feel." It is always a good idea to evaluate the feel and action of the brake pedal before starting any brake job. Then when you deliver the finished job, pedal feel should be noticeably improved. The biggest cause of spongy or low brake pedal action is air in the system, so careful bleeding of the system will do a lot to ensure customer confidence.

Brake Pedal Position Switch

Many late-model vehicles use a BPP sensor to inform the body control module (BCM) of the brake pedal position (Figure 4-13). The BPP sensor is a potentiometer. The BCM

Customer Care

This feature highlights those little things a technician can do or say to enhance customer relations.

Name ______________________ Date ____________ Drum Brake Service 425

DIAGNOSING DRUM BRAKE PROBLEMS

JOB SHEET 36

Upon completion of this job sheet, you will be able to diagnose poor stopping, noise, pulling, grabbing, dragging or pedal pulsation problems.

ASE Education Foundation Correlation

This job sheet addresses the following **MLR** task:

C.4. Inspect wheel cylinders for leaks and proper operation; remove and replace as needed. **(P-2)**

This job sheet addresses the following **AST/MAST** tasks:

C.1. Diagnose poor stopping, noise, vibration, pulling, grabbing, dragging or pedal pulsation concerns; determine necessary action. **(P-1)**

C.5. Inspect wheel cylinders for leaks and proper operation; remove and replace as needed. **(P-2)**

We Support ASE Education Foundation

Tools and Materials
- Basic hand tools

Protective Clothing
Goggles or safety glasses with side shields

Describe the vehicle being worked on:
Year ________ Make ____________ Model ____________ VIN ____________
Engine type and size ______________________

Procedure

1. Begin the inspection of the drum brake system by checking the tires for excessive or unusual wear or improper inflation. What did you find?
2. Wheels for bent or warped rims. What did you find?
3. Wheel bearings for looseness or wear. What did you find?
4. Suspension system for worn or broken components. What did you find?
5. Brake fluid level in the master cylinder. What did you find?
6. Signs of leakage at the master cylinder, in brake lines or hoses, at all connections, and at each wheel. What did you find?

Job Sheets

Located at the end of each chapter, the Job Sheets provide a format for students to perform procedures covered in the chapter. A reference to the ASE Education Foundation task addressed by the procedure is included on the Job Sheet.

ASE Challenge Questions

Each technical chapter ends with five ASE challenge questions. These are not mere review questions; rather, they test the students' ability to apply general knowledge to the contents of the chapter.

424 Chapter 8

ASE CHALLENGE QUESTIONS

1. While discussing mounting a drum on a lathe, Technician A says a two-piece drum mounts to the lathe arbor with tapered or spherical cones. Technician B says a one-piece drum is centered on the lathe arbor with a spring-loaded cone and clamped in place by two large cup-shaped adapters. Who is correct?
   A. A only
   B. B only
   C. Both A and B
   D. Neither A nor B
2. Brake drums are being discussed. ...

4. Technician A says a tire depth gauge can be used to measure lining thickness. Technician B says most carmakers specify a minimum lining thickness of 1/32 inch (0.030 in. or 0.75 mm) above the shoe table or above the closest rivet head. Who is correct?
   A. A only
   B. B only
   C. Both A and B
   D. Neither ...

Case Studies

Each chapter ends with a Case Study describing a particular vehicle problem and the logical steps a technician might use to solve the problem. These studies focus on system diagnosis skills and help students gain familiarity with the process.

Electrical Braking Systems Service 499

Warning Lamps

The instrument cluster on the majority of vehicles built within the last several years is not serviceable in the typical shop. The instrument panel has actually become another computer module on the automotive network. The gauges and lights are actually told to display based on serial data, not a direct reading from any sensor or switch. Most instrument clusters are sent out to specialty instrument panel repair shops. Many times a cluster that has already been rebuilt is exchanged, which helps reduce down time while a cluster is being repaired. Additionally, the instrument cluster will also need to be reprogrammed or reinitialized once it has been replaced.

SERVICE TIP Most of the latest systems in use share many common elements, but it is always necessary to verify that the right information is used. Always make sure to check service information for the specific repair for the vehicle concerned. As the newer systems begin to show up for repairs, consult the latest service information, recalls, and technical service bulletins for up-to-date diagnostic testing.

CASE STUDY

A vehicle came into the shop with the owner complaining of the ABS activating at a very low speed. All the normal checks were made, brakes were inspected, everything appeared to be okay, except for the fact that the ABS activated with the vehicle on dry pavement and coming to a slow stop. The technician took an assistant out to drive while the technician read the scan tool and everything looked normal. The technician finally decided that the BPMV had to be the source of the problem. After replacing the expensive modulator, the vehicle still had the same problem. At this point he decided to go to the shop foreman for help. The foreman told the technician to replace the rear shoes and resurface the drums, even though they did not appear to be contaminated in any way or worn to the point of needing replacement. When the shoes were replaced and the drums were machined, the vehicle performed normally. There are two lessons here: Start with the basics, and do not be afraid to ask a more experienced technician when you need help.

ASE-Style Review Questions

Each chapter contains ASE-style review questions that reflect the performance objectives listed at the beginning of the chapter. These questions can be used to review the chapter as well as to prepare for the ASE certification exam.

Drum Brake Service 423

ASE-STYLE REVIEW QUESTIONS

1. Before trying to remove a brake drum for service, Technician A backs off the brake shoe adjuster. Technician B takes up all slack in the parking brake cable. Who is correct?
   A. A only
   B. B only
   C. Both A and B
   D. Neither A nor B
2. When inspecting a wheel cylinder, Technician A finds liquid brake fluid behind the piston boot and rebuilds the wheel cylinder based on this fact. Technician B does not rebuild the wheel cylinder if only dampness is found in the boot. Who is correct?
   A. A only
   B. B only
   C. Both A and B
   D. Neither A nor B
3. When adjusting the brake shoes on a car with self-adjusting brakes, Technician A moves the self-adjusting lever away from the star wheel. Technician B says it is best policy to just force the star wheel against the self-adjuster without disengaging it. Who is correct?
   A. A only
   B. B only
   C. Both A and B
   D. Neither A nor B
4. Drum linings are badly worn at the toe and heel areas of the linings. Technician A says that the problem is an out-of-round drum. Technician B says that the problem is a tapered drum. Who is correct?
   A. A only
   B. B only
   C. Both A and B
   D. Neither A nor B
5. Technician A says that the drum discard dimension is the maximum diameter to which the drums can be refinished. Technician B says that the drum discard diameter is the maximum allowable wear dimension and not the allowable machining diameter. Who is correct?
   A. A only
   B. B only
   C. Both A and B
   D. Neither A nor B
6. When machining a drum on a brake lathe, Technician A uses a spindle speed of approximately 150 rpm. Technician B makes a series of shallow cuts to obtain the final drum diameter. Who is correct?
   A. A only
   B. B only
   C. Both A and B
   D. Neither A nor B
7. Technician A says that if the drum-to-lining adjustment is correct, the diameters of the two drums on an axle set do not matter as long as they do not exceed the discard dimension. Technician B says that the drum diameters on given axles must be exactly the same. Who is correct?
   A. A only
   B. B only
   C. Both A and B
   D. Neither A nor B
8. Technician A says that new drums must be cleaned to remove the rustproofing compound from the drum surface. Technician B says that refinished drums must be cleaned to remove all metal particles from the drum surface. Who is correct?
   A. A only
   B. B only
   C. Both A and B
   D. Neither A nor B
9. Technician A says that brakes with linings with more wear at the wheel cylinder end indicate a normal wear condition. Technician B says that if one lining on a duo-servo brake is worn more than the other, the shoes may be installed incorrectly. Who is correct?
   A. A only
   B. B only
   C. Both A and B
   D. Neither A nor B
10. Technician A says that weak or broken return springs can cause brake drag or pulling to one side. Technician B says that the same problems can be caused by a loose backing plate or an inoperative self-adjuster. Who is correct?
    A. A only
    B. B only
    C. Both A and B
    D. Neither A nor B

Name ____________ Date ____________

JOB SHEET 30

DIAGNOSING DISC BRAKE PROBLEMS

Upon completion of this job sheet, you will be able to diagnose poor stopping, noise, pulling, grabbing, dragging, or pedal pulsation problems.

ASE Education Foundation Correlation

This job sheet addresses the following **AST/MAST** task:

D.1. Diagnose poor stopping, noise, vibration, pulling, grabbing, dragging, or pulsation concerns; determine necessary action. **(P-1)**

We Support ASE Education Foundation

Tools and Materials

Basic hand tools

Protective Clothing

Goggles or safety glasses with side shields

Describe the vehicle being worked on:

Year ________ Make ________ Model ________ VIN ________

Engine type and size ________

Procedure

1. Begin the inspection of the disc brake system by checking the tires for excessive or unusual wear or improper inflation. What did you find?
2. Wheels for bent or warped wheels. What did you find?
3. Wheel bearings for looseness or wear. What did you find?
4. Suspension system for worn or broken components. What did you find?

Job Sheets

Located at the end of each chapter, the Job Sheets provide a format for students to perform procedures covered in the chapter. A reference to the ASE Education Foundation task addressed by the procedure is included on the Job Sheet.

APPENDIX

ASE PRACTICE EXAMINATION

1. *Technician A* says that if the master cylinder pushrod is adjusted too long, the brakes might not be able to fully apply. *Technician B* says that if the master cylinder pushrod is adjusted too short, the brakes might drag. Who is correct?
   A. A only C. Both A and B
   B. B only D. Neither A nor B
2. While discussing master cylinders, *Technician A* says normal brake lining wear causes a slight drop in fluid level. *Technician B* says a sure sign of brake fluid contamination with mineral oil is the swelling of the master cylinder cover diaphragm. Who is correct?
   A. A only C. Both A and B
   B. B only D. Neither A nor B
3. *Technician A* says that master cylinder leaks can be internal or external. *Technician B* says that a leaking master cylinder will remove paint from the area below the master cylinder. Who is correct?
   A. A only C. Both A and B
   B. B only D. Neither A nor B
4. While discussing brake lines, *Technician A* says that copper tubing can be used for brake lines. *Technician B* says that brake lines can use double-flare or an ISO flare fittings. Who is correct?
   A. A only C. Both A and B
   B. B only D. Neither A nor B
5. *Technician A* says to replace a double-flare fitting with an ISO-type fitting as new brake lines are required. *Technician B* says that flexible brake hoses allow movement of components. Who is correct?
   A. A only C. Both A and B
   B. B only D. Neither A nor B
6. A vehicle drifts to the right while driving. *Technician A* says that a crimped line to the left wheel could be the cause. *Technician B* says that the interior of the right brake hose could be damaged. Who is correct?
   A. A only C. Both A and B
   B. B only D. Neither A nor B
7. *Technician A* says service information circuit diagrams or schematics make it easy to identify common circuit problems. *Technician B* says if several circuits fail at the same time, check for a common power or ground connection. Who is correct?
   A. A only C. Both A and B
   B. B only D. Neither A nor B
8. *Technician A* says that there is a vacuum check valve in line between manifold vacuum source and the booster. *Technician B* says this check valve is to allow air pressure into the booster during wide-open throttle operation of the engine. Who is correct?
   A. A only C. Both A and B
   B. B only D. Neither A nor B
9. Drum brakes are being discussed. *Technician A* says that a grabbing brake could be traced to a leaking axle seal. *Technician B* says that a leaking wheel cylinder can also cause drum brake grabbing. Who is correct?
   A. A only C. Both A and B
   B. B only D. Neither A nor B
10. Before trying to remove a brake drum for service, *Technician A* uses the self-adjuster to back off the brake shoes. *Technician B* adjusts the parking brake cable to remove the slack. Who is correct?
    A. A only C. Both A and B
    B. B only D. Neither A nor B

539

ASE Practice Examination

A 50-question ASE practice exam, located in the Appendix, is included to test students on the content of the complete Shop Manual.

SUPPLEMENTS

Instructor Resources

The *Today's Technician* series offers a robust set of instructor resources, available online at Cengage's Instructor Resource Center and on DVD. The following tools have been provided to meet any instructor's classroom preparation needs:

- An Instructor's Guide provides lecture outlines, teaching tips, and complete answers to end-of-chapter questions.
- Power Point presentations include images, videos, and animations that coincide with each chapter's content coverage.
- Cengage Learning Testing Powered by Cognero® delivers hundreds of test questions in a flexible, online system. You can choose to author, edit, and manage test bank content from multiple Cengage Learning solutions and deliver tests from your LMS, or you can simply download editable Word documents from the DVD or Instructor Resource Center.
- An Image Gallery includes photos and illustrations from the text.
- The Job Sheets from the Shop Manual are provided in Word format.
- End-of-Chapter Review Questions are also provided in Word format, with a separate set of text rejoinders available for instructors' reference.
- To complete this powerful suite of planning tools, a pair of correlation guides map this edition's content to the NATEF tasks and to the previous edition.

MindTap for Today's Technician: Automotive Brake Systems, 7e

MindTap is a personalized teaching experience with relevant assignments that guide students to analyze, apply, and improve thinking, allowing you to measure skills and outcomes with ease.

- Personalized Teaching: Becomes yours with a Learning Path that is built with key student objectives. Control what students see and when they see it. Use it as-is or match to your syllabus exactly—hide, rearrange, add, and create your own content.
- Guide Students: A unique learning path of relevant readings, multimedia, and activities that move students up the learning taxonomy from basic knowledge and comprehension to analysis and application.
- Promote Better Outcomes: Empower instructors and motivate students with analytics and reports that provide a snapshot of class progress, time in course, engagement and completion rates.

REVIEWERS

The author and publisher would like to extend special thanks to the following instructors for reviewing the draft manuscript:

Rodney Batch
University of Northwestern Ohio
Lima, OH

Christopher J. Marker
University of Northwestern Ohio
Lima, OH

Tim Pifer
Midlands Technical College
Columbia, SC

Larry Stanley
Arizona Western College
Yuma, AZ

Claude F. Townsend
Oakland Community College
Bloomfield Hills, MI

CHAPTER 1
BRAKE SYSTEM FUNDAMENTALS

Upon completion and review of this chapter, you should be able to:

- List and describe the operation of the basic parts of a brake system.
- Describe the operation of the brake system during and after pedal application.
- Discuss the increasing use of disc brakes instead of drum brakes.
- Describe a typical brake hydraulic system.
- Describe the use of valves and lines to direct and control the hydraulic fluid.
- Discuss the purpose of brake power boosters and the parking brake.
- Discuss the general operation of electronic and active braking systems.
- Discuss the general operation of trailer brakes and air brakes.

Terms To Know

Active braking	Friction	Service brakes
Actuators	Fulcrum	Steering wheel position sensor
Air brakes	Lateral acccelerometer	Stroke sensor
Antilock brake system (ABS)	Leverage	Stroke simulator
Automatic ride control (ARC)	Lockup	Traction-control system (TCS)
Bulkhead	Master cylinder	Vehicle stability control (VSC)
Caliper	Negative wheel slip	Wheel cylinder
Disc brake	Parking brakes	Wheel speed sensors
Drum brake	Positive wheel spin	Yaw
Force	Pressure	
	Regenerative braking	

INTRODUCTION

The brake system is one of the most important systems on a vehicle. It has four basic functions:

1. It must slow a moving vehicle.
2. It must bring a vehicle to a stop.
3. It must hold a vehicle stationary when stopped.
4. It allows directional control during maximum braking.

If the brake system does not operate properly, the driver and passengers could be injured or killed in an accident. Technicians who service the brake system must be highly skilled experts because the work they do can save lives. In this chapter, we start our study of the brake system by presenting the basic concepts and parts of all brake systems.

Anti-lock brake system (ABS) is a braking system that is designed to prevent the brakes from locking up on hard stops so that the driver can maintain control of the vehicle.

This chapter also highlights some of the dynamics associated with braking and controlling a vehicle. If all the various dynamics are not considered during the design stage, most braking systems will under brake or over brake. When the brake system is not designed or operating correctly, it will be up to the driver to compensate, usually with poor results. In many cases, the human response is either too slow or too quick to react to a braking situation. In both cases, a loss of vehicle control is probably unavoidable. To prevent this, **antilock brake system (ABS)** and stability control have been added to help the driver maintain control.

Service brakes: The brakes that are used to stop the vehicle.

Parking brakes: The braking system that is used to hold the vehicle stationary while parked.

Regenerative braking recapture some of the lost energy during braking on hybrid vehicles.

BRAKE SYSTEM OVERVIEW

The complete brake system consists of the major components shown in **Figure 1-1**. These are the **service brakes**, which slow and stop the moving vehicle, and the **parking brakes**, which hold the vehicle stationary. On late-model vehicles, the ABS is a third major subsystem; and new cars now also include traction control and stability control as part of the brake system functions. Hybrid and electric vehicles make use of **regenerative braking**, which captures some of the energy normally lost as heat on the pads and rotors while stopping. Regenerative braking systems use electrical generators to help slow the vehicle during gradual stops and help recharge the electric batteries. Electric and hybrid vehicles still have conventional hydraulic brakes to stop the vehicle quickly when necessary. Regenerative brakes are a blending of the generators' ability to help slow the vehicle and conserve energy and the hydraulic systems' ability to stop the vehicle quickly.

Leverage and the Brake Pedal Design

AUTHOR'S NOTE A fulcrum is the point at which one lever pivots or sits to apply force to another lever or device. A seesaw pivots on a fulcrum.

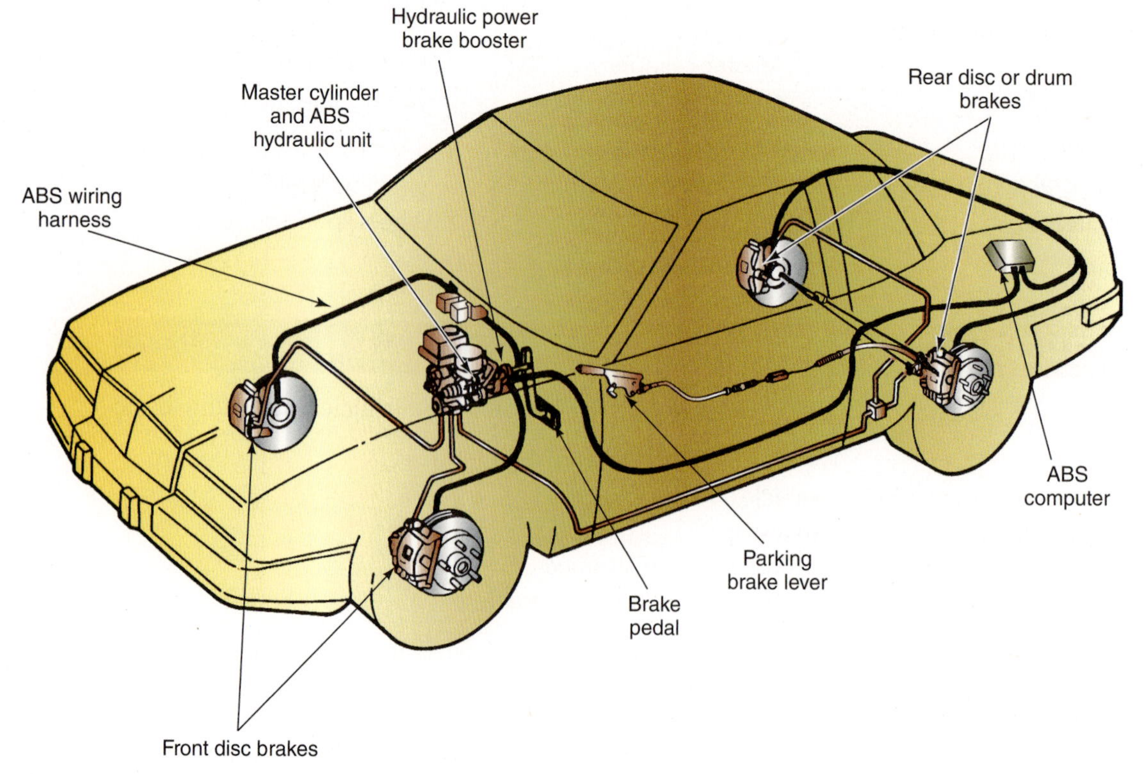

Figure 1-1 A typical automotive brake system comprises these major components and subsystems.

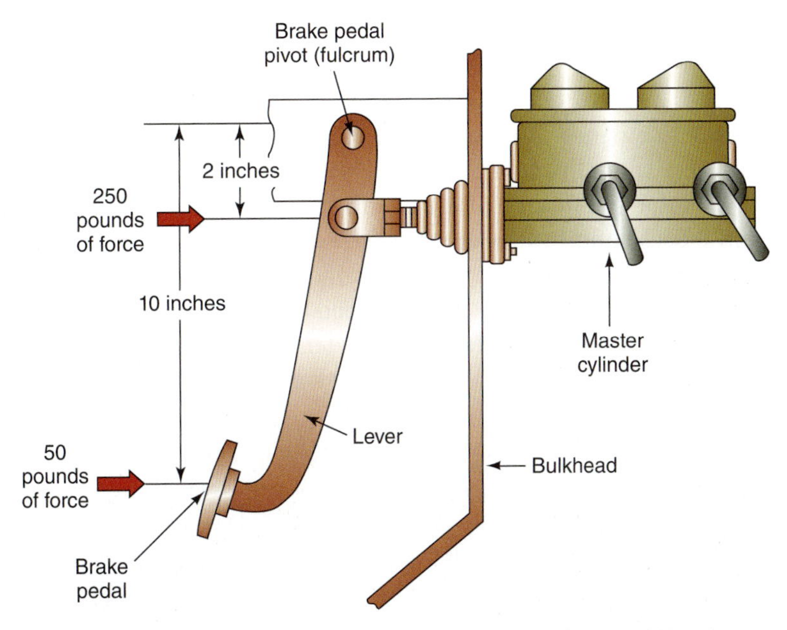

Figure 1-2 The brake pedal assembly uses leverage to increase force applied to the master cylinder.

Braking action on an automobile begins with the driver's foot on the brake pedal. The driver applies **force** to the pedal (which we learn more about later), and the pedal transfers that force to the master cylinder pistons. The brake pedal also multiplies the force of the driver's foot through leverage.

The brake pedal is mounted on a lever with a pivot near the top of the lever. The movement of the pedal causes a pushrod to move against a master cylinder. The master cylinder is mounted inside the engine compartment on the rear **bulkhead**. The master cylinder is a hydraulic pump that is operated by the driver through the brake pedal.

The **bulkhead** separates the engine compartment from the passenger compartment.

Most brake pedal installations are an example of what is called a second-class lever. In the science of physics, a second-class lever has a pivot point (or **fulcrum**) at one end and force applied to the other end. A second-class lever transfers the output force in the same direction as the input force, and multiplies the input force, depending on where the output load is placed. The brake pedal installation shown in **Figure 1-2** has a 10-inch lever, and the load (the master cylinder pushrod) is 2 inches from the fulcrum (8 inches from the pedal). The pedal ratio, or the force multiplying factor, is the length of the lever divided by the distance of the load from the fulcrum. In this case, it is:

The **fulcrum** is the support for a lever to pivot on.

$$\frac{10}{2} = 5:1$$

If the driver applies 50 pounds of force to the pedal, the lever increases the force to 250 pounds at the master cylinder. When the driver applies 50 pounds of input force, the pedal may travel about 2.5 inches. When the lever applies 250 pounds of output force, the pushrod moves only 0.5 inch. Thus, as **leverage** in a second-class lever increases force, it reduces distance by the same factor:

Leverage is the use of a lever and fulcrum to create a mechanical advantage, usually to increase force applied to an object.

$$\frac{2.5 \text{ inches}}{5} = 0.5 \text{ inch}$$

Service Brake History and Design

Modern automobile brakes evolved from the relatively crude brakes of horse-drawn vehicles. The earliest motor vehicle brakes were pads or blocks applied by levers and linkage to the outside of a solid tire on a wooden-spoked wheel. The same principles of leverage that work in modern brake pedal installations increased the force of the brake pad applied to the solid tire. These brakes worked well with speeds of 10 mph to 20 mph and little traffic. Higher performance (30 mph and beyond) and pneumatic tires meant that early wagon brakes were short-lived on automobiles.

By the end of the first decade of the twentieth century, automobiles were using either external-contracting band brakes or internal-expanding drum brakes. A few internal-expanding band brakes were tried on some early motor vehicles. External-contracting brakes have a band lined with **friction** material wrapped around a drum located on the driveline or on the wheels. The band is anchored at one end or at the center; levers and linkage tighten the band around the drum for braking force. The service brakes on Ford's famous Model T were a single contracting band applied to a drum inside the transmission.

Friction is the force that resists motion between the surfaces of two objects or forms of matter.

Band brakes, either internal or external, lose their effectiveness when higher braking force is needed. When you study **drum brakes**, you will learn about the mechanical servo action of brake shoes. It is very difficult to develop servo action with an internal band brake, and higher brake force is thus needed. Servo action on an external band brake tends to make the brake grab at high brake forces and high drum speed. Other problems associated with band brakes include dirt and water damage and loss of friction with external bands and the tendency of these brakes to lock if the drum overheated and expanded too much. Internal band brakes also suffer from band and drum overheating and reduced braking force.

A **drum brake** is a brake in which friction is generated by brake shoes rubbing against the inside surface of a brake drum attached to the wheel.

As drum brakes evolved, internal-expanding shoe-and-drum brakes became the standard. External-contracting band brakes were used as parking brakes until the late 1950s, but their days as service brakes were over by the late 1920s.

Drum Brakes. By the mid-1920s, drum brakes with internal-expanding shoes were the general rule. Early drum brakes were operated mechanically by levers and linkage (**Figure 1-3**). Expensive luxury cars such as the 1921 Duesenberg Model A were among the first to have hydraulic drum brakes. Hydraulic brakes started to appear on lower-priced cars in the mid-1920s with Chrysler's Light Six, which became the Plymouth. Ford Motor Company, however, used mechanical brakes through the 1938 model year.

AUTHOR'S NOTE There were two major reasons for the increased use of hydraulically applied brakes over the mechanically-applied ones: (1) The four brakes never seemed to apply the same amount of braking force at the same time because (2) the brake linkages required almost constant re-adjustment to make the brake work at all. The only reason mechanical brakes were ever practical was the fact that roads were rough and couldn't be traveled at high speeds.

The rigid brake shoes used with drum brakes could be made stronger than the flexible bands of earlier brake designs. This eliminated breakage problems that occurred with greater braking forces that were required as automobiles got more powerful and faster. With hydraulic actuation, four-wheel drum brakes remained the standard braking system for most cars into the middle and late 1960s. With the coming of Federal Motor Vehicle Safety Standards (FMVSS) 105 in 1967, brake systems had to pass specific performance tests that made front disc brakes the general rule in the 1970s. Even at the beginning of the twenty-first century, however, drum brakes are still used on the rear wheels of many cars and light trucks.

Disc Brakes: A braking system that forces two brake pads on opposite sides of a spinning rotor to stop the vehicle

Disc Brakes. Modern automotive **disc brakes** were developed from aircraft brakes of World War II. Known originally as "spot" brakes, disc brakes work by applying pressure to two brake pads on opposite sides of a spinning rotor attached to the wheel hub

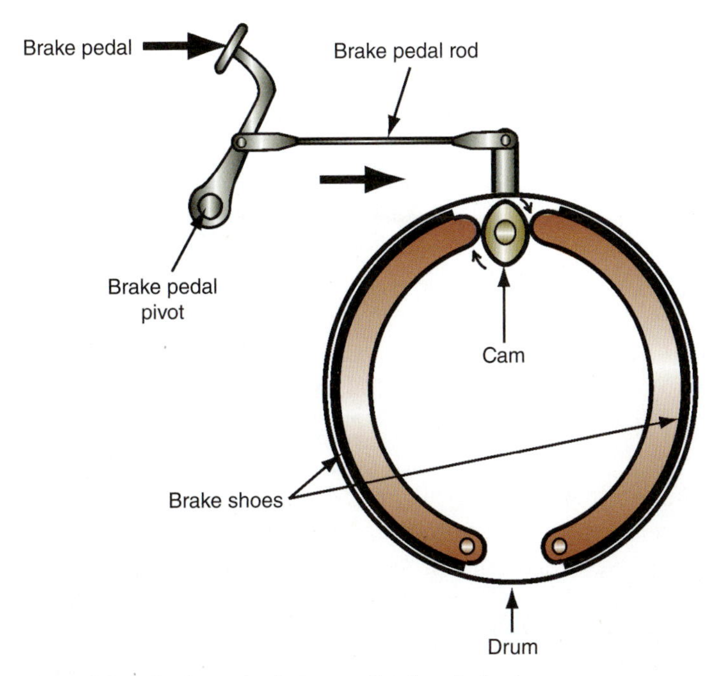

Figure 1-3 A simple mechanical expanding drum brake.

(**Figure 1-4**). Disc brake pads are mounted in a **caliper** that sits above the spinning rotor. The caliper is either fixed or movable on its mounting. With a fixed caliper, hydraulic pressure is applied to pistons on both sides to force the pads against the rotor (**Figure 1-5**). With a movable caliper, pressure is applied to a piston on the inboard side only. This forces the inboard pad against the rotor, and the reaction force moves the outboard side of the caliper inward so that both pads grip the rotor (**Figure 1-6**).

A **caliper** is a major component of a disc brake system that houses the piston(s) and supports the brake pads.

All the friction components of a disc brake are exposed to the airstream, which helps to cool the brake parts and maintain braking effectiveness during repeated hard stops from high speeds. This, in turn, leads to longer pad life and faster recovery from brake fade. Disc brakes do not develop the mechanical servo action that you will learn about as you study drum brakes. Therefore, disc brakes require higher hydraulic pressure and greater force to achieve the same stopping power as a comparable drum brake. These pressure and force requirements for disc brakes are met easily, however, with large caliper pistons

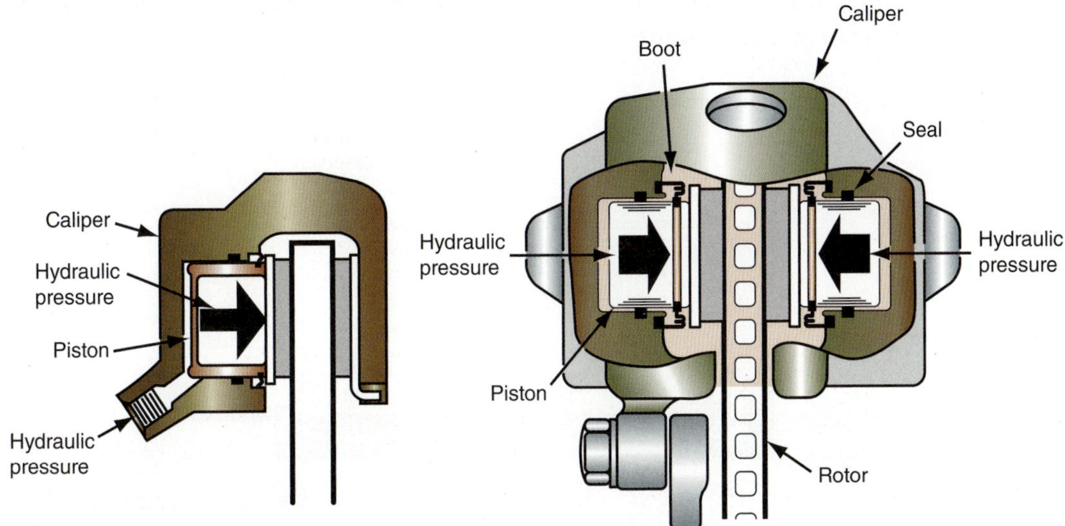

Figure 1-4 Hydraulic pressure in the caliper forces the disc brake pads against the spinning motor.

Figure 1-5 Hydraulic pressure is applied equally to pistons on both sides of a fixed caliper.

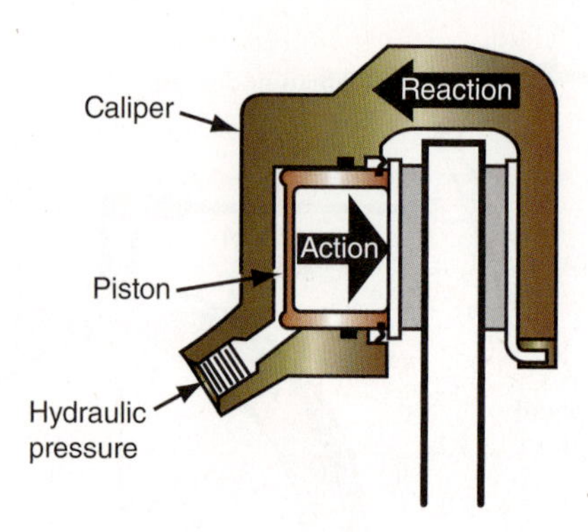

Figure 1-6 Hydraulic pressure in a movable caliper forces the piston in one direction and the caliper body in the other. The resulting action and reaction force the pads against the rotor.

A BIT OF HISTORY

Hydraulic brakes were invented in 1918 in the California shop of Malcolm Loughead. He later changed the spelling of his name to Lockheed, and he and his brother founded the aircraft company of that name. The Lockheed hydraulic brake first appeared on the 1921 Duesenberg Model A.

and power brake boosters. Because their advantages far outweigh any disadvantage, disc brakes have become the universal choice as the front brakes on all cars and light trucks built since the 1970s. Additionally, four-wheel disc brakes became standard equipment on high-performance automobiles, SUVs, and some trucks.

Brake Hydraulic Systems

The **master cylinder** is the liquid-filled cylinder in the hydraulic brake system in which hydraulic pressure is developed when the driver depresses a foot pedal.

Although brake systems have been changing recently, hydraulic operation of service brakes has been the universal design for more than 70 years. The complete hydraulic system consists of the **master cylinder**, steel lines, rubber hoses, various pressure-control valves, and brake apply devices at each wheel.

Master Cylinder. The master cylinder is the start of the brake hydraulic system. It actually is a cylindrical pump. The cylinder is sealed at one end, and the movable pushrod extends from the other end (**Figure 1-7**). The pushrod moves a pair of in-line pistons that

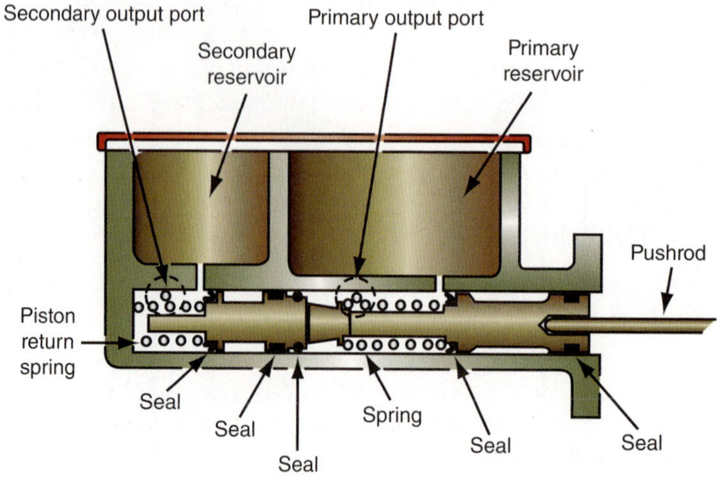

Figure 1-7 The master cylinder is a cylindrical pump with two pistons that develop pressure in the hydraulic lines to the front and rear brakes.

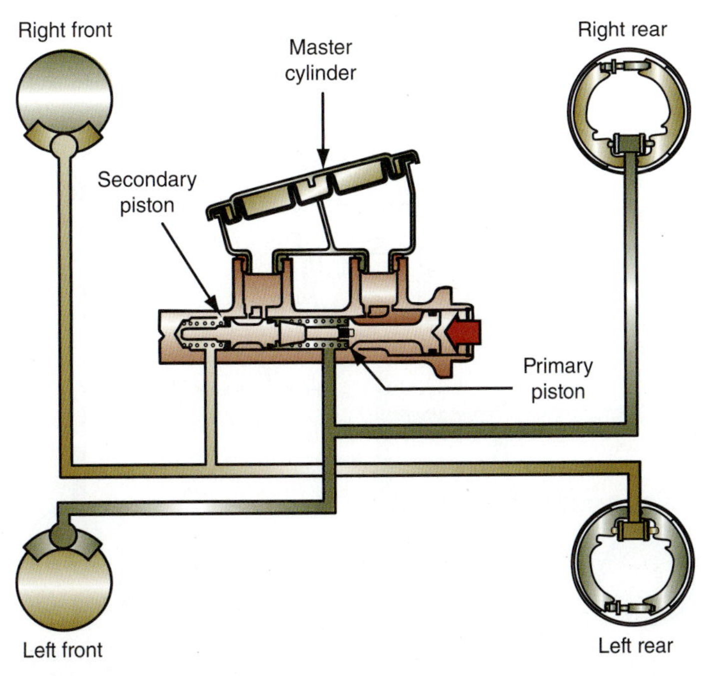

Figure 1-8 Each master cylinder piston feeds one system of a split hydraulic brake system. Shown is a diagonally split system.

produce the pumping action. When the brake pedal lever moves the pushrod, it moves the pistons to draw fluid from a reservoir on top of the master cylinder. Piston action then forces the fluid under **pressure** through outlet ports to the brake lines.

Pressure is the force exerted on a given unit of surface area—force divided by area—measured in pounds per square inch (psi) or kilopascals (kPa).

All master cylinders for vehicles built since 1967 have two pistons and pumping chambers, as shown in **Figure 1-7**. Motor vehicle safety standards require this dual-brake system to provide hydraulic system operation in case one hose, line, or wheel brake assembly loses fluid. Because the brake hydraulic system is sealed, all the lines and cylinders are full of fluid at all times. When the master cylinder develops system pressure, the amount of fluid moved is only a few ounces.

Split Systems. Modern-day vehicles have split brake systems. The pre-1970s vehicle had a single hydraulic system serving all four wheels. A leak anywhere in the system resulted in a complete braking failure. The split system was designed to prevent a total system failure. This required the use of a dual-piston master cylinder and the inclusion of various valves. A split system is fed by one piston in the master cylinder and feeds two wheel brakes of the vehicle. There are two types of split systems: diagonal and front/rear. The diagonal has one system feeding a front-wheel brake and the rear, opposing-side-wheel brake, that is, left front and right rear (**Figure 1-8**). The second diagonal split is to the other wheel brakes. The front/rear split is exactly as it sounds. One side or split feeds the rear-wheel brakes, and the other feeds the front wheels (**Figure 1-9**). Both types of split have advantages and disadvantages, but each prevents complete system failure from a single leak.

Hybrid Master Cylinders

Hybrid vehicles braking systems have some significant differences as compared to most conventional automobiles. One reason for this is regenerative braking used during light to moderate braking. Regenerative braking can slow the vehicle and recover much of the energy used by using the drive motor as a generator. Another reason is that the engine may not be running to provide vacuum to the conventional brake booster. These two factors produce a master cylinder used somewhat conventionally to produce pressure, but this pressure is amplified by the brake actuator (**Figure 1-10**) to apply the appropriate pressure to the wheel calipers. The brake **stroke sensor** (**Figure 1-11**) informs the brake

Stroke sensor informs the brake controller how fast and how much pressure the driver applied to the brakes.

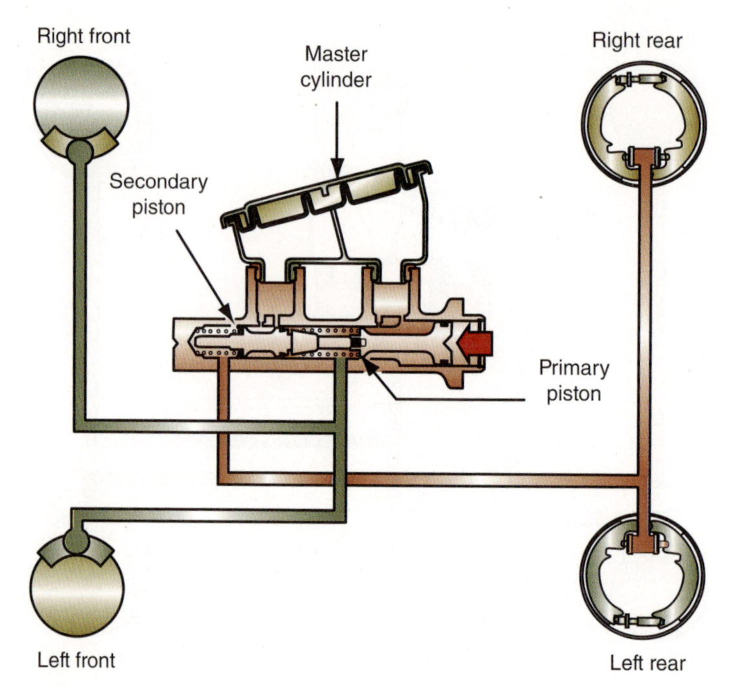

Figure 1-9 A front/rear split dual hydraulic brake system.

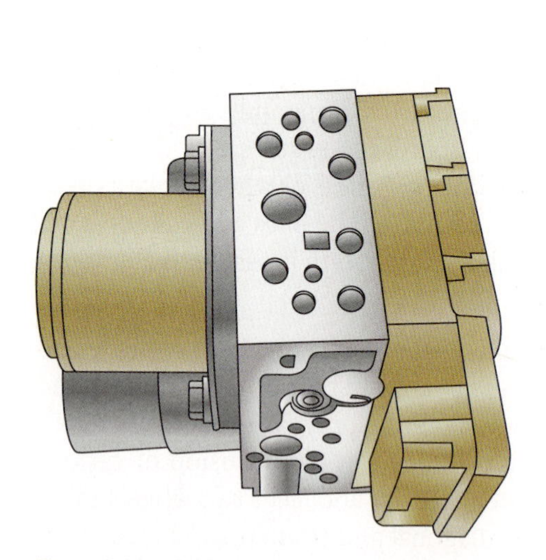

Figure 1-10 A hybrid brake actuator for a Toyota Camry contains a pump to boost brake pressure instead of using a vacuum booster.

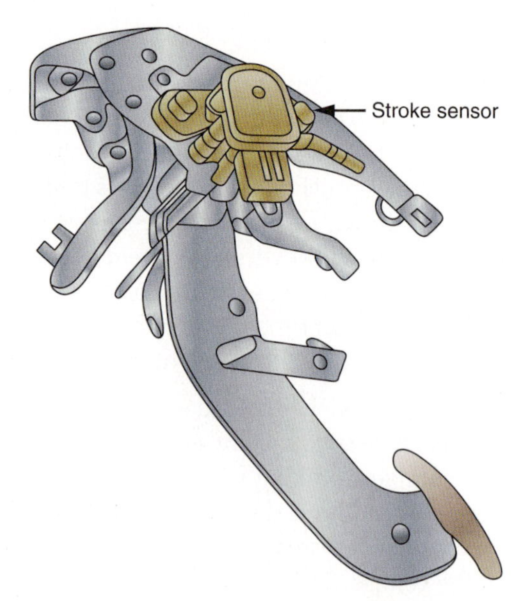

Figure 1-11 A brake pedal stroke sensor tells the braking system how far and how fast the brake pedal is being depressed.

controller how fast and how much pressure the driver applied to the brakes; the brake controller then decides on what ratio of regenerative braking to hydraulic braking needs to be applied. If the braking is light, then the controller may decide to use only the regenerative braking system to stop the vehicle. If the stop is moderate, the controller may command regenerative and hydraulic braking. If the stop is panic in nature, then the hydraulic braking system will be the dominant choice. In these systems, the **stroke simulator** (**Figure 1-12**) provides appropriate brake pedal feedback to the driver; in other words, the brakes should feel normal to the driver.

Stroke simulator provides brake pedal feedback to the driver.

Brake Lines and Hoses. The rigid lines or pipes of a brake hydraulic system are made of double-walled steel tubing for system safety. Flexible rubber hoses connect

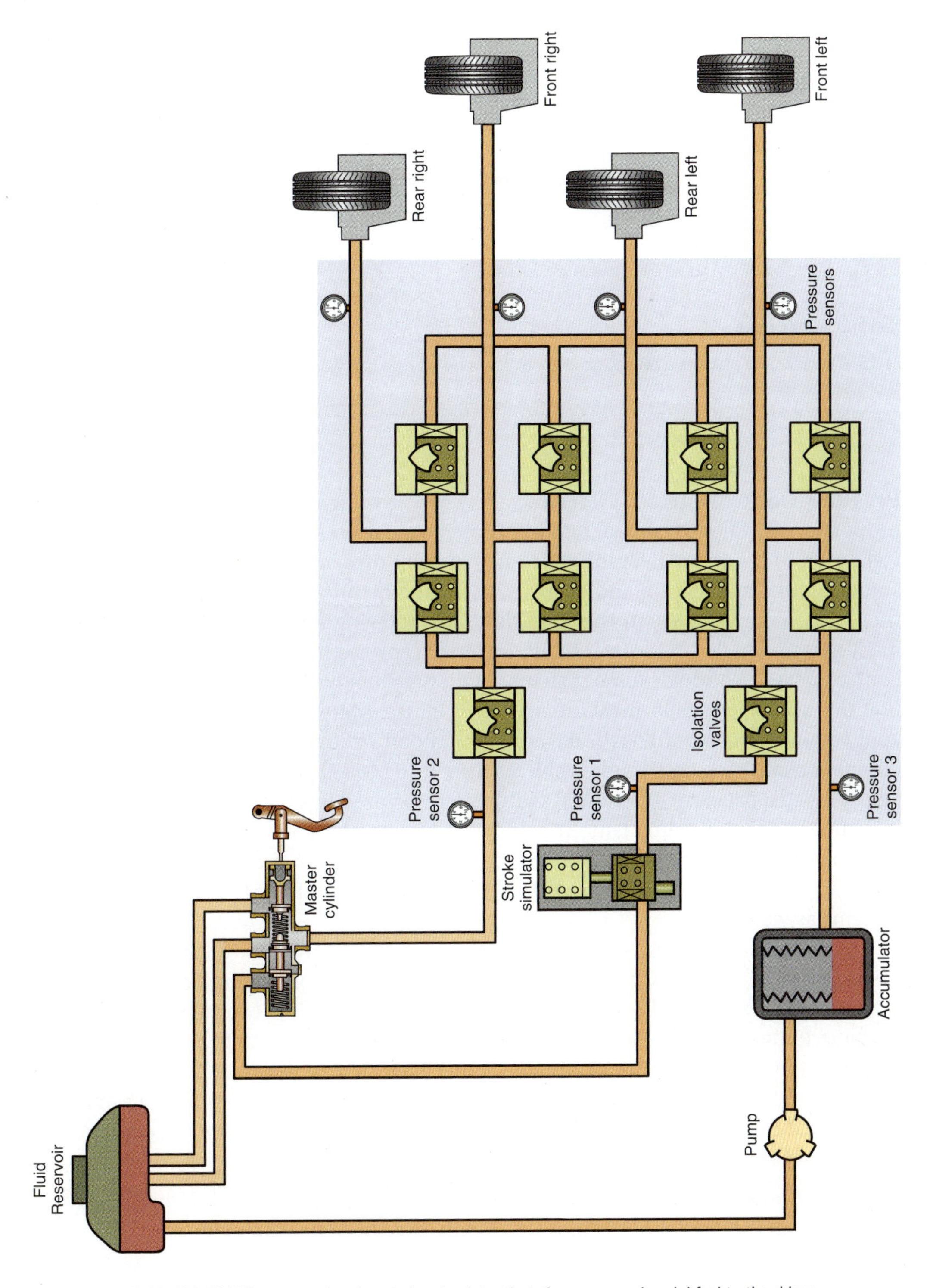

Figure 1-12 Hybrid ABS system showing stroke simulator that gives a normal pedal feel to the driver.

the wheel brakes to the rigid lines on the vehicle body or frame (**Figure 1-13**). The front brakes have a rubber hose at each wheel to allow for steering movement. Rear brakes may have separate hoses at each wheel or a single hose connected to a line on the body or frame if the vehicle has a rigid rear axle. Brake lines and hoses contain the high-pressure fluid, and the fluid acts as a solid rod to transmit force to the wheel cylinders and caliper pistons.

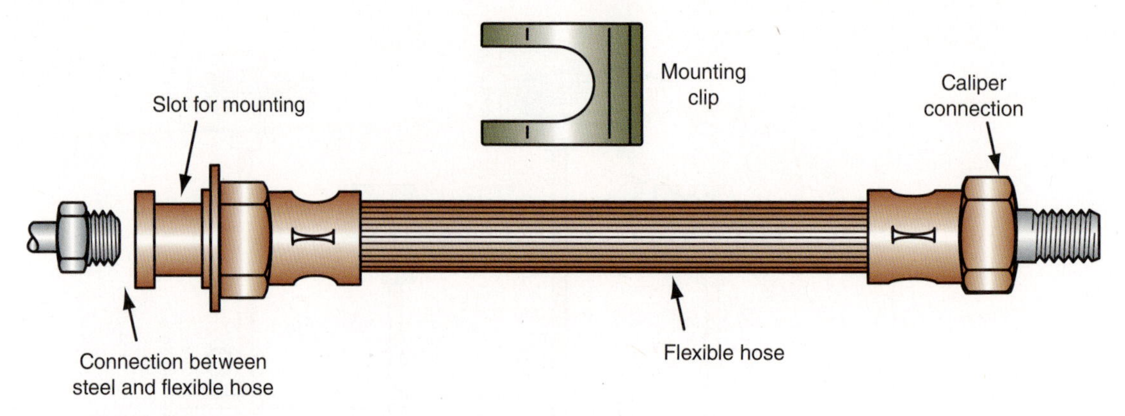

Figure 1-13 A flexible hose provides a connection between the vehicle's rigid frame and the movement of the wheel and suspension assemblies.

An **ABS** is a service brake system that modulates hydraulic pressure to one or more wheels as needed to keep those wheels from locking during braking. If the wheel locks during braking, steering control becomes very difficult. ABS allow the driver to maintain control of the vehicle during a panic stop.

Dynamic rear proportioning and **electronic brake distribution** are names for the electronic function of the proportioning valve by the ABS braking system.

The increasing use of four-wheel disc brake systems instead of the front disc/rear drum systems has also reduced the need for metering valves.

A **wheel cylinder** is the hydraulic cylinder mounted on the backing plate of a drum brake assembly.

Pressure-Control Valves. Older and a few newer vehicles depended on metering and proportioning vales to control braking application. Metering and proportioning valves were used to modulate hydraulic pressure to front disc or rear drum brakes to provide smooth brake application and reduce the tendency to lock the rear brakes. Although there are many vehicles on the road that use metering and proportioning valves, on most newer vehicles, ABS and stability control replaces the need for using mechanical control valves. Although some systems retain proportioning valves for use in the event that the ABS system fails, most are no longer in use. Metering valves, which slowed the application of the front disc brakes to allow the drum brakes to apply, have been less common since most late model vehicles have disc brakes in both front and rear. Older vehicles had a pressure differential switch that was used in some systems to turn on the instrument panel warning lamp if half of the hydraulic system loses pressure. Today's systems use a fluid level switch to detect the loss of fluid in the system, in addition to the ABS system, which can be utilized to detect problems in the hydraulic braking systems as well.

As was noted earlier, even though pressure-control valves have been part of brake systems for more than 60 years, **ABS** and **stability-control systems** are beginning to make some valves obsolete. An ABS electronic control module can modulate hydraulic pressure for normal braking better than metering and proportioning valves can. As ABS installations became standard equipment, some older hydraulic functions have been given over to the electronic modulator. The electronic function of the proportioning valve has been called **"dynamic rear proportioning"** and **"electronic brake distribution"** by various manufacturers.

Wheel Cylinders and Caliper Pistons. The **wheel cylinders** of drum brakes and the caliper pistons of disc brakes operate in response to the master cylinder. These hydraulic cylinders at the wheels change hydraulic pressure back into mechanical force to apply the brakes.

Most systems with drum brakes have a single, two-piston cylinder at each wheel (**Figure 1-14**). Hydraulic pressure enters the cylinder between the two pistons and forces them outward to act on the brake shoes. As the shoes move outward, the lining contacts the drums to stop the car. Wheel cylinder construction and operation of drum brakes are covered in Chapter 8 of this manual.

The caliper pistons for disc brakes also act in response to hydraulic pressure that enters a fluid chamber in the caliper. Hydraulic pressure in a stationary caliper is applied to one or two pistons on each side of the caliper to force the pads against the rotor, as shown in Figure 1-5. Pressure is applied to a single piston in a movable caliper on the

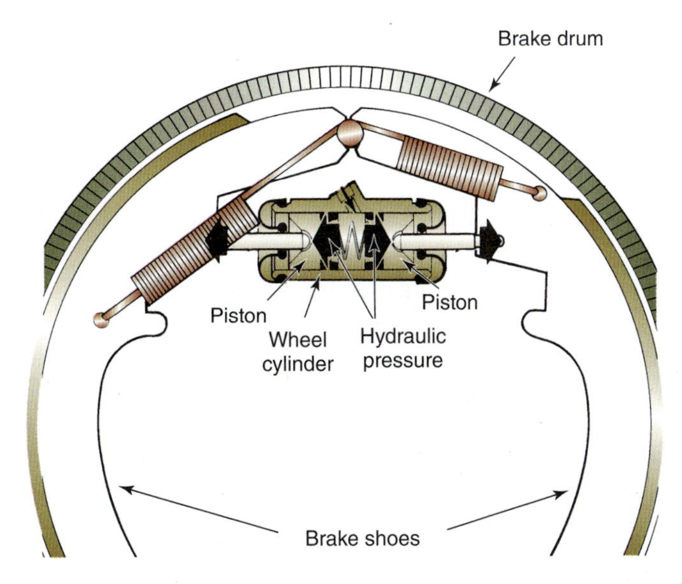

Figure 1-14 Hydraulic pressure in the wheel cylinder moves the two pistons outward to force the shoes against the drum.

inboard side to force the inboard pad against the rotor. As explained later, in Chapter 2, hydraulic pressure is equal in all directions in a sealed chamber. This equal pressure creates a reaction force that moves the outboard side of the caliper inward so that both pads grip the rotor, as shown in Figure 1-6. More details about caliper construction and operation of disc brakes are found in Chapter 7 of this manual.

Power Boosters

Almost all brake systems have a power booster that increases the force of the driver's foot on the pedal (**Figure 1-15**). Most cars and light trucks use a vacuum booster that uses the combined effects of engine vacuum and atmospheric pressure to increase pedal force. Some

Figure 1-15 The power brake booster increases the brake pedal force applied to the master cylinder.

vehicles without a ready vacuum source may have a hydraulic power booster that may be supplied with fluid by the power steering system or that may be a part of the brake system and driven by an electric motor. Alternatively, as was discussed earlier, the brake actuator can amplify the brake pressure applied to the wheel calipers based on input from the brake stroke sensor. Chapter 6 in this *Classroom Manual* explains power boosters in detail.

Parking Brakes

After the service brakes stop the moving car, the parking brakes help to hold it stationary. Parking brakes are mistakenly called "emergency" brakes, but their purpose is not to stop the vehicle in an emergency. The amount of potential stopping power available from parking brakes is much less than from the service brakes. Because the parking brakes work on only two wheels or on the driveline, much less friction surface is available for braking energy. In the rare case of total hydraulic failure, the parking brakes can be used to stop a moving vehicle, but their application requires careful attention and skill to keep the vehicle from skidding or spinning.

Parking brakes on most cars and light trucks are applied with cables. On rear drum brakes, the brakes' shoes are applied by the cable pulling a shoe into the drum. On some rear disc brakes, a cable applies the caliper mechanically, while most late model disc parking brakes are applied by a small set of brake shoes that sit inside a special brake rotor on the rear wheels. In either case, the parking brake shares the components of the disc and/or drum braking systems. Some parking brakes are applied through the use of an electric motor on the calipers instead of a cable, and some parking brakes use an electric motor to apply tension to the brake cables, but the basic concept is still the same.

traction control systems are designed to prevent wheel slip in low-traction situations by applying the brakes to the free spinning wheel.

vehicle stability control is an electronic assisted braking system that can selectively apply and release brakes during critical maneuvers to help the driver maintain control.

Lockup or **negative wheel slip** is a condition in which a wheel stops rotating and skids on the road surface.

ABS prevents tires from skidding during braking.

Positive wheel spin happens with no traction and the wheel spins but does not move the vehicle.

Electronic Braking Systems

Electronic braking covers all systems from ABS and **traction-control systems** (TCSs) up to and including **vehicle stability control** (VSCs). While there are different terms used by the various manufacturers for the same operation, all operate with electronic sensors and actuators with little or no input from the vehicle operator. Some systems such as VSC are in the second and third stages of development and application. The National Highway Traffic Safety Administration (NHTSA) mandated VSC for all models built after 2012. Now we are seeing additional developments, such as cruise control that reacts to traffic conditions, blind spot detection, lane departure alerts, and parking assist.

Whenever the brakes are applied with heavy pressure, the wheel may totally stop rotating. This condition is called wheel **lockup** or **negative wheel slip,** which does not help the car stop. Rather, the tire loses some frictional contact with the road and slides or skids. As the tires slide, the car is no longer stopping under control, and the driver is in a dangerous situation. Experienced drivers try to prevent wheel lockup by pumping the brake pedal up and down rapidly. This stops and starts hydraulic pressure to the brakes and gives the driver control during hard braking.

All late-model cars have an ABS. The ABS does the same thing as an experienced driver would, only it does it faster and more precisely. It senses when a wheel is about to lock up or skid. It then rapidly interrupts the braking pressure to the brakes at that wheel. Speed sensors at the wheels monitor the speed of the wheels and send this information to an on-board computer. The computer then directs the ABS unit to pulse the pressure going to the wheel that is starting to lock up.

Traction-Control Systems (TCS)

The TCS controls wheel spin or **positive wheel spin** during hard acceleration or slippery road conditions. The TCS has some options to control wheel spin. It can reduce engine power via the throttle actuator control (TAC), reduce engine timing through the ECM, and apply the brakes to the spinning wheel. This allows the torque to be transferred to the wheel, which has traction, allowing the vehicle to move.

Automatic Ride Control

Automatic Ride Control (ARC) is an electronic suspension system designed to provide a more comfortable ride to the passengers and allow better control of the vehicle during cornering. To some extent, it can command reduced engine power, thereby slowing the vehicle. The major component shared between ARC and active braking is the yaw sensor.

Active braking systems take the data from shared components, perform certain calculations, and apply the brakes without any driver input at all. There are more details on active braking and stability control in Chapter 10.

Vehicle Stability Control

Vehicle stability control (VSC) will be covered more in depth in later chapters, but a brief overview is given here to help you understand how all these systems interact. VSC is the result of advances in technology that is expected to save many lives, in particular by preventing single vehicle crashes due to loss of driver control. VSC systems depend on a vehicle responding to the application of the brakes to help control the vehicle that is in a skid. Like steering a bulldozer by locking the treads on the right or left sides, a vehicle with the brakes applied individually on one side can help steer the vehicle into control from a skid. The **yaw** or **lateral accelerometer** determines the direction and speed at which the vehicle is skidding so the VSC can apply the appropriate wheel brake(s) to correct the skid. The direction the driver is trying to steer is determined by the **steering wheel position sensor**, which is also a factor in the VSC decision on brake(s) application. The VSC can also determine if a wheel is slowing down too quickly or speeding up too fast by comparing it to the other wheels by use of the **wheel speed sensors** (WSS) and apply TCS or ABS braking at the appropriate times.

Active Braking

Mass-produced **active braking** systems are systems that can act to prevent a collision without input from the driver. In the context here, "active" means the brake system will perform some functions without input from the operator. The ABS could be considered as the first active braking system, but it functioned only if the driver had applied the brakes and if certain conditions, such as wheel skid, were present. The active braking systems of today use components from the ABS, TCS, VSC, and radar or cameras to determine how close the vehicle is to another vehicle or an obstruction. Most systems warn the driver of an impending collision. If the driver does not respond, then the braking system acts on its own to bring the vehicle to a stop, or at least reduce the severity of a collision if it cannot come to a complete stop. Each of the systems listed normally has its own controller or computer module.

TRAILER BRAKES

Normally a textbook of this type does not deal with trailer brakes. However, with the advent of active braking systems and the increasing sales and usage of ¾- and 1-ton pickups, it is important that some information is provided. Consult local and state laws for specific requirements.

Older trucks used to tow trailers heavy enough to warrant a brake system always used an add-on system. This system was not efficient in some cases, and its efficiency in many cases was based on "you got what you paid for." If the trailer was lightly loaded or the road was slick, the trailer brakes could lock and the trailer would probably jack-knife. Heavy trucks over 1 ton did have a trailer brake system that was either "hard-wired" at the factory

Throttle actuator control is electronic control of the throttle by electric motor instead of a conventional throttle cable.

Automatic ride control is a suspension system that can be electrically tailored to supply optimum passenger comfort and cornering ability action in a wide range of road surfaces.

Shared components among ABS and TCS are the wheel speed sensors and hydraulic modulator.

TCS is an outgrowth of the ABS.

Lateral accelerometer determines the direction and speed at which the vehicle is skidding.

Steering wheel position sensor determines which direction the driver is trying to steer.

Wheel speed sensors are mounted at selected (or all) wheels to monitor wheel speed during vehicle operation.

Yaw is the deviation in the line of travel commonly referred to as roll or lean during cornering.

Active braking systems can apply the brakes without the driver touching the brake pedal. Meant to be used during an emergency stop.

or at least had to meet stringent Department of Transportation (DOT) standards. The newest ¾- and 1-ton truck models offer a factory-installed trailer brake system. This system borrows from the ABS and active braking systems. Trailer brake systems are usually divided into two types: hydraulic surge and electric.

Hydraulic Surge

A hydraulic surge brake system is completely mounted on the trailer. It may be a disc or drum type and operates hydraulically and mechanically in a very similar manner to the drum brakes on the tow vehicle. There is a pressure differential valve mounted somewhere in the tow vehicle's rear brake system, usually at or near the vehicle's master cylinder. This valve operates so the pressure supplied to the trailer's master cylinder is in direct proportion to the pressure being directed to the tow vehicle's rear brake. The valve works very similarly to the regular proportioning valve found on almost every roadworthy vehicle. The pressure delivered to the trailer's master cylinder applies a mechanical pushrod, which, in turn, creates the correct pressure to apply the trailer brake. Trailer drum brakes may be uni-servo with one wheel cylinder and one pushrod or duo-servo with one wheel cylinder and two pushrods. The uni-servo type is the most common because of lower cost compared to the duo-servo system. The disc-brake type operates almost exactly like the disc brakes on a standard vehicle.

Electric Brakes

Electric trailer brakes are most commonly used on utility and RV units, whereas boat trailers or others that are designed to be submerged underwater use the hydraulic surge. This is because water severely shortens the life expectation of electric brake components.

A trailer electric brake system requires a brake controller mounted in the tow vehicle within hand control of the vehicle driver. It is usually mounted on the dash near the steering column (**Figure 1-16**). The driver may reduce or increase the power of the trailer brakes or lock them down as a kind of parking brake. The controller is not readily switched between different tow vehicles, so each tow unit must be equipped with its own controller. There are three types of controllers, but each performs basically the same function. For detailed information on the controllers and troubleshooting trailer brakes, investigate Champion Trailers at its website at http://www.championtrailers.com/techsup.html. The regulated electric power from the controller is sent to magnets within the wheel drum. The energized magnets are pulled toward a specially machined flat surface of the drum. As the magnets move, they, in turn, move a lever or levers attached to the brake shoe. The

Figure 1-16 This is a typical add-on trailer brake controller.

brake shoe is applied against the rotating drum in direct proportion to the braking action of the rear brakes of the tow vehicle.

Trailer Breakaway Condition

The DOT requires any trailer with its own braking system to have a method to apply its brakes in case the trailer disconnects from its tow vehicle. On an electric system, an emergency power battery is mounted on the trailer. If a breakaway condition occurs, the pull pin at the trailer/vehicle hitch is pulled loose, triggering a switch that connects the full power of the emergency battery to the wheel magnets. This effectively locks the wheels. The hydraulic surge system uses a slightly different method to accomplish the same result. A chain or cable runs from the tow vehicle to a lever positioned to apply force to the piston in the trailer's master cylinder. When a disconnect situation occurs, the lever is moved by the chain/cable and applies the trailer brakes. In either case, both breakaway systems may be used as parking brakes when the trailer is stored or being loaded/unloaded.

Reverse Braking

Trailer brakes do not recognize reverse braking from forward braking. Usually it is best if the trailer does not have braking during reverse operation. To prevent this, the hydraulic surge brake may be standard or free backing (**Figure 1-17**). The standard type requires a reversing solenoid that is triggered when the tow vehicle backup lights are activated (**Figure 1-18**). The solenoid is plumbed into the system at the outlet of the trailer's master cylinder. With the solenoid energized, pressurized fluid from the master cylinder is blocked from the wheels and is directed back into the master cylinder's reservoir. Free backing operation recognizes that the wheels are rotating backward and deactivates the brakes. The brake system reactivates when the wheels rotate forward. This is the system of choice for most trailers because it can be manufactured into the system. It should be noted that disc brakes on trailers must have a free backing (reversing) solenoid. Electric brake systems use only a reverse solenoid to prevent the trailer brakes from functioning in reverse. The solenoid is triggered by the backup lights circuit.

AUTHOR'S NOTE Technicians or do-it-yourselfers can get into difficulties if they do not understand how to connect an add-on trailer brake system. Also, if the add-on system is on the less expensive side, the rule "you get what you pay for" applies. There are some add-on units that meet DOT standards but have no leeway if the trailer is loaded or too heavy for the brake system.

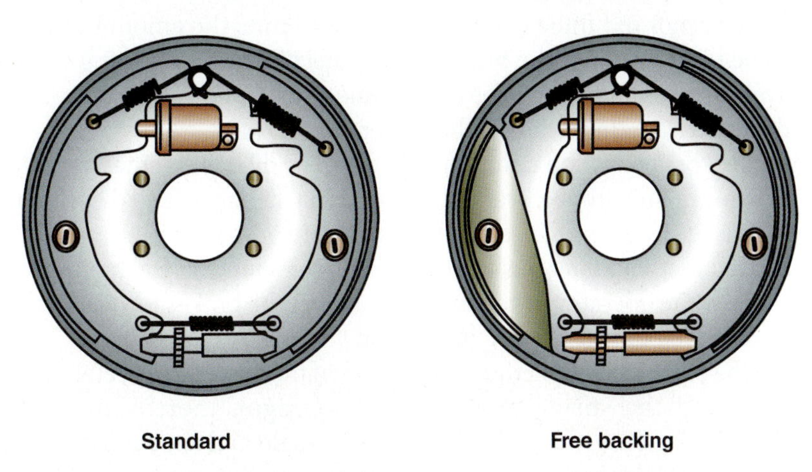

Figure 1-17 The free backing brake on the right recognizes reverse movement and the brake cannot be applied.

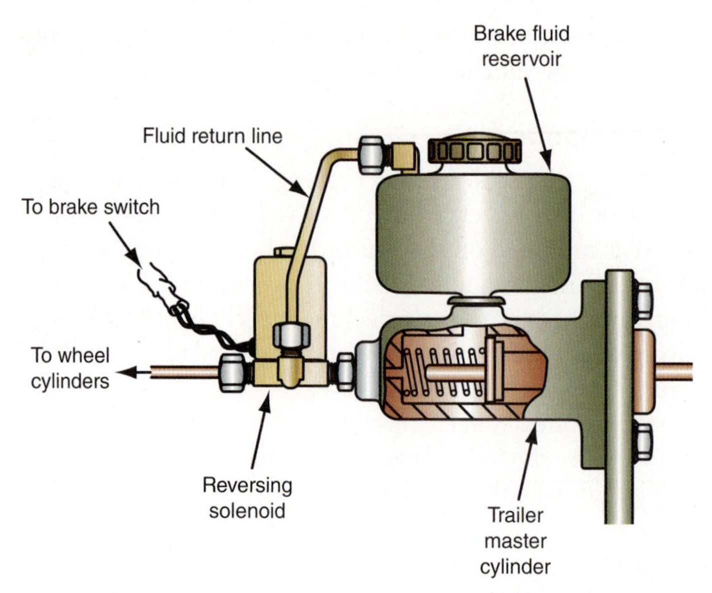

Figure 1-18 The reversing solenoid is needed with the standard brakes in Figure 1-15 to prevent braking in reverse.

Before installing an add-on system, read the specifications carefully, and talk to the customer about the weight and number of axles on the trailer. On multiple-axle trailers, braking components may be installed at each wheel or only on one axle. The total weight of the loaded trailer, not just the payload of one axle, must be considered when selecting the capacity of the braking system.

AUTHOR'S NOTE This short section on air brakes is to give an overview to the reader should he or she end up working for a fleet or having to provide some type of emergency assistance to an air brake–equipped vehicle. Remember that more in-depth training will be required to provide air brake maintenance and repair. Air brake systems must comply with DOT requirements and inspection procedures.

Air Brakes

Air brakes are brakes applied by using compressed air. Only used on heavy-duty trucks.

Air brakes require at least 100 psi to operate correctly. This pressure is provided by a belt-driven air compressor, and the compressed air is held in one or two air reservoirs (tanks). A governor mounted on the compressor limits the amount of pressure to about 125 psi. The reservoirs and the brakes are connected via steel tubing to a manifold valve (foot valve) usually mounted on the engine side of the bulkhead. A three-way valve directs the air dependent on the action of the driver.

The wheel brake friction components, drum or disc, are actually applied by a spring-operated diaphragm within a brake chamber at each wheel. The diaphragms are held off (brakes released) by air pressure on the brake side of the diaphragm. Slack adjusters are placed between the chamber pushrod and the S-cam in the wheel brake mechanism. Slack adjusters allow the operator or technician to adjust the brakes for wear and are one of the first items checked during a DOT inspection.

When the brakes are applied, some portion of the air pressure retaining the diaphragm is released, and the spring pushes the diaphragm, thereby moving the pushrod (**Figure 1-19**). The pushrod in turn rotates the S-cam and applies a proportional amount of movement to the brake shoes in relation to the amount of air released. A red button on the dash labeled PARK applies the parking brakes by releasing all of the air in all of the

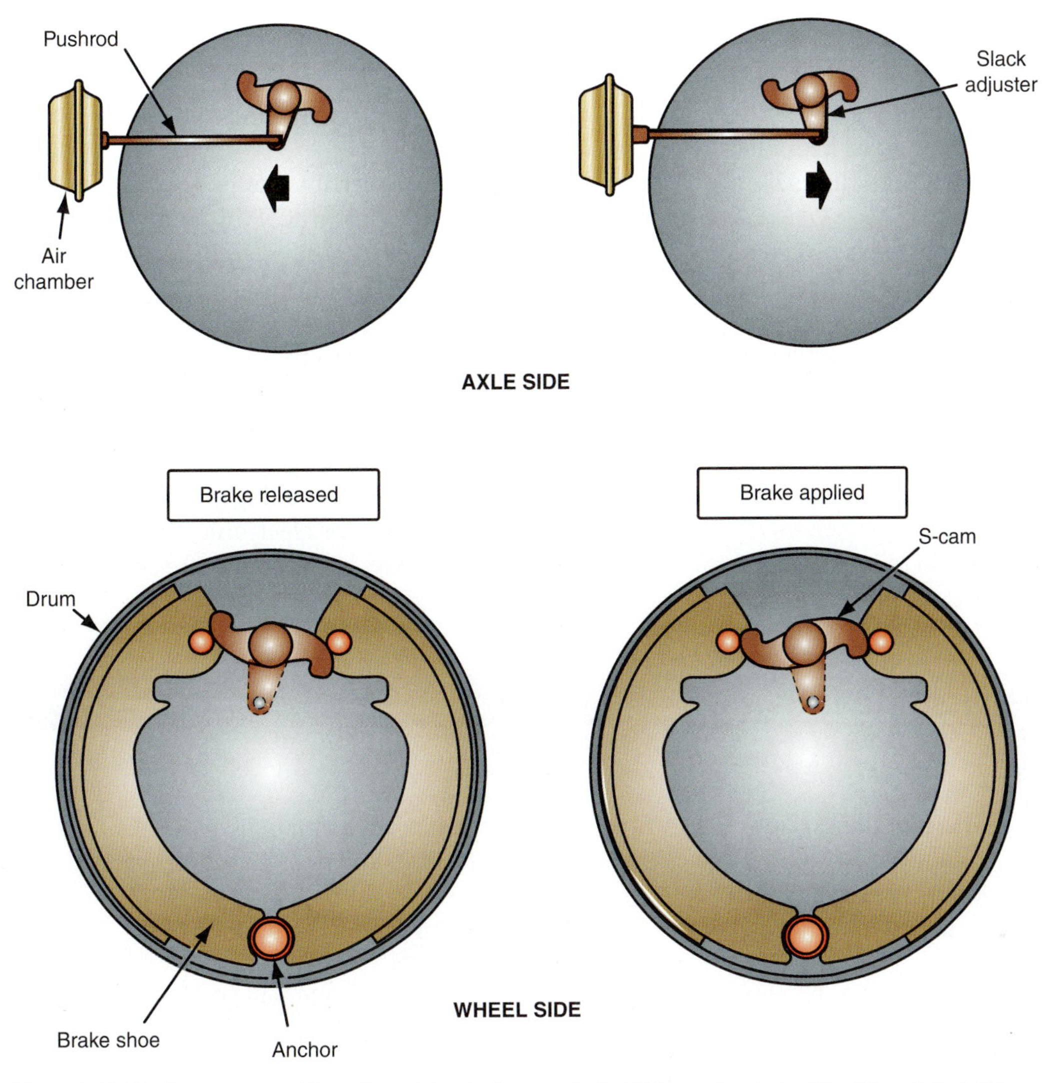

Figure 1-19 The S-cam is rotated by actions of the air chamber. As the "S" turns, the wings of the "S" push the brake shoe outward.

brake chambers. Anytime the vehicle is parked, the parking brakes should be engaged. On the steering column is a lever that allows the driver to apply only the trailer brakes from full lock to a moderated braking effect. The hissing noise commonly heard around air brake vehicles is from brakes being applied, parking brakes being applied, or a release of excess pressure from the air reservoirs.

When the tractor is attached to a trailer, two flexible air lines connect the tractor and trailer brakes. A tractor protection valve is located on the tractor to prevent a loss of brakes on the tractor if the trailer brake system develops a leak or becomes disconnected. The lines extending from the tractor are self-sealing while the connections on the trailer side are not. When the hoses are disconnected, the trailer brakes lock down (park mode).

Although this overview does not provide sufficient information to become an air brake technician, there is still enough information for the average automotive technician to make a quick safety inspection. With the system fully charged at 100 psi or more, shut down the engine and perform a walk-around inspection. Any hissing sound at this time usually denotes an air leak, and the leak can be traced by following the sound. At this point in your training, do not attempt a repair. Notify a trained heavy truck technician of your findings. If the air pressure does not reach 100 psi, there are several devices that can cause this problem, and repairs are best left to a properly trained technician.

SUMMARY

- An automotive brake system consists of a master cylinder and control valves connected hydraulically through lines to disc and drum brake units that stop the wheels.
- A hydraulic or vacuum power assist is used on most cars to decrease the braking effort required from the driver.
- A stroke sensor is used to tell the brake actuator how fast and how far the driver applied the brake pedal.
- A stroke simulator is used to give the driver a normal pedal feel.
- A mechanical brake, operated by levers and cables, is used for parking.
- All late model cars have an ABS to improve brake operation during emergency stopping.
- Active braking systems may function with or without driver input.
- Vehicle stability control systems share components with the ABS, TCS, and ARC to provide better vehicle control and comfort.
- Trailer brakes may now be controlled or regulated by components similar to those used in active braking systems.
- Air brake systems use compressed air to control the movement of a diaphragm.
- Regardless of the type of braking system, mechanical, hydraulic, or electronic, the final braking action occurs between the tire and the road.

REVIEW QUESTIONS

Short-Answer Essays

1. Name the four basic functions of the braking system.
2. Briefly describe the modern master cylinder.
3. Briefly describe why the split system is needed on modern vehicles.
4. Describe the purpose of the brake stroke sensor on a hybrid vehicle.
5. Describe the two types of split hydraulic systems.
6. Describe regenerative braking.
7. Describe the lines and hoses used on vehicle braking systems.
8. Why is it wrong to call the parking brake an "emergency brake."
9. Describe what is meant by negative wheel slip.
10. Explain how the traction control system works to control wheel spin.

Fill in the Blanks

1. ______________ braking systems use electrical generators to help slow the vehicle during gradual stops and help recharge the electric batteries.
2. The ______________ ______________ is mounted on a lever with a pivot near the top of the lever.
3. Modern automotive ______________ ______________ were developed from aircraft brakes of World War II.
4. Disc ______________ ______________ are mounted in a caliper that sits above the spinning rotor.
5. The ______________ ______________ is the start of the brake hydraulic system.
6. All master cylinders for vehicles built since 1967 have ______________ ______________ and pumping chambers
7. There are two types of split systems: ______________ and front/rear.
8. The electronic function of the proportioning valve has been called "______________ ______________ ______________" and "______________ ______________ ______________" by various manufacturers.
9. Most systems with drum brakes have a single, ______________ cylinder at each wheel.
10. The ABS does the same thing as an ______________ ______________ would, only it does it faster and more precisely.

Multiple Choice

1. Technician A says all the friction components of a disc brake are exposed to the airstream. Technician B says this helps to cool the brake parts and maintain braking effectiveness during repeated hard stops from high speeds. Who is correct?

   A. A only
   B. B only
   C. Both
   D. Neither

2. Technician A says that with a fixed caliper, hydraulic pressure is applied to pistons on both sides to force the pads against the rotor. Technician B says that with a movable caliper, pressure is applied to a piston on the outboard side only. Who is correct?

   A. A only
   B. B only
   C. Both
   D. Neither

3. Technician A says drum brakes do not develop the mechanical servo action that you will learn about as you study drum brakes. Technician B says drum brakes require higher hydraulic pressure and greater force to achieve the same stopping power as a comparable disc brake. Who is correct?

   A. A only
   B. B only
   C. Both
   D. Neither

4. While discussing the brake stroke sensor, Technician A says the brake stroke sensor informs the brake controller how fast the brake pedal is being pushed down. Technician B says the stroke sensor also informs the controller how much pressure the driver is applying to the brakes.

   A. A only
   B. B only
   C. Both
   D. Neither

5. Technician A says the stroke simulator is used to make hybrid brake application feel normal to the driver. Technician B says the stroke simulator is an electrical switch for the brake lamps. Who is correct?

   A. A only
   B. B only
   C. Both
   D. Neither

6. Technician A says that all new vehicles have metering and proportioning valves. Technician B says that the function of these valves has been taken over by the ABS system on most vehicles. Who is correct?

   A. A only
   B. B only
   C. Both
   D. Neither

7. While discussing wheel cylinders and caliper pistons, Technician A says wheel cylinders of drum brakes operate in response to the master cylinder. Technician B says the caliper pistons of disc brakes operate in response to the wheel cylinders. Who is correct?

   A. A only
   B. B only
   C. Both
   D. Neither

8. Technician A says almost all brake systems have a power booster that increases the force of the driver's foot on the pedal. Technician B says most cars and light trucks use a vacuum booster that uses the combined effects of engine vacuum and atmospheric pressure to increase pedal force. Who is correct?

   A. A only
   B. B only
   C. Both
   D. Neither

9. Technician A says some vehicles without a ready vacuum source may have a hydraulic power booster. Technician B says the hydraulic booster can be supplied with fluid by the power steering system or may be a part of the brake system and driven by an electric motor. Who is correct?

   A. A only
   B. B only
   C. Both
   D. Neither

10. Technician A says automatic ride control (ARC) is an electronic braking system designed to provide a more comfortable ride. Technician B says ARC also allows betters control of the vehicle during cornering. Who is correct?

    A. A only
    B. B only
    C. Both
    D. Neither

CHAPTER 2

PRINCIPLES AND THEORIES OF OPERATION

Upon completion and review of this chapter, you should be able to:

- Discuss the conversion of energy from one type to another.
- Discuss braking dynamics.
- Explain the importance of kinetic and static frictions in brake system.
- Describe the effects of pressure, surface area, and friction material on producing friction during braking.
- Discuss coefficient of friction.
- Explain why heat dissipation is important in a braking system.
- Explain how work is accomplished.
- Explain how the laws of motion affect the design and operation of a vehicle.
- Discuss hydraulic principles and how they may be applied in a vehicle.
- Explain the method for calculating mechanical advantage.
- Discuss how hydraulics can be used to transmit force.
- Discuss how a hydraulic brake system can produce mechanical advantage.
- Define and explain the basic electrical terms: amperes, voltage, and resistance.
- Explain how to use Ohm's law to calculate the amount of resistance, voltage, and current in a circuit.

Terms To Know

Adsorption
Ampere (A)
Brake fade
Ceramic
Coefficient of friction
Force
Gas fade
Inertia
Kinetic energy
Kinetic friction
Lining fade
Mass
Mechanical advantage
Mechanical fade
Momentum
Ohms (Ω)
Ohm's law
Perpetual energy
Rolling resistance
Static friction
Tensile force
Thermal energy
Vacuum
Voltage (volt)
Weight

INTRODUCTION

No vehicle or any other device will function correctly if the laws of physics are not considered during design and manufacturing. On the other hand, there is no current technology to make *all* designs meet *all* factors of physics. The automotive industry has performed well in making compromises between different physics laws to design and manufacture vehicles that work and are affordable. In this chapter, various laws of physics are discussed so that the reader will have a basic understanding of the physical environment that the designers, the manufacturers, the driver, and even you must overcome to operate any type of vehicle.

BRAKE OPERATION/CONVENTIONAL SYSTEM

Before discussing the energy and dynamics associated with braking, a review of brake system operation is essential. This may help when trying to understand the next sections.

When the driver decides to slow or stop the vehicle, the first action is to remove force from the accelerator pedal. The slowing engine, in most cases, begins to slow the vehicle. The driver then applies force to the brake pedal.

The movement of the brake pedal activates the brake power booster, which, in turn, boosts and transmits the driver's input force to the master cylinder pistons via pushrods (**Figure 2-1**). The master cylinder pistons turn mechanical action into hydraulic action by pressurizing the brake fluid and forcing it out through the hoses and lines to the wheel braking components (see Chapter 5). Older vehicles use proportioning valves and metering valves. Valves between the master cylinder and wheels control the pressure and volume of brake fluid reaching the wheels. These valves do not control actual braking force, but they mainly divide the force between the front and rear brakes for a smooth stop without allowing the rear wheels to lock up on hard activation. At the wheels, hydraulic pressure is converted into mechanical action and applies the last segment of the braking mechanism. In drum brakes, this action is accomplished by the wheel cylinders, brakes shoes, and brake drums. On disc brakes, the same action is performed by the calipers, brake pads, and brake rotors. The antilock brake system (ABS) becomes involved only if it senses one (or more) wheels about to lock up. The ABS can then modulate hydraulic pressure to that wheel(s) to reduce braking effects. Most late-model vehicles use the ABS to replace proportioning valves with hydraulic modulation of the brake pressure to prevent wheel lockup. Everything told, the braking system is a straightforward and simple concept and operation, but it is one of the more critical systems on a vehicle.

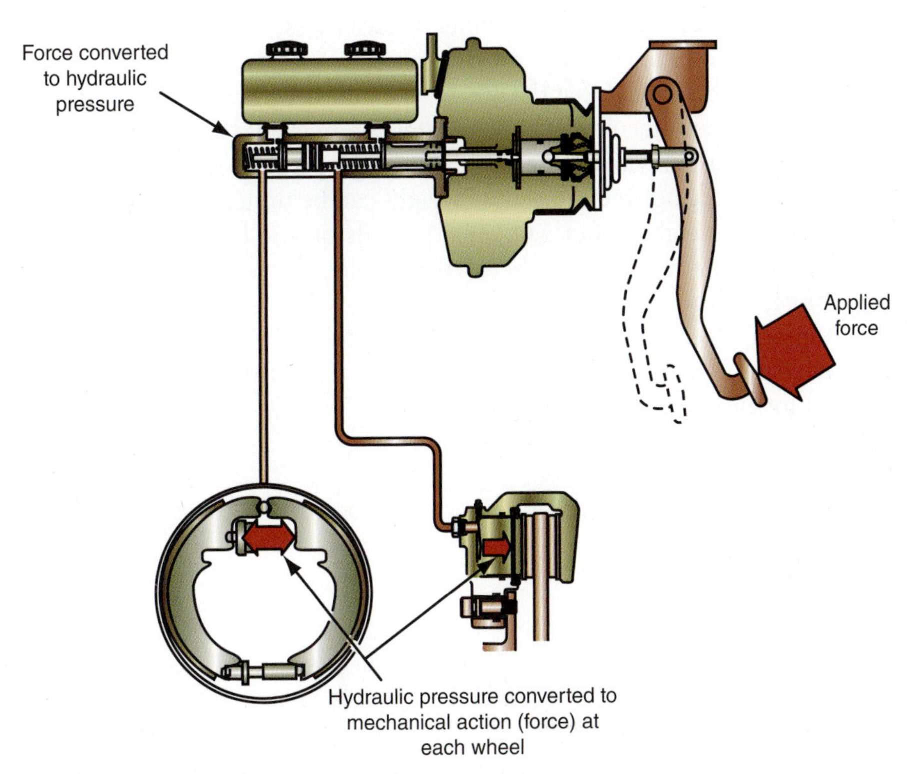

Figure 2-1 The energy exerted to brake a vehicle is converted from mechanical to hydraulic and back to mechanical.

BRAKE SYSTEM ENERGY

All brake systems work according to a few principles or "laws" of physics, and the concept of energy is a basic part of physical science. Energy is the ability to do work and comes in many familiar forms: chemical energy, mechanical energy, heat energy, and electrical energy are among the most obvious forms in all automotive systems.

Kinetic energy is the energy of mechanical work or motion.

A brake system converts one form of physical energy to another. To slow and stop a moving vehicle, the brakes change the **kinetic energy** of motion to heat energy through the application of friction. When the brakes change one form of energy to another, they are doing work. Work is the result of releasing or using energy.

AUTHOR'S NOTE It is impossible at this time to create or destroy energy. However, it can be converted from one form to another. The master cylinder is one place this happens: the mechanical energy of the brake pedal is converted into hydraulic energy in the master cylinder bore. It is later converted back to mechanical energy at the wheels.

Kinetic Energy, Mass, Weight, and Speed

Kinetic energy is the energy of mechanical work or motion. When an automobile starts, accelerates, decelerates, and stops, kinetic energy is at work. The amount of kinetic energy at work at any moment is determined by a vehicle's mass (weight), speed, and the rate at which speed is changing.

Mass is the measure of the inertia of an object or form of matter or its resistance to acceleration; it also is the molecular density of an object.

The terms "mass" and "weight" can be used interchangeably to describe objects on the surface of the Earth, but the two terms are not technically the same. **Mass** is a measurement of the number of molecules that make up an object. **Weight** is a measurement of the effect of gravity on that mass. All objects have mass, from a steel brake shoe to a quart of hydraulic fluid to the air in an air compressor. Without going too deeply into the science of physics, it can be said that the greater the number of molecules in an object and the more complex the molecules are, the greater the mass of that object and the more dense it is. The effect of gravity on the mass of an object is that object's weight.

The basic difference between mass and weight can be understood by thinking of the space shuttle, which weighs about 1,000,000 pounds on the launch pad, on the Earth. When the shuttle is in orbit, outside the Earth's gravity, it is weightless (**Figure 2-2**). Its mass stays the same, however.

The combined effects of weight and speed constitute kinetic energy, but speed has a much greater effect than weight. The kinetic energy of any moving object can be calculated with this formula, which is quite simple:

$$\frac{mv^2}{29.9} = Ek$$

where

$$m = \text{mass (weight) in pounds}$$
$$v = \text{velocity (speed) in miles per hour}$$
$$Ek = \text{kinetic energy in foot-pounds}$$

Consider two cars, both traveling at 30 miles per hour (mph). One weighs 2,000 pounds; the other weighs 4,000 pounds (**Figure 2-3**).

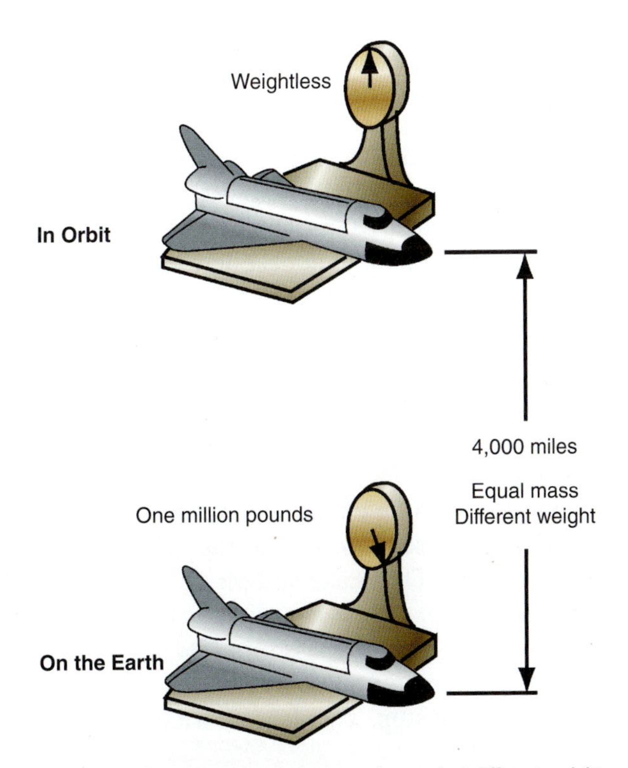

Figure 2-2 The space shuttle has equal mass but different weights on the Earth and in orbit.

$$\frac{2{,}000 \times 30^2}{29.9} = 60{,}200 \text{ foot - pounds of kinetic energy for the lighter car}$$

$$\frac{4{,}000 \times 30^2}{29.9} = 120{,}400 \text{ foot - pounds of kinetic energy for the heavier car}$$

Doubling the car weight doubles the kinetic energy when the speeds are equal. Therefore, kinetic energy increases and decreases proportionally with weight. Now let us accelerate the lighter car to 60 mph (**Figure 2-4**).

$$\frac{2{,}000 \times 60^2}{29.9} = 240{,}802 \text{ foot - pounds of kinetic energy for the lighter car}$$

When the speed is doubled, kinetic energy increases not two times, but four times. Kinetic energy increases as the square of the speed. If we accelerate the same 2,000-pound

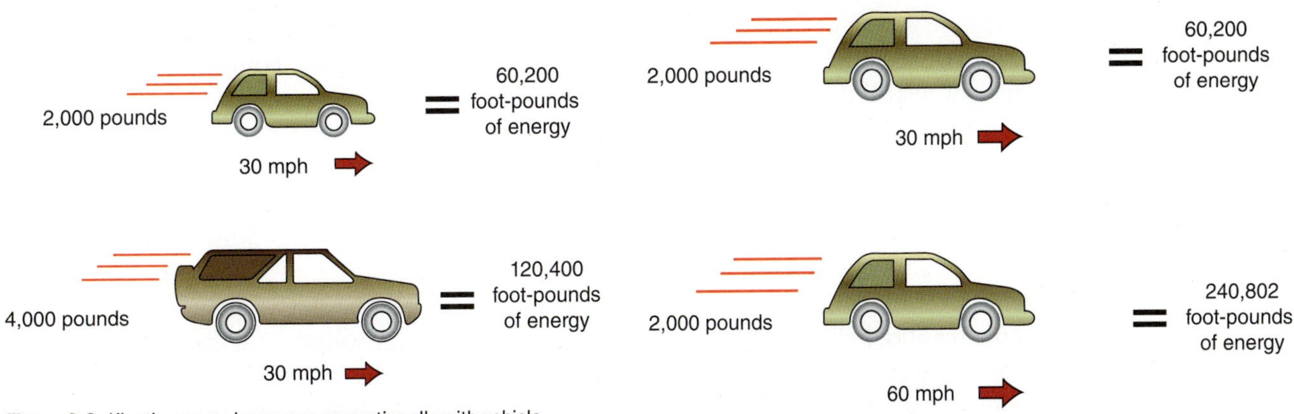

Figure 2-3 Kinetic energy increases proportionally with vehicle weight.

Figure 2-4 Kinetic energy increases exponentially with vehicle speed.

car to 120 mph, its kinetic energy is 16 times greater than it was at 30 mph—almost 1,000,000 foot-pounds.

Remember the brake performance test described in Chapter 1 of the *Shop Manual*, in which the required stopping distance at 30 mph was 57 feet, but increased to 216 feet at 60 mph. That is a practical example of kinetic energy at work. The effects of kinetic energy are also why a high-performance car needs a brake system with much greater stopping power than an economy car with only modest performance.

Inertia and Momentum

Inertia is the tendency of an object in motion to keep moving and the tendency of an object at rest to remain at rest.

When a car is accelerated and then decelerated and brought to a stop, two forms of **inertia** are in play. Inertia is simply the resistance to a change in motion. An object at rest tends to remain at rest; an object in motion tends to remain in motion. In both cases, the object resists any change in its motion.

Momentum is the force of continuing motion; the momentum of a moving object equals its mass times its speed.

Static inertia is the inertia of an object at rest; dynamic inertia is the inertia of an object in motion. When the brake system slows and stops a vehicle, it overcomes dynamic inertia and imposes static inertia. The brake system also must overcome the vehicle's momentum. **Momentum** is another way to view kinetic energy at work because it, too, is the mathematical product of an object's mass times its speed. Physical force starts an object in motion and gives it momentum. Another kind of force must overcome the momentum to bring the object to a stop. That force is friction.

BRAKING DYNAMICS

One important brake system dynamic is called weight transfer. You have probably had the experience of applying the brakes hard for an emergency stop. During such a stop, the front of the car lowered and you could feel yourself being thrown forward against your seat belt. This is caused by the vehicle weight being transferred from the rear to the front during braking, which means that more of the braking must be done by the front wheels and less by the rear wheels.

AUTHOR'S NOTE When the driver applies the brakes, the vehicle—not the driver or the passengers—is being slowed. Without seat belts and air bags, the vehicle occupants tend to keep moving in the same direction and at the same speed of the vehicle as before the brakes were applied. Hence, the use of safety restraints is required to reduce injuries.

The weight of a car is not distributed evenly on all four wheels even when the vehicle is standing still. The position of the heavy engine and powertrain components determines weight distribution. During braking, the weight of these components transfers forward. On a rear-wheel-drive car, about 70 percent of the weight shifts to the front. On a front-wheel-drive car, as much as 90 percent of the car weight is shifted to the front during braking. The vehicle's momentum and weight combine to cause the rear wheels to lift (have less down force applied) and the front wheels to be forced down (**Figure 2-5**). Therefore, front brake systems are larger than rear units.

Another dynamic to think about is braking power. As the previous section of this chapter about kinetic energy explained, the more the car weighs and the faster it is moving, the greater the braking power must be to stop the car. The brake system is designed with more power than the engine. Consider a typical car with a 200-hp engine. This car can accelerate from 0 to 60 mph in 10 seconds. The brake system must be able to

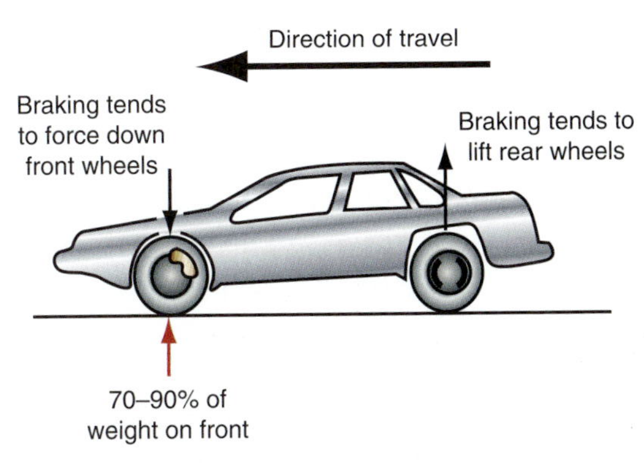

Figure 2-5 Weight transfer during braking can place as much as 90 percent of the vehicle weight on the front brakes.

decelerate the car from 60 to 0 mph in nearly one-fifth that time. This means that the comparable power to stop this car is about 1,000 hp.

This chapter focuses on the friction between the brake linings and the brake rotor or drum. However, remember that the vehicle is actually stopped by the friction between the tire and the road. Large, effective brake assemblies will stop the wheels from rotating, but the car will stop only if the tires maintain traction with the road.

When the brake system completely stops the tire so that it can no longer maintain traction with the road, the tire locks up or skids. A car with locked brakes takes much longer to stop because hot molten rubber can form between the tire and the road, causing a lubricating effect. Steering control can be lost, sending the car out of control. Even if the car stops before hitting something, lockup can damage the tires by causing flat spots on them.

FRICTION PRINCIPLES

Service brakes use friction to stop the car. The parking brakes use friction to hold the car stationary. Friction is used in the wheel brake units to stop the wheels. The friction between the tires and the road stops the car. Because friction is so important to the brake system, it is necessary to understand some of its principles.

Kinetic and Static Frictions

Two basic types of friction are at work in the brake system (**Figure 2-6**). The first is called kinetic, or moving, friction. The second is called static, or stationary, friction.

When the brakes are applied on a moving car, the frictional parts (brake shoes or pads) are forced against the rotating parts of the car (brake rotors or drums). The friction causes

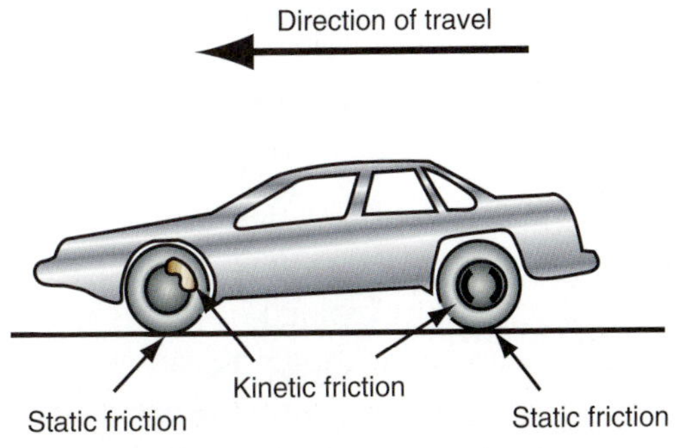

Figure 2-6 Both static friction and kinetic friction are at work during braking.

Kinetic friction is the friction between two moving objects or between one moving object and a stationary surface. It is used to stop, control, or reduce kinetic energy.

Thermal energy is the energy of heat.

Static friction is friction between two stationary objects or surfaces.

Rolling resistance is a combination of inertia of the mass to moving, including the friction between all moving parts of the vehicle and the outside environment.

the rotating parts to slow down and stop. Just as the energy of the rotating parts is called kinetic energy, the friction used to stop them is called **kinetic friction**.

Kinetic friction changes kinetic energy into **thermal energy** (heat). For a simple example of kinetic friction changing kinetic energy to heat, rub the hands together. Rub them together fast enough and the palms will feel warm.

Brake parts and tires get very hot during braking because of kinetic friction. For example, the temperature of the brake friction material in a typical car going 60 mph (95 km/h) can be more than 450°F (230°C) during an emergency stop.

Static friction holds the car in place when it is stopped. The friction between the applied brake components and between the tire and the road resists movement. To move the car, the brake components must be released. Then, the power of the engine must be great enough to overcome the **rolling resistance** between the tires and the road. The car can then begin to move.

Friction and Pressure

One important factor in the amount of friction developed in a braking system is the amount of pressure used to force the friction material against the rotating brake part. The more pressure applied to two frictional surfaces, the harder they grip each other and the harder they resist any movement between them.

Figure 2-7 shows the basic frictional parts of a disc brake system. The rotor is the part connected to and rotating with the car wheel. The friction pads are forced against the frictional surfaces or sides of the rotor. The friction causes the rotor to slow and stop, which stops the wheel from turning.

Braking force in the brake system is achieved by using mechanical leverage, hydraulic pressure, and different kinds of power-assist systems. The more force used to push the friction pads against the rotor, the more friction developed. The more friction developed, the more braking action—and heat—achieved.

Friction and Surface Area

Another important factor affecting the amount of friction produced is the amount of frictional surface being contacted. Two hands will stop a revolving shaft faster than one hand. The larger of the two brake units shown in **Figure 2-8** has more frictional contact area than the smaller unit. The larger unit will stop a car faster than the smaller unit, which is why large cars and trucks require larger brake components than do small cars.

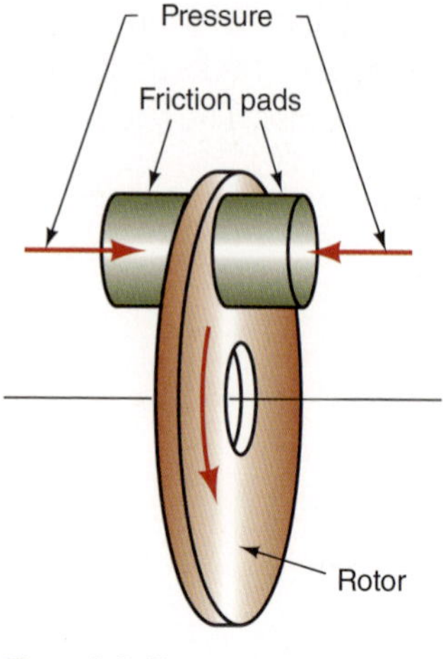

Figure 2-7 The greater the pressure against the friction pads, the greater the braking force.

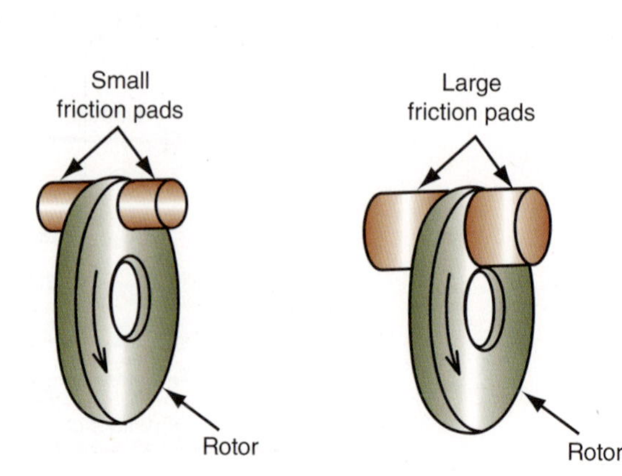

Figure 2-8 The larger the frictional contact area, the greater the braking force.

Coefficient of Friction

It takes more force to move some materials over a surface than others, even though the applied pressure and the amount of surface in contact are the same. Different materials have different frictional characteristics or coefficients of friction. It is easier to slide an ice cube over a surface than it is a piece of sandpaper.

To calculate the **coefficient of friction**, measure the force required to slide an object over a surface and then dividing it by the weight of the object. The moving force that slides an object over a surface is **tensile force**. The weight of the object is the force pushing the object against the surface.

Coefficient of friction is a numerical value that expresses the amount of friction between two objects, calculated by dividing tensile force (motion) by weight force; it can be either static or kinetic.

As an example, assume it takes about 100 pounds of force to slide a 100-pound block of iron over a concrete floor, but only about 2 pounds of force to slide a 100-pound block of ice over the same surface (**Figure 2-9**). When the amount of force is divided by the weight of the object, we find that the coefficient of friction for the metal block is 1.0, whereas the coefficient for the ice block is 0.02.

$$\frac{\text{Tensile force (motion)}}{\text{Weight force of object}} = \text{coefficient of friction}$$

$$\frac{\text{100 pounds of tensile force}}{\text{100-pound iron block}} = \text{coefficient of friction} : 1.0$$

$$\frac{\text{2 pounds of tensile force}}{\text{100-pound ice block}} = \text{coefficient of friction} : 0.02$$

Tensile force is the moving force that slides or pulls an object over a surface.

The results of these calculations provide the coefficient of friction between two materials. The coefficient of friction between brake friction components is always less than 1.0. A coefficient of friction greater than 1.0 means that material actually is transferred from one surface to another. Although brake friction surfaces wear, material is not transferred from pads to rotors or from shoe linings to drums. The coefficient of friction between a tire and the road can exceed 1.0, however. When material transfers from the tire to the road, we see a skid mark. That means that the coefficient of friction momentarily exceeded 1.0.

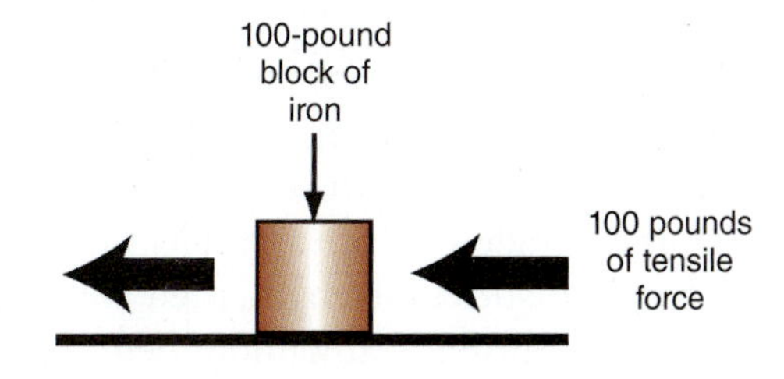

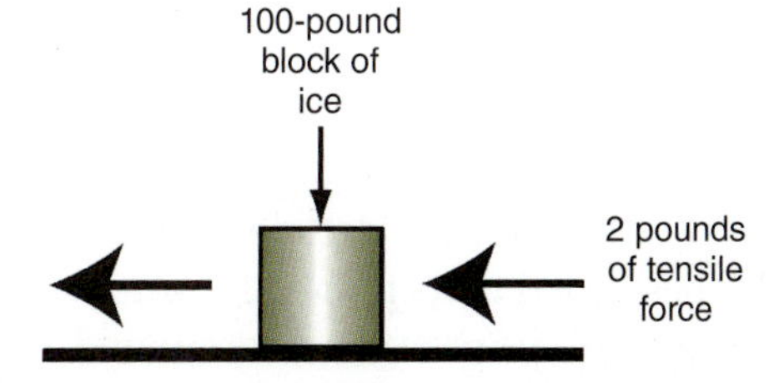

Figure 2-9 A high coefficient of friction requires greater force to move one surface against another.

Designing or selecting brake materials in relation to the coefficient of friction is not a simple matter of picking the highest number available. If the coefficient of friction of a material is too high, the brakes may grab and cause the wheels to lock up and slide. If the coefficient of friction is too low, excessive pressure on the brake pedal would be required to stop the car.

Automotive engineers carefully determine the required coefficient of friction for the best braking performance on a particular vehicle. The friction materials selected for brake pads and rotors or brake shoes and drums are designed to give the best cold and hot temperature performance. The best materials have a coefficient that stays within narrow limits over a wide range of temperatures. If the selected material increases or decreases its coefficient of friction as temperature changes, then the brakes can fade, grab, or work erratically. This illustrates an important reason to use approved replacement brake parts so that the brake performance will meet the system design requirements.

Three factors affect the coefficient of friction in a brake system:

1. The surface finish of both friction surfaces
2. The material: the metal of the rotors and drums and the friction material of the pads and linings
3. Temperature

Surface Finish. The 100-pound metal block previously used to calculate a basic coefficient of friction can be used to show how friction can be easily changed. By polishing the metal surface and waxing the concrete floor, the required tensile force to move the block may be changed from 100 pounds to 50 pounds. The metal block still weighs 100 pounds, but the coefficient of friction is now:

$$\frac{\text{50 pounds of tensile force}}{\text{100-pound metal block}} = \text{0.5 coefficient of friction}$$

In a brake system, the surface finishes of pads, rotors, linings, and drums must be a trade-off between smoothness for long life and roughness for good stopping power. Friction surfaces as rough as coarse sandpaper could provide excellent stopping action—for just a few brake applications before wear becomes excessive.

Materials. The kinds of materials being rubbed together also have a big effect on the coefficient of friction, as the earlier comparison of a 100-pound block of metal with a 100-pound block of ice shows. Just as it is with selecting a surface finish, selecting the kinds of brake friction materials is a trade-off. Iron and steel work best as rotors and drums because of their combined characteristics of strength, their ability to hold a surface finish for a long time, and their ability to transfer heat without being harmed by high temperatures. One of the newer materials used for brake rotors is carbon ceramic. Carbon ceramic rotors have low weight and good wear characteristics.

Friction materials for pads and linings can be allowed to wear but must have a reasonable lifespan. If pads and linings are soft and wear fast, the vehicle may have very good braking ability, but friction material life will be short. However, if friction material is hard and wears slowly for long life, it will create a low coefficient of friction and poor stopping ability.

Asbestos was used almost universally for decades as the basis of brake friction material. The health hazards of asbestos led to its removal from most pads and linings. Today, most friction materials are semi-metallic or organic compounds. All of them have different coefficients of friction, and engineers select material that best matches the intended use of a vehicle and its braking requirements.

Temperature. Temperature also affects the coefficient of friction, but it affects different materials in different ways. A moderate amount of heat actually increases the coefficient of friction of most brakes. The semi-metallic and carbon fiber materials of some racing brake linings must be heated quite a bit for maximum efficiency.

Too much heat, however, reduces the coefficient of friction (**Figure 2-10**), and as heat increases, the coefficient of friction continues to drop. This leads to brake fade, discussed later in this chapter, and braking efficiency is reduced.

Heat Dissipation. The weight and speed of the car during braking determine how much frictional energy is required to stop the vehicle. As mentioned previously, heat is a natural result of friction. The more frictional energy needed to stop the car, the greater the amount of heat generated during braking.

The heat caused by friction must be removed or it will damage the brake system. Heat is removed from the friction surfaces as it passes through the friction material and metal of the brake components into the surrounding air. This heat removal is called heat dissipation.

Brake Fade

The buildup of heat from continuous braking can cause the friction material to become glazed or polished and the metal surface of the rotors or drums to become hardened. As a result, the driver will need to push harder on the brake pedal to stop the car.

Brake drums and rotors must absorb heat faster than they can dissipate it to the surrounding air. Drum or rotor temperature can rise 100°F (55°C) in just a couple of seconds during a hard stop. Dissipating that same heat to the air can take 30 seconds or more. Repeated hard stops in a short period of time can overheat brake components severely and reduce braking effectiveness. This is called **brake fade**.

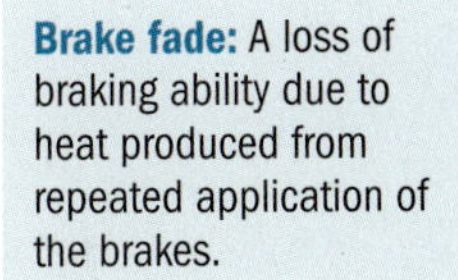
Brake fade: A loss of braking ability due to heat produced from repeated application of the brakes.

A brake system is designed to provide the best possible heat dissipation. Parts are often vented to allow maximum airflow around the hot surfaces. The sizes of the friction surfaces are carefully designed with heat dissipation in mind. The larger the frictional contact area of the brake system, the better the heat dissipation and its resistance to brake fade. Severe heat causes three kinds of brake fade:

Lining fade occurs when the brake linings are hot enough that the coefficient of friction changes.

1. **Lining fade**. Lining fade occurs when the pad or lining material overheats to a point at which its coefficient of friction declines severely. Pressing harder on the brake pedal may further overheat the lining and increase—not decrease—the fade. Lining fade affects both disc and drum brakes but is usually less severe with discs. The rotor friction surfaces are exposed directly to cooling airflow, whereas drum surfaces are not. Ventilated and slotted rotors also aid the cooling process and counteract lining fade.

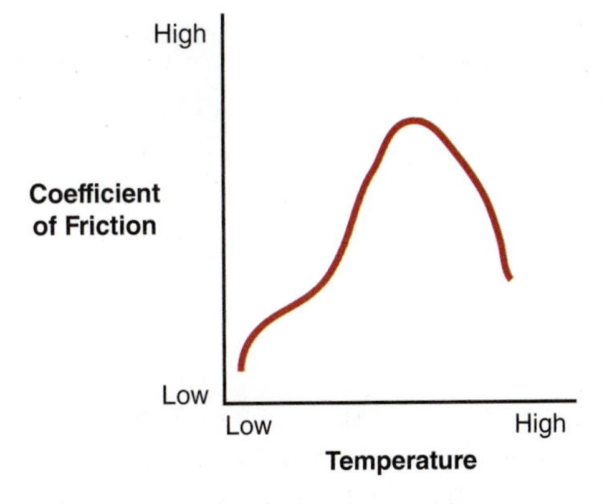

Figure 2-10 Initially, the coefficient of friction increases with heat, but very high temperatures cause it to drop off.

Mechanical fade: Brake fade resulting from the expansion of the brade drum.

2. **Mechanical fade**. Mechanical fade occurs with drum brakes when the drums overheat and expand away from the shoes. More pedal travel is then needed to move the shoes farther and let the linings contact the drum. Extreme mechanical fade can cause the pedal to travel to the floor before braking action starts. Pumping the brakes rapidly sometimes can force more fluid into the lines and move the shoes farther. Large and heavy drums and drums with cooling fins help to dissipate heat and reduce mechanical fade. Mechanical fade is not a factor in disc brake operation.

Gas fade results from adsorption, which is the condensation of a gas on the surface of a solid.

3. **Gas fade**. Gas fade is a rare condition that occurs under very hard or prolonged braking. A thin layer of hot gas develops between the pads or linings and the rotors or drums. The thin layer of gas—although only a few molecules thick—acts as a lubricant and an insulator and reduces friction. Gas fade is a greater problem with friction materials that have large surface areas that trap the gas. Slots cut in pads and linings provide airflow that aids cooling and a route for hot gases to escape. Gas fade is one of the reasons there are signs at the top of some hills requesting truck drivers to select a lower gear before heading down.

Brake Friction Materials

The friction material for a disc brake is bonded, riveted, or molded to a metal brake pad (**Figure 2-11**). The friction material for a drum brake is bonded or riveted to a metal brake shoe (**Figure 2-12**). Some brake pad manufacturers are also employing the use of hooks to hold the brake material onto the metal backing. Several different friction materials are

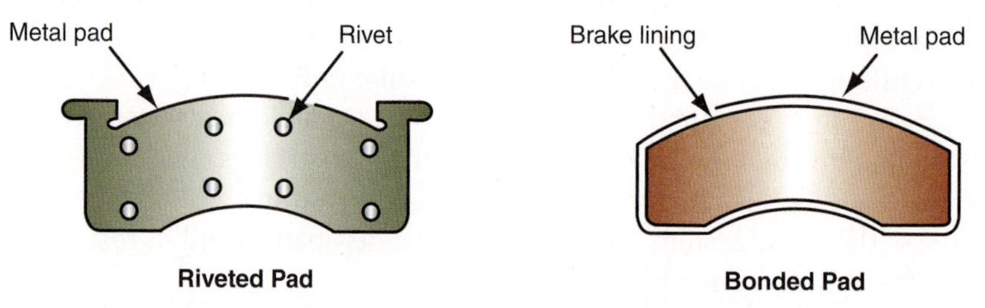

Figure 2-11 The friction material of a disc brake pad is bonded or riveted to the metal pad.

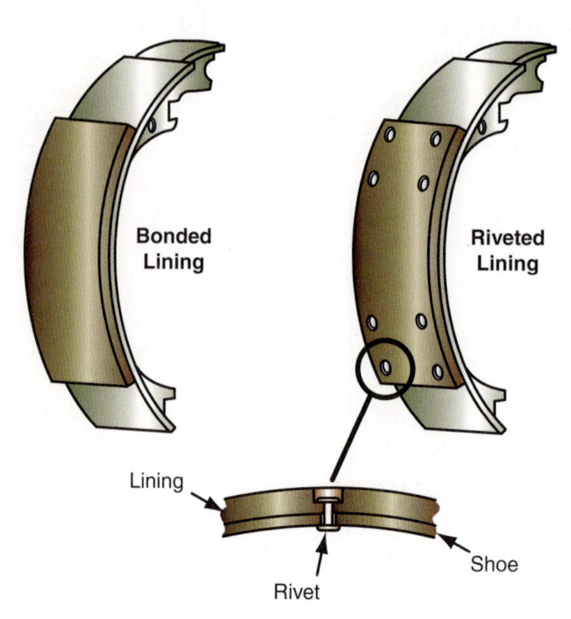

Figure 2-12 Drum brake lining also can be bonded or riveted to the metal shoe.

used for brake pads and shoes. Each of these has different characteristics. Five of the most important characteristics are:

1. The ability to resist fading when the brake system temperature increases
2. The ability to resist fading when the parts get wet
3. The ability to recover quickly from heat or water fading
4. The ability to wear gradually without causing wear to brake rotors or drum surfaces
5. The ability to provide a quiet, smooth frictional contact with a rotor or drum

Asbestos was the most common brake lining friction material for many years. Asbestos has very good friction qualities and provides long wear but its health hazards caused its removal from most brake friction materials. The most common types of non-asbestos lining materials can be divided into four general categories:

1. *Nonmetallic materials* are made from synthetic fibers bonded together to form a composite lining.
2. *Semi-metallic materials* are made from a mix of organic or synthetic fibers and metals molded together. Semi-metallic linings are harder and more fade-resistant than nonmetallic materials but require higher brake pedal effort. They are often blamed for wear to the rotors and drums.
3. *Fully metallic materials* have been used for many years in racing. The lining is made from powdered metal that is formed by heat and pressure in a process called sintering. These materials provide the best resistance to brake fade but require high brake pedal pressure and create the most wear on rotors and drums.
4. **Ceramic** pads are based on clay or porcelain material blended with copper fibers. They provide suitable life, provide good stopping, and reduce chances of fading, but they retain heat. This heat can distort or damage other adjacent brake components.

Chapter 7 in this *Classroom Manual* provides more detailed information on friction materials used in disc and drum brake linings.

ENERGY AND WORK

No matter how much a person may enjoy the task being performed, work is being done. Work is the expenditure of energy. Energy, in a loose definition, is the fuel for work. Energy cannot be created nor can it be destroyed. However, and this is the key, energy can be converted or changed from one form to another.

Many times, the phrase "that plant produces or makes electrical energy" is heard. Not really. The process is much simpler than making electrical energy, a task that no one can do at this time. It is actually using the energy of falling water or heat to operate a mechanical device rotating within a magnetic field and converting that input energy into electrical energy. Although the mechanics and engineering required to perform that process are complex, the actual concept is very simple.

Let's move a little closer to home. The human body uses a lot of energy just to function without seemingly doing any work. The heart, lungs, and other organs work all the time. The body changes the chemical and heat energy of food into electrical energy for the brain and muscle control that, in turn, allows most organs to perform some type of chemical, electrical, or mechanical function. When the energy supply is low, the organs may become exhausted or unable to work. It is feeding time, and the process continues. The same is true of vehicles.

The most common vehicle fuel of today is gasoline derived from crude oil. Crude oil is a by-product of all types of ancient animals and plants that died, decayed, and then heated under enormous pressure. This oil is recovered by the petroleum industry and eventually distilled into fuel for a vehicle. The original energy from the plants and animals has been converted into a chemical energy contained in the fuel. That chemical energy is compressed and burned in the engine and is converted into radiant heat and expanding gases. The gases push downward on a piston, and the last change in this process is the piston's mechanical action or energy. Run out of gas, and everything stops.

Hybrid-powered vehicles currently use an electric motor to drive the vehicle that is touted to save gasoline. But the electric power comes from a battery that must be recharged at intervals. The recharging can be accomplished using an on-board generator powered by the gasoline engine or by plugging into the electric grids where the electricity is produced by a power plant. Either way, some other source of power or energy is needed to achieve electric drive. This leads to a search for a never-ending energy source.

A BIT OF HISTORY

The first friction materials used in brake systems were wood, leather, and camel hair. These materials would burn or char easily, causing erratic brake action. In 1908, the Raymond Company of Bridgeport, Connecticut, developed an asbestos and copper wire mesh brake lining. This material worked much better than the other materials because it did not char and provided a smooth and grab-free pedal effort. The company later became known as the Raybestos Corporation.

Perpetual energy is everlasting, continuous energy.

There is one major drawback to all energy-conversion devices used today. It is a fact that a machine or device cannot produce more energy than it consumes, or, in other words, there is no **perpetual energy** device today. An example of a device that seems to contradict that statement is the alternating current generator or alternator on a vehicle. The magnetic fields in an alternator are both energized or charged by electricity from the vehicle battery or from the output current flow of the alternator. Typically, a fully charged battery has about 12.6 volts of direct current, 13.7 volts to 14.7 volts with the engine operating. However, if the alternator's regulator fails and goes to "full field," it is possible for the alternator to produce up to 20 volts. This seems to prove that the alternator is a perpetual-energy device until the engine power needed to drive the alternator is considered. The result is a large use of energy to produce a mere 20 volts of electrical current.

The preceding text is a simplistic overview of energy, but it shows that energy is neither destroyed nor created. It is just changed, and the changes can be used to do work. So, whether engaging in a fun-filled activity or working the graveyard shift at the local factory, energy is being used and work is being performed. In the next sections, the discussion highlights how various physics or theories of operation must be considered and how the engineers may use those theories to accomplish work with the amount of energy available at any given time.

NEWTON'S LAWS OF MOTION

Inertia is the tendency of an object to continue doing what it is already doing.

One of the biggest uses of energy in transportation is the energy to overcome the laws of motion. Every object on the Earth must deal with these basic theories of motion. Newton developed several basic laws about motion that apply directly to the automobile. The first one deals with inertia.

This unchanging action applies to an object in motion or an object at rest. Without an outside force being applied to the object, its actions will not change. A stationary object will remain stationary, whereas a moving object will continue to move. And it will move in a straight line. Obviously, this would not work well with a vehicle. Although natural forces such as wind, gravity, and terrain will affect the vehicle, the driver may have to wait a while before the vehicle begins to move, and the driver still would not have control of it.

An engine and a driveline are used to power the vehicle, with steering and suspension providing some means of control. The brake system is used to overcome the inertia of

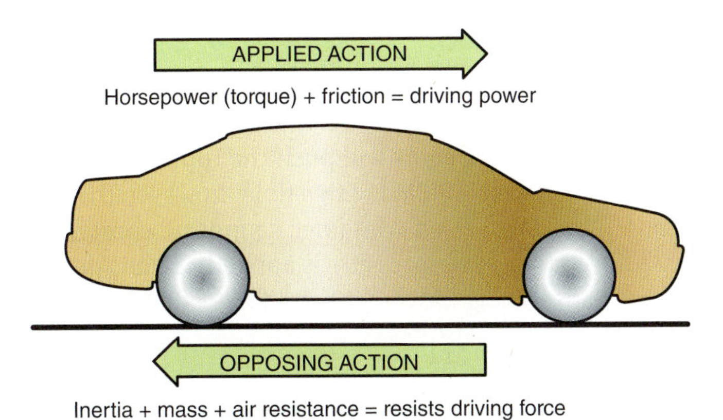

Figure 2-13 The force needed to move and accelerate a vehicle must be greater than the opposing force. When the two forces are equal, the vehicle cannot achieve more speed or torque.

motion and helps stop the vehicle. A typical vehicle uses large amounts of energy to move the vehicle from stationary and even more energy to stop it. The steering system also requires energy to move the vehicle out of a straight-line motion so it can follow the road. But there are other factors involved.

For every action, there is an opposite and equal reaction. This is generally the second law. The force applied to the drive wheels must create enough friction (see next paragraph) against the road to produce that equal, opposite reaction (**Figure 2-13**). A spinning tire wastes that reaction, and the vehicle does not move as it should. On the other end, the same wheels must grip the road so the applied braking force can provide the slowing action. The problem with stopping a vehicle is the speed at which it is traveling. Upon acceleration, the energy is used to move the mass of the vehicle. During braking, the speeding mass must be overcome and still retain driver control.

Inherent in this law is a rule regarding acceleration of an object. The acceleration of a vehicle is directly proportional to the driving forces applied to it. This applied force must overcome the friction of the various moving components, the tire against the road, and the total mass of the vehicle and its load. Mass is not the weight of the object but the amount of matter in that object. A vehicle weighing two tons on Earth will weigh about 670 pounds on the Moon, but its mass will be the same. When calculating the acceleration rate of a vehicle, other forces must be considered. As the drive wheels' force is applied against the road and begins to accelerate the vehicle, it is resisted by a force equal to the vehicle's mass multiplied by its acceleration rate. Aerodynamic forces act simultaneously against the vehicle in the opposite direction of the driving force, reducing the acceleration rate. In general, this reduces the amount of force available to keep the acceleration rate high. At some point, the driving forces equal the resisting forces, indicating that the vehicle has reached its maximum speed. Most light vehicles will accelerate quickly to a certain speed before the rate of acceleration drops off. This can be checked easily with a stopwatch and an open, clear stretch of road. Measure the time taken to accelerate from 0 to 25 mph. Record the time but continue running the clock until the speed reaches 50 mph. The acceleration time between 25 and 50 will be a little slower than the time between 0 and 25. The distance needed for that extra acceleration will also be a little greater.

HYDRAULIC PRINCIPLES

Automotive brake systems use the force of hydraulic pressure to apply the brakes. Because automotive brakes use hydraulic pressure, a study of some hydraulic principles used in brake systems is needed. These include the facts that fluids cannot be compressed; fluids can be used to transmit movement and force; and fluids can be used to increase force.

Fluids Cannot Be Compressed

Fluid resistance to compression allows it to transmit motion and pressure.

Hydraulic devices such as the brake system work because fluids, unlike gases, do not compress. If a container with a top on it is filled with a gas, as shown in **Figure 2-14**, and a weight is placed on the top, the weight will push the top down. The top moves down because the gas can be compressed. If filled with a fluid, however, the weight will not push the top down, because fluid in a sealed container cannot be compressed. If air is introduced into the hydraulic system when pressure is applied, the air (or gas) in the system is compressed instead of transmitting movement and force effectively (**Figure 2-15**).

Fluids Can Transmit Movement

Fluid trapped in a sealed system such as the brake system will act like a steel rod when force is applied to it. That force will be transmitted to every interior surface of the system equally. It is measured in pounds per square inch of pressure.

Fluids can be used to transmit movement. Two cylinders of the same diameter are filled with a fluid and connected by a pipe, as shown in **Figure 2-16**. If piston A is forced downward, fluid will push piston B upward. Because piston A starts the movement, it is called the apply piston. Piston B is called the output piston. If the apply piston moves 10 inches, the output piston also will move 10 inches (**Figure 2-17**).

The principle that motion can be transmitted by a liquid is used in hydraulic brake systems. A master cylinder piston is pushed when the driver applies the brakes. The

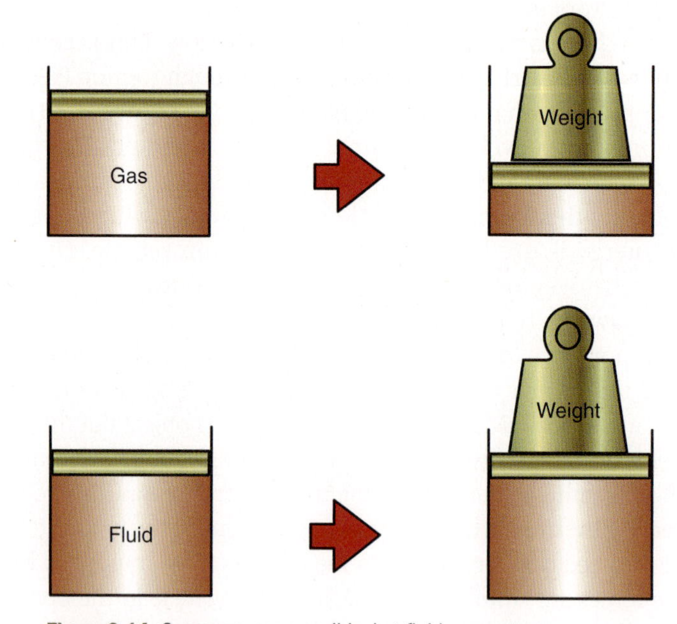

Figure 2-14 Gases are compressible, but fluids are not.

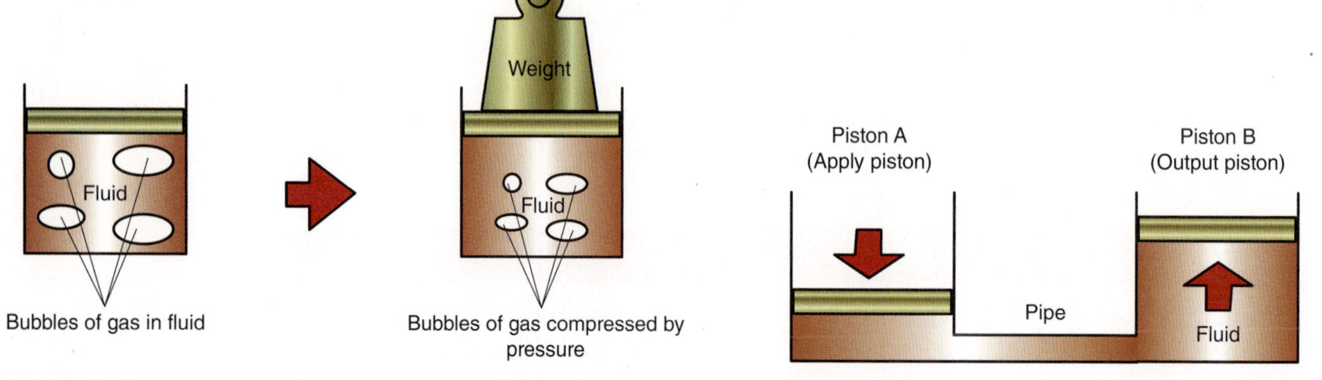

Figure 2-15 When air is in the brake system hydraulics, the force applied at the master cylinder simply compresses the gas instead of doing work.

Figure 2-16 Fluid can transmit motion through a closed system.

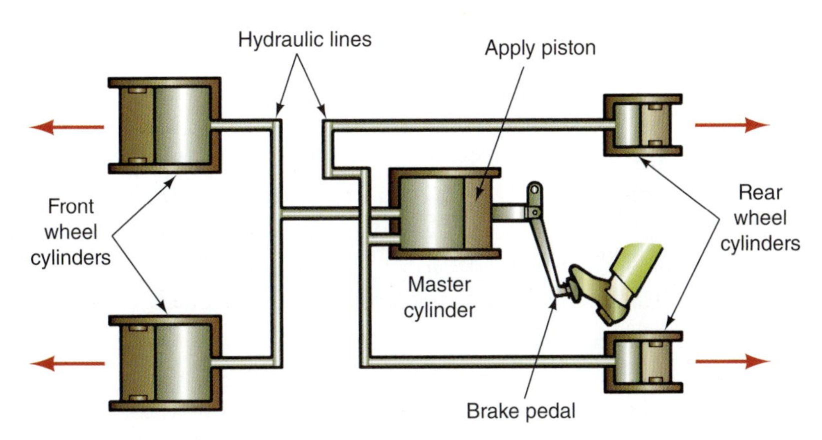

Figure 2-17 The master cylinder is an apply piston, working as a pump, to provide hydraulic pressure to the output pistons at the wheel brakes.

master cylinder piston is the apply piston. The fluid in the master cylinder is connected by tubing to pistons in each of the car's front and rear wheel brake units. Each of the wheel brake pistons is an output piston. They move whenever the master cylinder input piston moves.

Fluids Apply Pressure to Transmit and Increase Force

Fluids are not just used to transmit movement; they are also used to transmit **force**. Force is power working against resistance; it is the amount of push or pull exerted on an object needed to cause motion. Force is usually measured in the same units that are used to measure weight: pounds or kilograms.

Piston A and piston B (**Figure 2-18**) are the same size. If a 100-pound force is applied to piston A, the same force will be applied to piston B. This demonstrates that force is transmitted from the apply piston to the output piston.

This principle can be used to increase the output force of a hydraulic system. To do this, change the sizes of the input and output pistons and calculate the amount of pressure developed in the system.

Output Force Creates System Pressure. Pressure is the force exerted on a given unit of surface area. Therefore, pressure equals force divided by surface area:

$$\frac{F}{A} = P$$

where

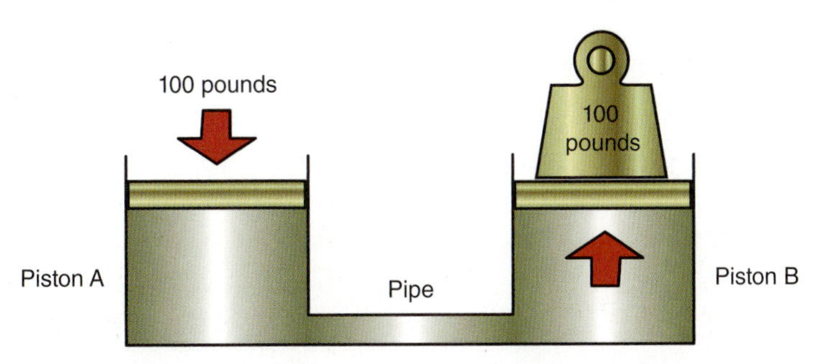

Figure 2-18 As hydraulic fluid transmits motion through a closed system, it also transmits force.

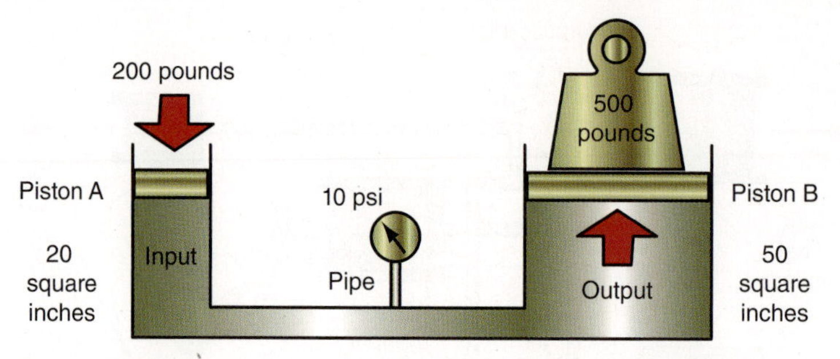

Figure 2-19 A hydraulic system also can increase force.

F = force in pounds (or kilograms)

A = area in square inches (or centimeters)

P = pressure

The pressure of a liquid in a closed system such as a brake hydraulic system is the force exerted against the inner surface of its container, which is the surface of all the lines, hoses, valves, and pistons in the system. In customary English units, pressure is measured in pounds per square inch (psi). In the metric system, pressure can be measured in kilograms per square centimeter, but the preferred metric pressure measurement unit is the Pascal. Regardless of how it is measured, pressure applied to a liquid exerts force equally in all directions. When pressure is applied to a movable output piston, it creates output force.

In **Figure 2-19**, input piston A is smaller than output piston B. Piston A has an area of 20 square inches, and in the example, 200 pounds of force are applied. Therefore,

$$\frac{200\text{ pounds }(F)}{20\text{ square inches }(A)} = 10\text{ psi }(P)$$

If that same 200 pounds of force is applied to a piston of 10 square inches, system pressure is 20 psi, because

$$\frac{200\text{ pounds }(F)}{10\text{ square inches }(A)} = 20\text{ psi }(P)$$

Therefore, pressure is inversely related to piston area. The smaller the piston, the greater the pressure that is developed with the same amount of input force.

Pressure and Output Piston Area Determine Force. Apply the 10 psi of pressure in the first example to an output piston with an area of 50 square inches. In this case, output force equals pressure times the surface area:

$$P \times A = F$$

where

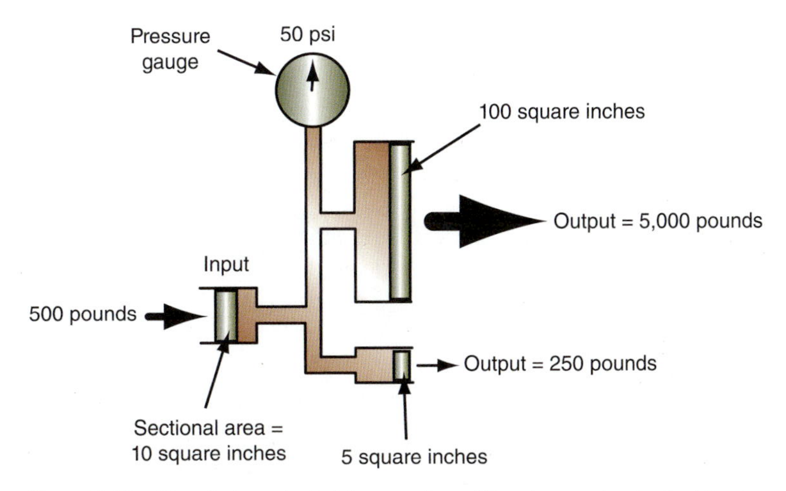

Figure 2-20 Different-size output pistons produce different amounts of output force from the same hydraulic pressure.

F = force in pounds (or kilograms)

A = area in square inches (or centimeters)

P = pressure

Therefore, 10 psi of pressure on a 50-square-inch piston develops 500 pounds of output force: $10 \times 50 = 500$. Brake systems use hydraulics to increase force for brake application. This is called **mechanical advantage**.

Mechanical advantage: With hydraulics, the multiplication of force by using a larger output piston than input piston.

Figure 2-20 shows a hydraulic system with an input piston of 10 square inches. A force of 500 pounds is pushing on the piston. The pressure throughout the system is 50 psi (force $500 \div$ area $10 = 50$ psi). A pressure gauge in the system shows the 50-psi pressure.

There are two output pistons in the system. One has 100 square inches of area. The 50-psi pressure in the system is transmitted equally everywhere in the system. This means that the large output piston has 50 psi applied to 100 square inches to deliver an output force of 5,000 pounds (100 square inches $\times$ 50 psi = 5,000 pounds).

The other output piston in Figure 2-20 is smaller than the input piston with a 5-square-inch area. The 5-square-inch area of this piston has 50-psi pressure acting on it to develop an output force of 250 pounds (5 square inches $\times$ 50 psi = 250 pounds).

In a brake system, a small master cylinder piston is used to apply pressure to larger pistons at the wheel brake units to increase braking force. It is important that the pistons in the front brakes (now almost exclusively caliper pistons) have a larger surface area than the pistons in the rear brakes. This creates greater braking force at the front wheels to overcome the weight transfer that momentum creates during braking.

Hydraulic Pressure, Force, and Motion. Hydraulic pressure acting on pistons of different surface areas creates different output forces. As the output pistons move, another effect of pressure and force appears. When output force increases, output motion decreases. If the 10-square-inch input piston moves 2 inches as it applies 50 psi to the 100-square-inch output piston, that output piston will move only 0.2 inch as it applies 5,000 pounds of output force (**Figure 2-21**). The ratio of input motion to output motion is the ratio of the input piston area to the output piston area, and you can use this simple equation to calculate it:

$$\frac{A_1}{A_2} \times S = M$$

where

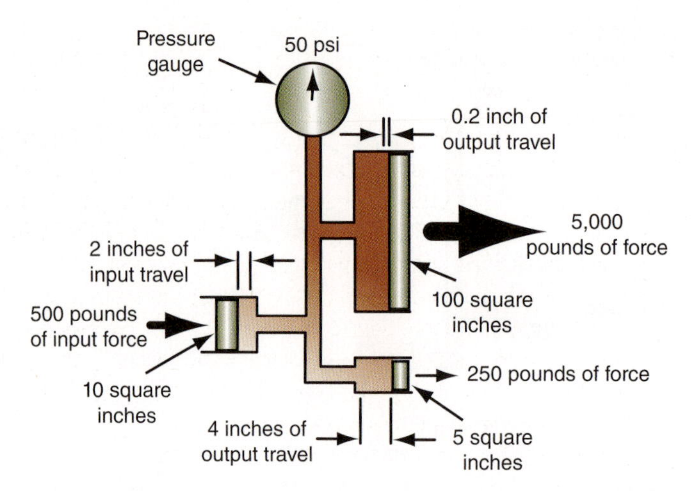

Figure 2-21 The output pistons' movement and their created force will be proportional to their size in relationship to the input piston size.

$$A_1 = \text{input piston area}$$
$$A_2 = \text{output piston area}$$
$$S = \text{input piston stroke (motion)}$$
$$M = \text{output piston stroke (motion)}$$

$$\frac{A_1}{A_2} \times S = M$$

$$\frac{\text{10 square inches (input piston)}}{\text{100 square inches (output piston)}} = \frac{1}{10} \times \text{2 inches (input stroke)}$$
$$= \text{0.2 inch of output motion}$$

If the output piston is larger than the input piston, it exerts more force but travels a shorter distance. The opposite also is true. If the output piston is smaller than the input piston, it exerts less force but travels a longer distance. Apply the equation to the 5-square-inch output piston in Figure 2-21:

$$\frac{\text{10 square inches (input piston)}}{\text{5 square inches (output piston)}} = \frac{2}{1} \times \text{2 inches (input stroke)}$$
$$= \text{4.0 inches of output motion}$$

In this case, the smaller output piston applies only half the force of the input piston, but its stroke (motion) is twice as long.

This relationship of force, pressure, and motion in a brake system is shown when the force applied to the master cylinder pistons and the resulting brake force and piston movement at the wheels is considered. Wheel cylinder pistons move only a fraction of an inch to apply hundreds of pounds of force to the brake shoes, but the wheel cylinder piston travel is quite a bit less than the movement of the master cylinder piston. Disc brake caliper pistons move only a few thousandths of an inch but apply great force to the brake rotors.

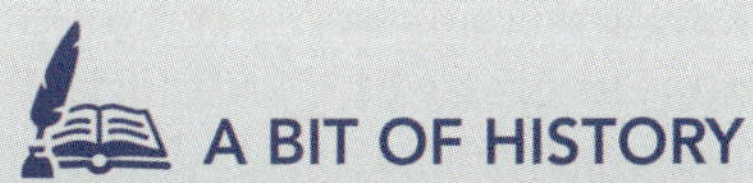

A BIT OF HISTORY

All the hydraulic principles that are applied in a brake system are based on the work of a seventeenth-century scientist named Blaise Pascal. Pascal's work is known as Pascal's law. Pascal's law says that pressure at any one point in a confined liquid is the same in every direction and applies equal force on equal areas. Additionally, liquids take the path of least resistance. One of the most important results of Pascal's work was the discovery that fluids can be used to increase force. Pascal was the first person to demonstrate the relationships of pressure, force, and motion and the inverse relationship of motion and force. On an automobile, Pascal's laws are applied not just to the brake system. These same hydraulic principles are at work in the hydraulic system of an automatic transmission and other systems. Pascal's laws are even at work in the movement of liquid fuel from a tank to the fuel-injection system on the engine.

Hydraulic Principles and Brake System Engineering

Engineers must consider these principles of force, pressure, and motion to design a brake system for any vehicle that will give maximum stopping efficiency but still be easy to control. If the engineer chooses a master cylinder with relatively small piston area, the brake system can develop very high hydraulic pressure, but the pedal travel will be extreme. Moreover, if the master cylinder piston travel is not long enough, this high-pressure system will not move enough fluid to apply the large-area caliper pistons regardless of pressure. If, however, the engineer selects a large-area master cylinder piston, it can move a large volume of fluid but may not develop enough pressure to exert adequate braking force at the wheels.

The overall size relationships of master cylinder pistons, caliper pistons, and wheel cylinder pistons are balanced to achieve maximum braking force without grabbing or fading. Most brake systems with front discs and rear drums have large-diameter master cylinders (large piston area) to move enough fluid and a power booster to increase the input force.

VACUUM AND AIR PRESSURE PRINCIPLES

Vacuum is another force used in most brake systems. Many power brake systems use vacuum to provide a power assist for the driver. Because the most significant use of atmospheric pressure and vacuum in a brake system is in the operation of a power booster, these principles are covered in Chapter 6 of this manual on power brake systems.

Vacuum is generally considered to be air pressure lower than atmospheric pressure (a true vacuum is a complete absence of air).

ELECTRICAL PRINCIPLES

Many of the brake system components requiring repairs are controlled or powered by electricity. Examples are brake system warning lamps, stop lamp switches, brake fluid level sensors, and ABS components. Therefore, a basic understanding of some of the electrical principles, including amperage, voltage, and resistance, is needed.

Amperage, Voltage, and Resistance

Think of electricity in terms of the same principles that work in hydraulic systems. The flow of electricity through a circuit is similar to the flow of fluid through a hydraulic line.

Ampere: A measurement of the number of electrons passing a point in one second, which is 6.28 x 10 (need to place the 18th power on the ten as it is in the text).

Current is the movement of free electrons, under pressure, in a conductor. A flow of current through a conductor requires a source of free electrons to supply the demand, just as fluid in a tank or reservoir is a source of flow in a hydraulic system. The rate of fluid moving through a line often is measured in gallons per minute. The rate of current flowing in a conductor is measured in amperes (A). One **ampere (A)** equals 6.28×10^{18} electrons passing a given point in a circuit per second.

Voltage is the electromotive force that causes current to flow; the potential force that exists between two points when one is positively charged and the other is negatively charged.

Just as pressure is necessary to move fluid through hydraulic lines, there must be pressure to move electrons through a conductor. The pressure pushing the electrons through an electrical circuit is called the **voltage**, measured in **volts (V)**.

Electrical resistance is measured in **ohms**.

Friction between the walls of a hydraulic line and the fluid will cause some resistance to the flow of fluid. Similarly, some resistance to electron flow through a circuit is offered by any material. Electrical resistance is measured in **ohms (Ω)**.

To summarize the comparison of electrical and hydraulic systems, the following apply:

- Voltage is the pressure (or electrical force) that moves electrons (current or amperes) through a wire just as pressure moves fluid through a pipe.
- Amperage, or current, is similar to the amount of fluid flowing in a line.
- Electrical resistance is a load on the moving current that must be present to do any useful work, just as a hydraulic system must have the load of an output piston or motor to do work.

Ohm's Law

The amount of fluid flowing through a pipe will increase if the fluid pressure is increased. Likewise, the amount of fluid flowing in the pipe will decrease if the resistance offered by the hydraulic system increases. In the same way, the amount of current flowing in an electrical circuit will increase if the voltage is increased. It will decrease if the resistance of the circuit increases even when the voltage is constant.

Ohm's Law: Ohm's law states that it takes one volt of electrical pressure to push one ampere of current through a resistance of one Ohm.

The relationship of these electrical factors (voltage, current, and resistance) is summarized by a mathematical equation known as **Ohm's law** (**Figure 2-22**). Ohm's law states that it takes 1 volt of electrical pressure to push 1 ampere of current through a resistance

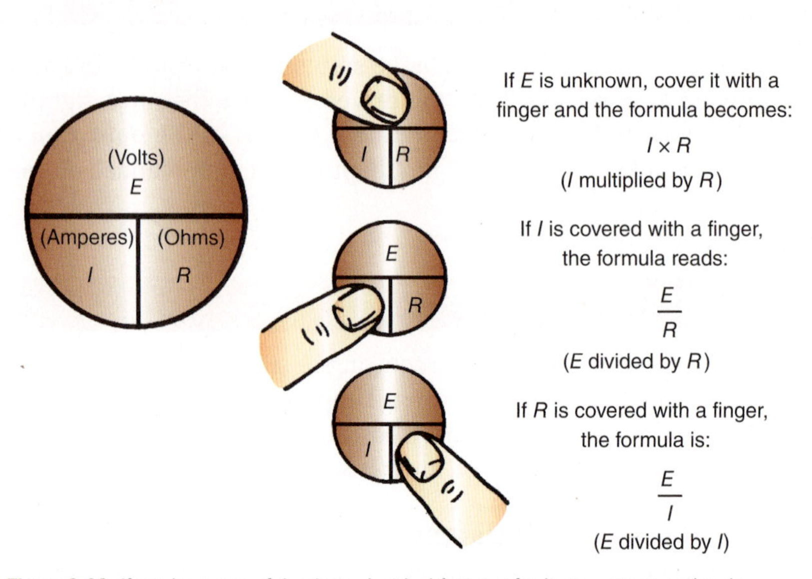

Figure 2-22 If you know two of the three electrical factors of voltage, current, and resistance, you can calculate the third as shown here.

of 1 ohm. Mathematically, Ohm's law is expressed by the following equations, using the symbols E for voltage, I for current, and R for resistance:

$E = I \times R$ to find voltage

$I = E \div R$ to find current

$R = E \div I$ to find resistance

Georg Simon Ohm (1787–1854) wrote the law regarding electrical theory and measurement. It came to be known as Ohm's law.

For example, if a 12-volt circuit has 2 amperes of current flowing, you can use Ohm's law to determine the resistance as follows:

$$R = E \div I$$
$$= 12 \div 2$$
$$= 6 \text{ ohms}$$

The resistance of the circuit is 6 ohms.

These equations can be used to calculate the voltage, current, or resistance of a circuit. In most cases, you can use electrical testing instruments to measure these values. There are times, however, when this is not possible. Then, if you know two of the values, you can find the third value by using the problem-solving wheel, which is based on Ohm's law (Figure 2-22).

A BIT OF HISTORY

Early experiments with electricity by Volta and Ampere led to names for the new phenomenon that remains today. The term *electromotive force* is now called volts. The term *intensity* is now called amps.

SUMMARY

- The energy used by the braking system is based on the kinetic energy, mass, weight, and speed of the vehicle and its braking system.
- Inertia is an object's resistance to change.
- Vehicle subsystems are used to overcome inertia.
- The acceleration rate of a vehicle is the force applied against the opposing forces.
- Kinetic and static frictions are at work in the brake system.
- The amount of friction in a brake system depends on the pressure (force) exerted on the friction components and the surface area.
- Friction materials may be made from organic, metallic, or ceramic compounds.
- Energy provides the fuel for work.
- Work is accomplished any time energy is expended.
- The laws of motion affect the vehicle and all its subsystems.
- Hydraulic fluid can transmit motion and force because a fluid cannot be compressed.
- A small hydraulic input piston and a large output piston can increase force at the wheel braking components.
- Mechanical advantage in a hydraulic system is proportional to the size of the input piston versus the output piston.
- Electrical circuits have an electrical pressure (force) that is used to move electrons through a circuit.
- In Ohm's law, E = voltage, I = current (ampere), and R = resistance (ohm).
- In electrical terms, a change in E, I, or R will cause a change in one or both of the other two.
- Resistance in a circuit is needed to change electrical energy into mechanical energy (action).

REVIEW QUESTIONS

Short-Answer Essays

1. Briefly explain how a braking system converts one form of physical energy to another.
2. Explain the brake system dynamic called weight transfer.
3. Explain the difference between static and dynamic inertia.
4. Explain the relationship between pressure and friction.
5. Explain what the term coefficient of friction means.
6. Explain why heat dissipation is important in a braking system and how it is removed.
7. Briefly describe inertia.
8. Explain what happens if air is introduced into a hydraulic system.
9. Explain how the force from a small master cylinder piston is multiplied by the caliper piston.
10. Briefly describe amperage, resistance, and voltage.

Fill in the Blanks

1. _______________ is the ability to do work.
2. The brake system is used to overcome the _______________ of motion and helps stop the vehicle.
3. Mass is not the weight of the object but the amount of _______________ in that object.
4. For every _______________, there is an opposite and equal _______________.
5. If the _______________ piston is larger than the _______________ piston, it exerts more force but travels a shorter distance.
6. _______________ is the mathematical product of an object's mass times its speed.
7. The rate of current flowing in a conductor is measured in _______________.
8. The pressure pushing the electrons through an electrical circuit is called _______________.
9. Electrical resistance is measured in _______________.
10. The _______________ _______________ _______________ becomes involved only if it senses one (or more) wheels about to lock up.

Multiple Choice

1. Technician A says kinetic energy is the energy of mechanical work or motion. Technician B says when an automobile starts, accelerates, decelerates, and stops, kinetic energy is at work. Who is correct?

   A. A only
   B. B only
   C. Both A and B
   D. Neither A nor B

2. Technician A says weight is a measurement of the number of molecules that make up an object. Technician B says mass is a measurement of the effect of gravity on that mass. Who is correct?

   A. A only
   B. B only
   C. Both A and B
   D. Neither A nor B

3. Technician A says the weight of a car is not distributed evenly on all four wheels even when the vehicle is standing still. Technician B says the position of the heavy engine and powertrain components determines weight distribution. Who is correct?

   A. A only
   B. B only
   C. Both A and B
   D. Neither A nor B

4. Technician A says on a rear-wheel-drive car, about 70 percent of the weight shifts to the front. Technician B says on a front-wheel-drive car, as much as 60 percent of the car weight is shifted to the front during braking. Who is correct?

   A. A only
   B. B only
   C. Both A and B
   D. Neither A nor B

5. Technician A says a car can accelerate from 0 to 60 mph in 10 seconds. Technician B says the brake system must be able to decelerate the car from 60 to 0 mph in nearly one-fifth that time. Who is correct?

   A. A only
   B. B only
   C. Both A and B
   D. Neither A nor B

6. Technician A says this chapter focuses on the friction between the brake linings and the brake rotor or drum. Technician B says to remember that the vehicle is actually stopped by the friction between the tire and the road. Who is correct?

   A. A only
   B. B only
   C. Both A and B
   D. Neither A nor B

7. Technician A says a car with locked brakes has a shorter stopping distance. Technician B says this is because hot molten rubber can form between the tire and the road, causing a grabbing effect.

   A. A only
   B. B only
   C. Both A and B
   D. Neither A nor B

8. Two basic types of friction are at work in the brake system. These are:

   A. Static and dynamic
   B. Disc and drum
   C. Moving and stopping
   D. Kinetic and static

9. Technician A says it takes more force to move some materials over a surface than others, even though the applied pressure and the amount of surface in contact are the same. Technician B says different materials have the same frictional characteristics or coefficients of friction.

   A. A only
   B. B only
   C. Both A and B
   D. Neither A nor B

10. Technician A says if pads and linings are soft and wear fast, the vehicle may have very good braking ability, but friction material life will be short. Technician B says if friction material is hard and wears slowly for long life, it will create a high coefficient of friction and good stopping ability.

    A. A only
    B. B only
    C. Both A and B
    D. Neither A nor B

CHAPTER 3

RELATED SYSTEMS: TIRES, WHEELS, BEARINGS, AND SUSPENSIONS

Upon completion and review of this chapter, you should be able to:

- Describe the basic kinds of tire construction and identify the most common construction method for modern tires.
- Identify and explain the various letters and numbers used in tire size designations and other tire specifications.
- Explain the basic effects of tire tread design on vehicle handling and braking.
- Explain the most important effects of tire design and condition on brake performance.
- Explain how wheel and tire runout and wheel rim width and offset affect braking.
- Identify the common types of wheel and axle bearings used on cars and light trucks.
- Identify the basic wheel alignment and steering angles.
- Explain how wheel alignment and steering angles can affect braking.
- Explain how the condition of steering and suspension parts can affect braking.

Terms To Know

Aspect ratio
Belted bias ply tire
Bias ply tire
Camber
Carcass
Casing
Caster
Cold inflation pressure
Geometric centerline
Gross vehicle weight rating (GVWR)
Hydroplane
Lateral runout
National Highway Transportation and Safety Agency (NHTSA)
P-metric system
Pressure-sensor base (PSB)
Radial ply tire
Radial runout
Run-flat
Scrub radius
Section width
Setback
Side-to-side
Steering axis inclination (SAI)
Thrust angle
Thrust line
Tire load range
Tire pressure monitoring system (TPMS)
Toe angle
Toe-out on turns (turning radius)
Tread
Tread contact patch
Tread wear indicator
Unidirectional tread pattern
Uniform Tire Quality Grading (UTQG) indicators
Wheel offset
Wheel-speed base (WSB)

INTRODUCTION

The brake shoes or pads apply friction to the wheels, but it is the friction between the tires and the road that actually stops the car. Tire design, condition, and inflation pressure can affect braking, and attention to these factors often can solve braking problems.

The tires are mounted on wheels, which ride on bearings on steering knuckle spindles and axles. Steering and axle components, in turn, are supported by suspension struts and springs. Any of these components can create braking problems if they are not in proper working order. This chapter outlines the key relationships between brake systems and the related systems of wheels, tires, wheel bearings, and suspensions.

TIRE FUNDAMENTALS

Brake systems are engineered in relation to many vehicle factors of weight, size, and performance. Among these factors are the construction, size, and tread design of the tires and the amount of traction or friction expected to be available between the tires and the road. For the best and most reliable brake performance, tires at all four wheels should be identical in construction, size, and tread pattern.

Shop Manual
page 98

Carmakers' Recommendations

Most passenger cars and light trucks built since 1968 have a tire information placard on a door, on a door pillar, or inside the glove compartment (**Figure 3-1**). The tire information placard lists the manufacturer's original equipment tire size and any recommended optional sizes. It also lists the recommended cold front and rear inflation pressures, and maximum front and rear **gross vehicle weight rating (GVWR).** Brake systems are engineered to work most efficiently with the tire sizes and pressures listed on the placard.

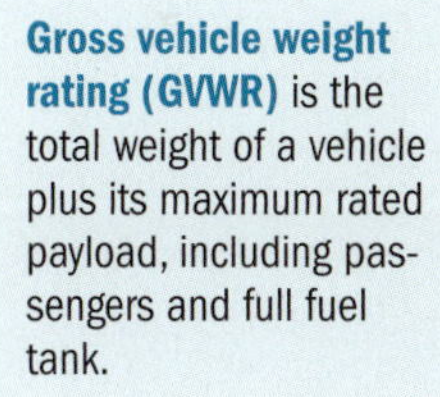

Gross vehicle weight rating (GVWR) is the total weight of a vehicle plus its maximum rated payload, including passengers and full fuel tank.

A few carmakers install different sized wheels and tires at the front and rear of some vehicles, but this practice is reserved for a small percentage of high-performance sports cars like the Porsche 911. More than 99 percent of the vehicles on the road are originally fitted with wheels and tires of the same size at each corner. Although manufacturers may recommend one or two optional tire sizes at the rear that are larger than the front original equipment size, a large variation from the carmaker's recommendation can lead to braking problems, as well as problems with other vehicle systems.

For example, an extreme difference in tire diameters from front to rear may produce unequal speed signals from the wheel speed sensors of ABSs. Tires much larger than those recommended by the vehicle maker may produce inaccurate vehicle speed-sensor signals to the PCM or the ABS control module. This same problem exists if all four tires are larger or smaller than the manufacturer's recommendations.

A BIT OF HISTORY

When radial tires were first introduced in the 70s, there was a lot of resistance by drivers to using the new design. Complaints ranged from "feels funny when driving" to "they don't have enough air in them." Some drivers even went so far as to remove radial tires from a brand-new vehicle and to install bias tires. Two major characteristics of the radial tire overcame this die-hard resistance: a much smoother ride and increased fuel mileage. Lower-profile tires of today have also eliminated most of the comments about the tires "appearing underinflated."

TIRE AND LOADING INFORMATION | RENSEIGNEMENTS SUR LES PNEUS ET LE CHARGEMENT

SEATING CAPACITY: TOTAL 5 FRONT 2: REAR 3 | NOMBRE DE PLACES: TOTAL 5 AVANT 2: ARRIÈRE 3

The combined weight of occupants and cargo should never exceed 410 kg or 900 lbs. | Le poids total des occupants et du chargement ne doit jamais dépasser 410 kg ou 900 lb.

TIRE	SIZE	COLD TIRE PRESSURE	PNEU	DIMENSIONS	PRESSION DES PNEUS A FROID
FRONT	225/65R17	230kPa, 33PSI	AVANT	225/65R17	230kPa, 33PSI
REAR	225/65R17	230kPa, 33PSI	ARRIÈRE	225/65R17	230kPa, 33PSI
SPARE	T165/80R17	420kPa, 60PSI	DE SECOURS	T165/80R17	420kPa, 60PSI

SEE OWNER'S MANUAL FOR ADDITIONAL INFORMATION | VOIR LE MANUEL DE L'USAGER POUR PLUS DE RENSEIGNEMENTS

4 9 42590

Figure 3-1 This placard is located on the driver door and lists recommended tire size and cold inflation pressure.

AUTHOR'S NOTE Changing the size of the driving tires/wheels may result in an incorrect speedometer reading. This could lead to two traffic tickets: exceeding the speed limit and improper equipment.

Tire Construction

Every tire is constructed of these basic parts (**Figure 3-2**):

Carcass: Steel beads around the rim and layers of cords or plies that are bonded together to give the tire its shape and strength.
Casing: Additional layers of sidewall and undertread rubber added to the carcass.
Tread: The layer of rubber that contacts the road and contains a distinctive pattern to provide the desired traction.

Before radial ply tires became the standard of the industry, tires were made with the plies laid at an angle or bias to the tire beads. This construction was called bias ply (**Figure 3-3**). The cords in a **bias ply tire** run at an angle closer to the tread, giving a much stiffer sidewall but creating a rougher ride than a radial tire. A second ply is added to the first so that its cords run in the opposite direction across the first ply. Additional plies can be added for more strength. Bias ply tires can carry more load and are still used on large trucks and most trailers. Most passenger car and light truck tires were made with two, four, six, or eight plies.

The cords of bias ply tires were usually made of nylon or polyester. This construction provided decent stiffness to prevent unnecessary twisting while still allowing enough flexibility to absorb some road shock and cushion the ride.

Belted bias ply tires incorporate the belts used in radial ply tires with the older bias ply construction. Two or more belts of nylon, fiberglass, or steel cords are wrapped around two or more bias plies (**Figure 3-4**). The belts keep the tire from deforming out of round at high speeds and thus improve tire stability. Belted bias ply tires were basically an interim design that combined some of the features of radial ply tires with the older bias ply construction. They are still used on some vehicles today.

Since the mid-1970s, almost all original equipment and replacement tires for cars and light trucks have been the radial ply type. The cords in a **radial ply tire** run at 90 degrees to the tread, creating a more flexible sidewall and creating the impression that the tire is low on air pressure (**Figure 3-5**). Each cord is parallel to the radius of the tire circle, which

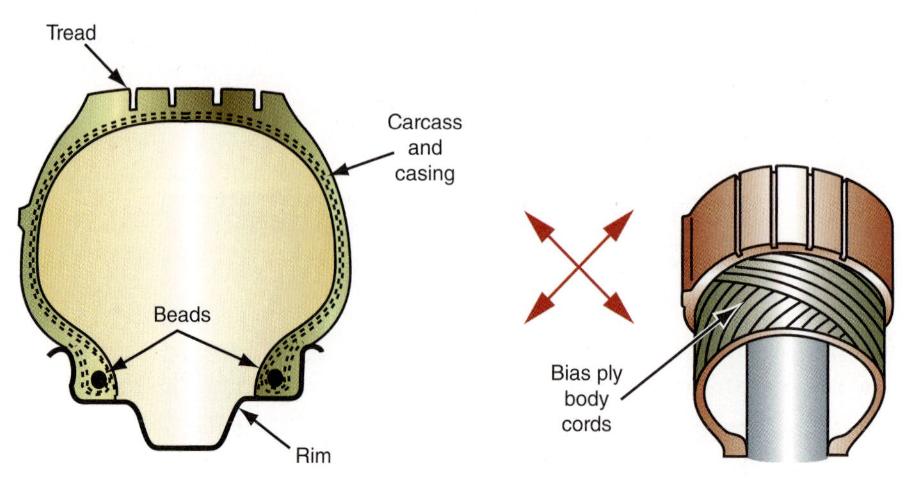

Figure 3-2 Major parts of a tire.

Figure 3-3 Bias ply tire construction.

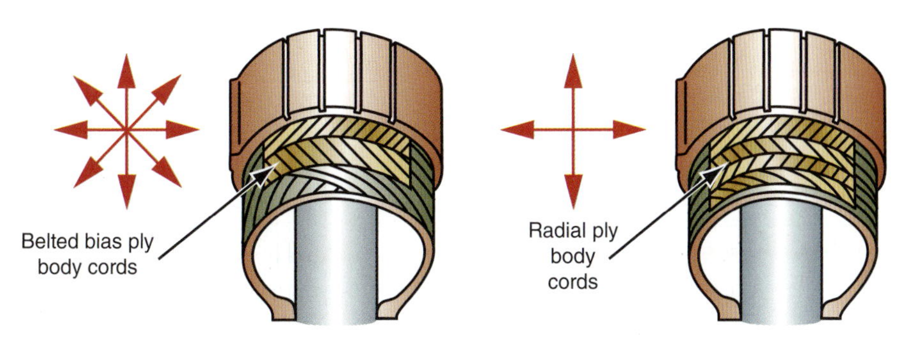

Figure 3-4 Belted bias ply tire construction.

Figure 3-5 Radial ply tire construction.

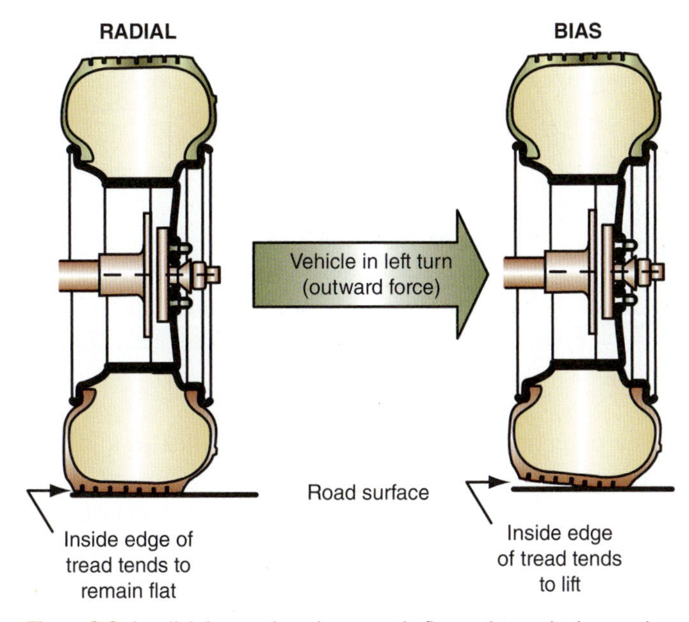

Figure 3-6 A radial tire tread tends to remain flat on the road, whereas the bias tread tends to lift the inner edge from contact with the road.

gives the tire the name radial ply. Two or more additional belts of cords are added around the tire circumference between the radial plies and the tread for more strength and stability. These belts may be made of nylon, rayon, fiberglass, or steel. The radial plies provide maximum flexibility in the tire sidewall, and the belts ensure excellent tread stiffness for good road contact.

Of the three types of tire construction, radial ply tires have the best braking performance. The flexible sidewalls of a radial ply tire provide a larger tread contact patch than a comparable size bias ply or belted bias ply tire. Radial ply construction also has greater flexibility on the wheel during cornering without decreasing tread contact with the road (**Figure 3-6**). The circumferential belts give a radial ply tire maximum stability and protect against deforming at high speed.

Tire Size

All tire sizes are based on three dimensions: the inside diameter, the section width, and the aspect ratio. The inside diameter is the same as the wheel size, or rim diameter, on which the tire fits: for example, 15 inches, 17 inches, or 21 inches. This diameter is measured across the tire beads or across the bead seat of the wheel rim. The **section width** is the width of the tire across the widest point of its cross section, usually measured in millimeters. The **aspect ratio** is the ratio of the cross-sectional height to the cross-sectional

The **aspect ratio** is the ratio of the cross-sectional height to the cross-sectional width.

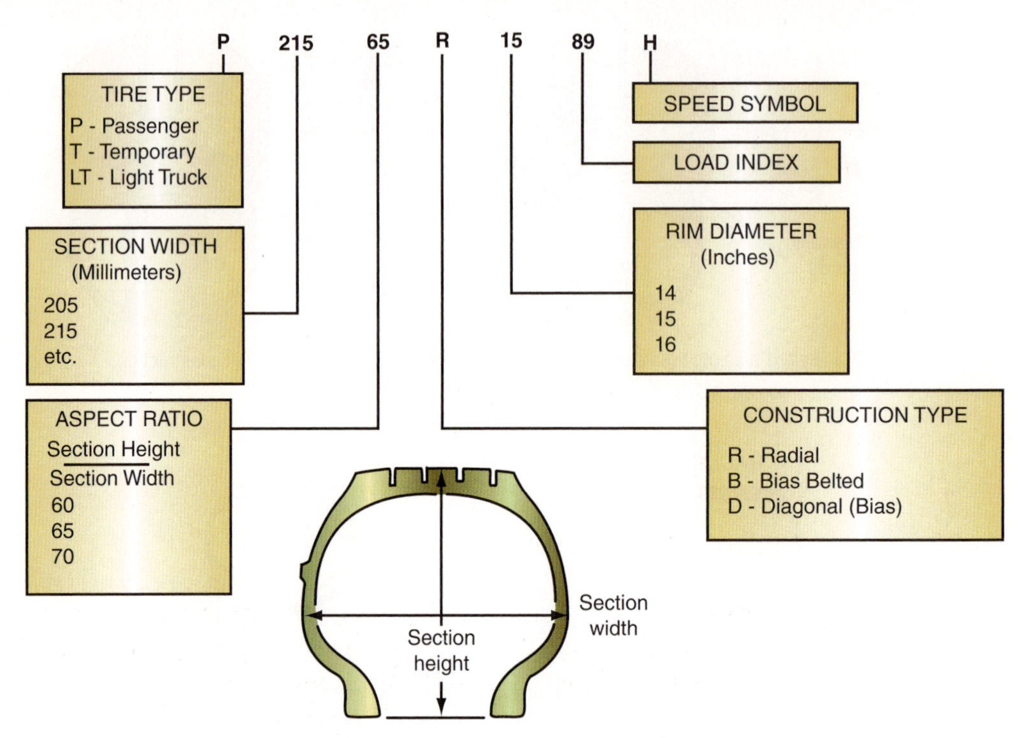

Figure 3-7 Typical tire markings.

width, expressed as a percentage. A high aspect ratio indicates a tall tire. Most modern tires have aspect ratios from 45 to 80.

The **P-metric system** of measurement is an alpha-numeric size identification for tires.

Modern tires are sized by the **P-metric system**. The first character indicates the type of service for which the tire is designed (**Figure 3-7**). The most common designation is "P" for passenger car, which gives the size identification system its name. The numbers after the first letter indicate the tire section width in millimeters, and the numbers after the slash indicate the aspect ratio. An optional speed-rating letter may follow the aspect ratio, and the final letter indicates the construction type. The last numbers indicate the wheel diameter. The European metric variation of this size specification system omits the first letter, such as *"P"* or "C," but includes the speed-rating letter.

Other Tire Specifications

Tire load range is the load-carrying capacity of a tire, expressed by the letters "A" through "L."

Other tire specifications appear on tire information placards, in owner's manuals, and on the tire sidewall itself. The **tire load range** is indicated by the letters "A" through "L," which often appear after the size specification on the tire sidewall. The load range letters replace the older method of rating tire strength by the number of plies used in its construction. Load range "A" corresponds to a two-ply rating, and load range "L" corresponds to a 20-ply rating. Most tires for cars and light trucks are load ranges "B," "C," and "D." As an alternative, tires may be marked with load index numbers that range from 65 to 104. These numbers indicate the load capacity of the tire from 639 pounds to 1,984 pounds. Most passenger car tires have a load index from 75 to 100. Many tires have a maximum inflation pressure molded into the sidewall. This is *not* the recommended operating pressure but the pressure that may cause the tire to burst if exceeded.

Low-Profile Tires

The modern trend has been toward lower-profile tires. These tires have an aspect ratio that is lower than 50. The height of the tire sidewall is smaller than those of a more conventional tire (**Figure 3-8**). The advantages of a low-profile tire are handling and ease of

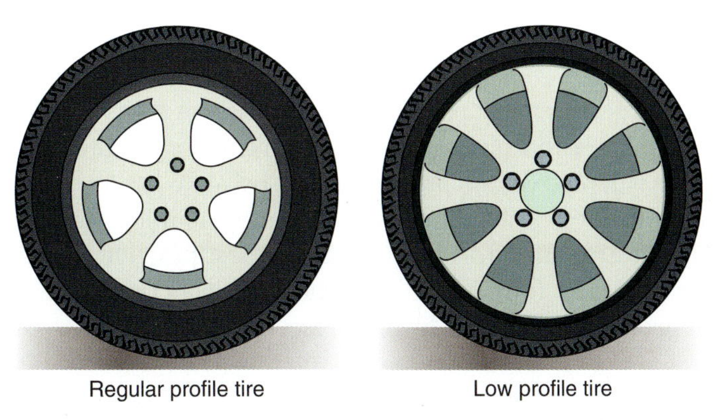

Figure 3-8 Comparison of low profile to regular profile tire and wheel combinations.

steering. Low-profile tires, like many of the advances in tire technology, were developed in racing. Low-profile tires do not flex as much as a higher profile tire during hard cornering (**Figure 3-9**). The larger wheel size can allow for the use of larger brakes. The disadvantage of low-profile tires is the fact that there is less tire sidewall to absorb shocks from hitting irregularities in the road, which can lead to a rougher ride, and can also lead to wheel damage. How far will the trend towards lower-profile tires? Kumho and Nexen have shown tires with a profile of 15 at the SEMA show!

The maximum **cold inflation pressure** for any tire is molded into the tire sidewall (**Figure 3-10**). This is the inflation pressure after the tire has been standing for 3 hours or driven less than 1 mile after standing for 3 hours. The vehicle manufacturer's recommended inflation pressures are usually slightly less than the maximum cold inflation pressure.

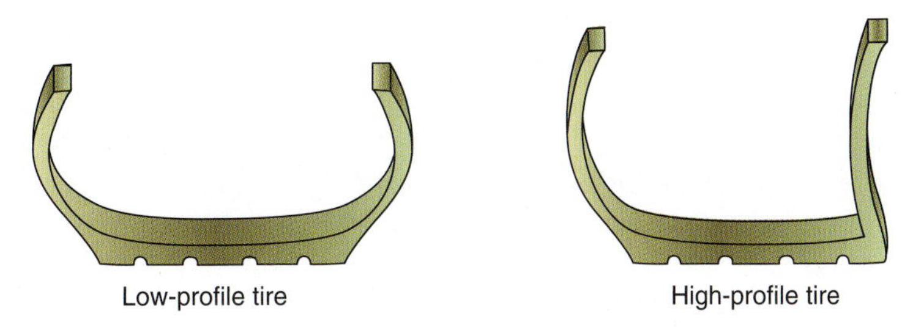

Figure 3-9 During cornering, the high-profile tire tends to distort more than the low-profile tire.

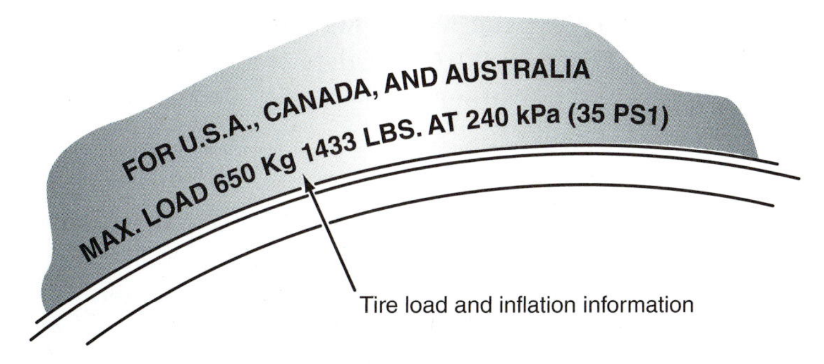

Figure 3-10 The maximum cold inflation pressure specification is molded into the sidewall of every tire.

As tire aspect ratios or profiles have gotten lower, inflation pressures have gotten higher. Until the early 1980s, most passenger car tires were equivalent to the old load range "B" with a maximum pressure of 32 psi. Maximum pressures then moved upward to about 36 psi. Today, a popular ¾-ton pick-up-trucks list front tire inflation at 55 psi front and 80 psi rear. With the older maximum pressures of 32 psi to 36 psi, it was common to inflate tires to the maximum for best handling, tire life, and gas mileage. With today's higher pressures and tire pressure monitoring systems, the best practice is to follow the manufacturer's recommendations found on a late-model vehicle's tire placard, and remember that the inflation pressure listed is with cold tires.

AUTHOR'S NOTE Starting in 1996, there were many accidents involving a brand of SUVs equipped with a particular brand of tire. There were two opposing arguments concerning these accidents. First, the vehicle manufacturer claimed that the tires came apart during normal operation because of poor tire manufacturing. Second, the tire manufacturer claimed that the vehicle manufacturer's recommended tire pressure could critically damage the tire during operation. Both the SUVs and tires were recalled at a huge cost to both manufacturers and the general public. Regardless of the root cause, it is obvious that tire pressure can affect overall operational durability, *particularly* if there is a problem with the tire's manufacturing process.

Uniform Tire Quality Grading

The Uniform Tire Quality Grading Standards were developed by an association of tire manufacturers, vehicle manufacturers, and governmental agencies in the early 1980s. The **Uniform Tire Quality Grading (UTQG) indicators** have been molded into the sidewall of every tire sold in the United States since 1985. The UTQG indicators are a number to indicate relative tread life, a letter to indicate wet weather traction, and a second letter to indicate heat resistance. The treadwear rating starts at 100. A tire with a treadwear rating of 200 should last twice as long as a tire rated at 100, on the test track set up by the NHTSA. The wet traction rating is designated by an AA, A, B, or C; the best traction rating is AA; and the worst acceptable is C. A tire's resistance to heat is A, B, or C. The UTQG indicators are intended to be a way that consumers can compare tires when making a purchasing decision.

Uniform tire quality grading indicators gives consumers a standard with which to compare the performance of tires.

In addition to the UTQG markings, the U.S. National Highway Traffic Safety Administration (NHTSA) requires all tires to have a tire identification number. This data set is prefixed by the letters DOT followed by 10, 11, or 12 numbers, which give the place of manufacturer, tire size, manufacturer code, and week and year of manufacture. The date of manufacture is the last four digits in the identification number. The first two digits represents the week and the last two the year. In **Figure 3-11**, the last four digits are 5116 representing the 51st week of the year 2016. The date of manufacture is of importance to the owner because tires can dry rot sitting on the shelf.

Tire Tread Design

The best possible traction and braking performance on dry pavement can be had with the treadless, slick tires used on race cars. Unfortunately, all of the great traction of a slick tire goes away on wet pavement. On wet pavement, a slick tire will **hydroplane** on a layer of water trapped between the tread and the pavement.

Hydroplaning is the sliding or skidding brought on by the vehicle riding on a thin film of water, risking a serious loss of traction. Hydroplaning is aggravated by having worn tires and excessive speed.

Therefore, all street tires have a tread pattern of ribs and grooves that lets water be displaced from under the tire while still maintaining good traction.

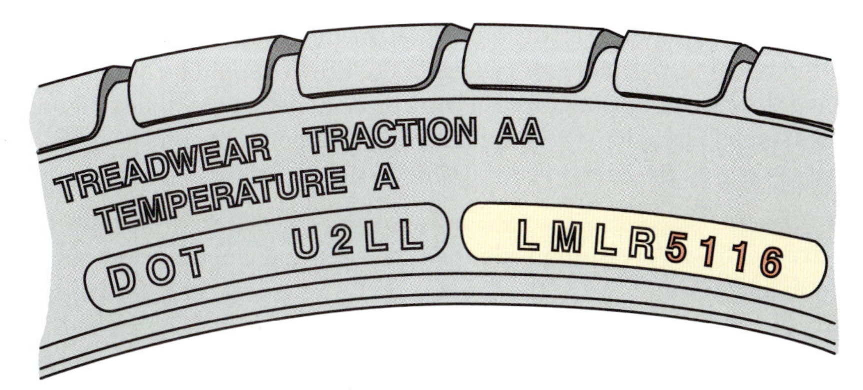

Figure 3-11 The number 5116 indicates that this tire was manufactured in the 51st week of year 2016.

Figure 3-12 When wear indicator bars appear across any two tread grooves, the tire is ready for replacement.

Tread designs vary widely, but most original equipment tires have some kind of mud-and-snow or all-weather tread pattern. In fact, the characters "M+S" or "M/S" often appear on a tire sidewall to indicate a mud-and-snow tread pattern. In addition to providing traction, the tread pattern acts as a radiator to expose more tire surface area to the air and improve tire cooling. Some high-performance tires have a **unidirectional tread pattern,** which means that each tire can be installed on only one side of the car.

Tires with a **unidirectional tread pattern** are specific to the side of the vehicle they are installed on.

Tread depth is measured from the top of the tread ribs to the bottom of the groove and usually is expressed in thirty-seconds of an inch. The tread on a new passenger car tire is usually in the range of 9/32 inch to 15/32 inch. The tread on truck tires and some mud-and-snow tires may be deeper.

All passenger car and light truck tires made since 1968 have **tread wear indicators** that appear as continuous bars across the tread when the tread wears down to the last 2/32 (1/16) inch (**Figure 3-12**). Note that the tire should be replaced when the indicator is visible across two or more grooves. However, if the vehicle is extremely out of alignment, the tire may wear completely to the steel cords before wearing the indicator in the second groove. The two-groove indication assumes that the vehicle is maintained and properly aligned.

Tread wear indicators are meant to make it obvious when a tire is worn out.

RUN-FLAT TIRES

Self-sealing tires. While not technically considered a run-flat tire, these tires can seal themselves against most minor punctures up to about 3/16 inch (4.8 mm). These tires are lined with a sealant layer under the tread. When an object punctures the tire, the sealant acts to seal the area around the intrusion. When the intrusion is removed, then the sealer flows into the opening, permanently repairing the tire. It should be noted that this does not mean the tire cannot be punctured; a large piece of debris can still result in a flat tire. Also, some manufacturers have a compressor and a can of sealant instead of a spare tire. Sealant that is sprayed into the tire at the time of the flat is not considered a permanent repair.

Run-flat tires prevent quick air loss during punctures or minor damage to the tire.

Run-flat tires. The air pressure inside the tire is normally needed to hold the vehicle weight and maintain the shape of the tire. Once the air is lost on a normal tire and it is driven even a very short distance, the tire, and possibly the rim, is destroyed. To prevent this from happening, some tires have an inner structure or are reinforced to help prevent damage from driving up to 50 miles at speeds of 50 mph. The advantage is that you can get the vehicle to a shop instead of fixing the tire on the road.

There are two types of run-flat tires, self-supporting (**Figure 3-13**) and auxiliary supported (**Figure 3-14**). In the self-supporting design, the tires are made stiffer by adding layers of rubber to the tire and the utilization of heat resistant cords to prevent breakage of the cord layers. These also utilize a bead design that allows a better grip on the wheel in case the tire goes flat. Self-supporting tires are best used with a tire pressure warning system because the driver may not realize that the tire is actually

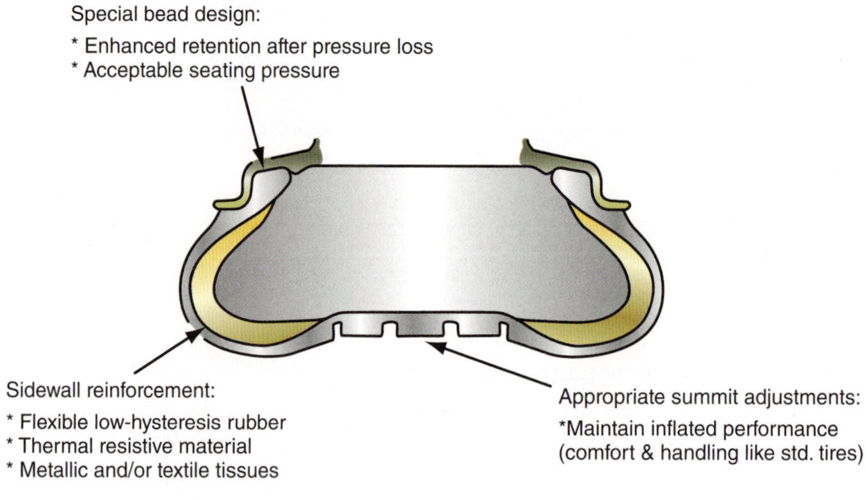

Figure 3-13 A run-flat tire.

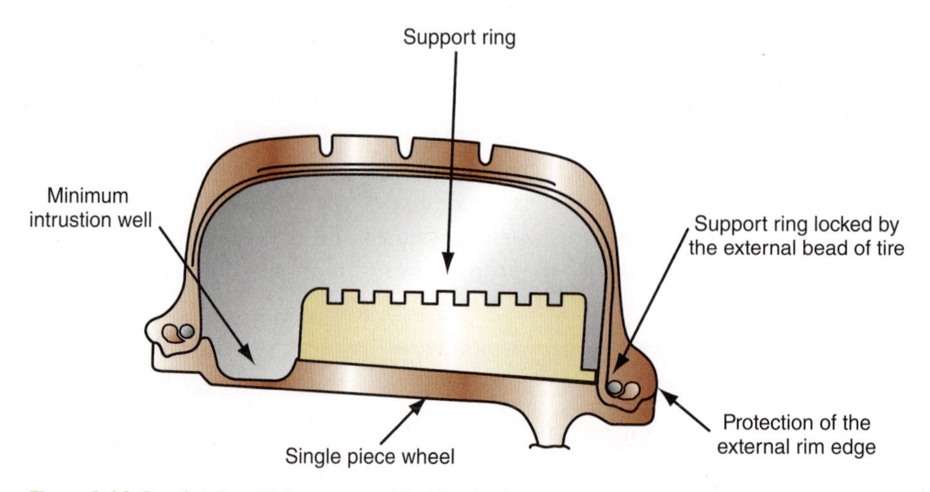

Figure 3-14 Run-flat tire with inner support inside wheel.

flat. As stated before, these tires are not intended to be driven over long periods or high speeds without being properly repaired. In the auxiliary supported design, the tire and the wheel are both specially designed to work together to support the vehicle when the tire is flat. In case of a flat tire, a supporting ring attached to the wheel itself holds the vehicle weight. The tires and wheels must made specifically for this purpose.

Some racing tires use race-proven technology by installing a tire-within-a-tire, but this is expensive and not suitable for mass-produced units. The most common design weaves the tire fabric in such a way that it will grip small penetrating objects in the tire. Another design weaves the fabric so it will attempt to close the hole if the penetrating object is withdrawn from the tire. In both cases, air inside the tire cannot escape quickly and, under the right conditions, will support the vehicle up to 50 miles at reasonable speeds and loads. Neither type will prevent the complete air loss because of a major incident such as a blowout. Leaking run-flat tires affect braking and steering of the vehicle the same as would any tire with insufficient air pressure.

TIRE PRESSURE MONITORING SYSTEM

Tires that are underinflated not only affect fuel mileage and tire life; they may also lead to problems in braking and steering. The Transportation Recall Enhancement, Accountability and Documentation (TREAD) Act of 2000 directed the **National Highway Transportation and Safety Agency (NHTSA)** to research and develop rules as a means of continuously monitoring tire pressure. After several years of testing, it was determined that two viable **tire pressure monitoring systems (TPMSs)** could perform the job. The **wheel-speed base (WSB)** system uses the wheel speed sensors that are used with ABS to detect the faster spinning or rotation caused by an underinflated tire. This illuminates a light in the dash, warning the driver of the condition. A second option, **pressure-sensor base (PSB),** has a pressure sensor located on each wheel and reads tire pressure directly. Currently, most vehicles have a direct pressure measurement of tire pressure. If the tire pressure drops 25 percent below the specified cold inflation pressure, the tire pressure warning light is illuminated on the dash. Maintenance of the TPMS system is discussed in Chapter 11 of the *Shop Manual.*

The **NHTSA** is a federal agency assigned to develop regulations for highway safety, including vehicle safety features.

The TPMS systems were phased in from 2003–2005. During 2005, the NHTSA developed and issued the final ruling based on any new data collected during the 2003–2005 time period.

The NHTSA mandated that vehicles manufactured after October 2006 (2007 model year) must have a TPMS (type determined by the vehicle manufacturer). A standard warning icon (**Figure 3-15**) is used in the dash to alert the driver if one or more tires have a pressure that is less than 25 percent of the cold inflation recommendation. The system also has to be able to indicate which tire(s) is underinflated.

The TPMS systems have been referred to as direct (pressure-based system) and indirect (wheel-speed-based system).

Based on current data, it appears that the PSB system is slightly more responsive, but advances in ABS technology may change that shortfall. A wheel-mounted pressure sensor uses wireless communication to send the pressure signal to a PSB computer, which, in turn, illuminates the warning light or icon in the dash. Factory sensors are usually installed in place of the regular tire valve stem. Specific repair and reset procedures are mandated by the manufacturer. Aftermarket units usually have a pressure sensor installed on the valve stem in lieu of the valve cap. If a tire, pressure sensor, or TPMS module is replaced or repaired or if the tires are rotated, the system must be reset so the sensor can determine the rotational speed of the new/ repaired/rotated tires and determine its own location on the vehicle. This reset and training procedure was one problem that the NHTSA found with the WSB and PSB systems.

Besides being dangerous and wasting fuel, underinflation of tires is the most common cause of premature wear.

Figure 3-15 The standard symbol for the tire pressure monitoring system.

WHEEL FUNDAMENTALS

Wheel offset and rim width are design factors of the wheel itself that affect brake performance. Their effects are most significant on the front wheels, which provide steering control as well as most of the braking force.

Rim Width

Rim width is measured between the inside surfaces of the rim flanges (**Figure 3-16**). Every tire has three or four rim widths on which it can be mounted. Similarly, every rim has

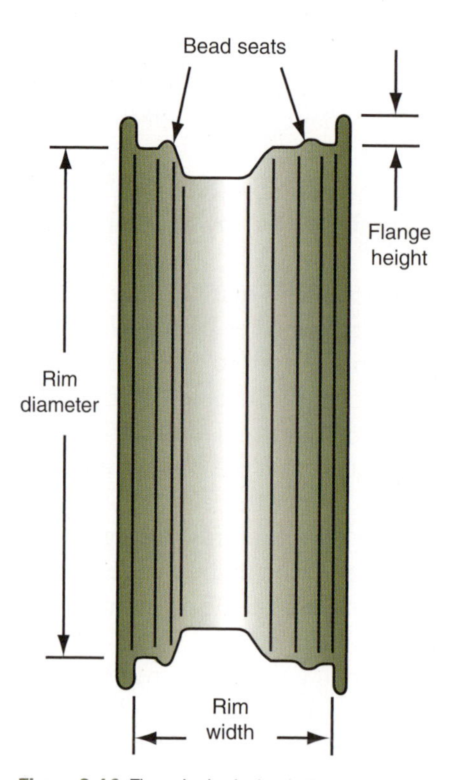

Figure 3-16 The principal wheel dimensions are diameter and width.

three or four tire sizes that it can safely hold. Although there is not a rigid one-to-one correlation between tire sizes and rim sizes, there are a limited number of combinations for any given wheel or tire.

If a tire is mounted on a wheel rim that is too wide or too narrow, the beads will not seat properly and the contact patch area will be affected. The tire may shift excessively on the rim and cause uneven braking. In the worst cases, the tire may lose air and be pulled off the rim under hard cornering or braking.

Wheel Offset

Wheel offset is the distance between the centerline of the rim and the mounting plane of the wheel (**Figure 3-17**). If the mounting plane is outboard of the centerline, the wheel has positive offset. This is the more common design, particularly for front-wheel drive (FWD) cars. If the mounting plane is inboard of the centerline, the wheel has negative offset. Negative offset is common on many aftermarket wheels, particularly for trucks and SUVs. Along with low-profile tires, reverse center rims were also used to change the appearance of the vehicle. This type of rim pushed the tire away from the centerline of the body. Some tires were almost completely out of the wheel well. Reverse center rims and low-profile tires completely change the steering and suspension geometry and affect the braking characteristics of the vehicle. This involves extra cost over the price of the tires and rims.

Wheel offset is the distance between the centerline of the rim and the mounting plane of the wheel.

The amount and direction of wheel offset are important design factors because they affect the load-carrying capacity of axles, spindles, hubs, and bearings. Wheel offset also affects the steering scrub radius and overstresses the wheel bearings.

The **scrub radius** is the distance from the tire contact patch centerline to the point where the steering axis intersects the road. If the tire contact patch centerline is outboard of the steering axis at the road, the car has positive scrub radius (**Figure 3-18**). If the tire contact patch centerline is inboard of the steering axis at the road, the car has negative scrub radius (**Figure 3-19**).

Scrub radius is the distance from the tire contact patch centerline to the point where the steering axis intersects the road.

Scrub radius is a design factor that is engineered for each vehicle. Negative scrub radius is the more common design because it tends to push the tires inward as they roll. In case of a blowout or a failure of half of a diagonally split brake system, negative scrub

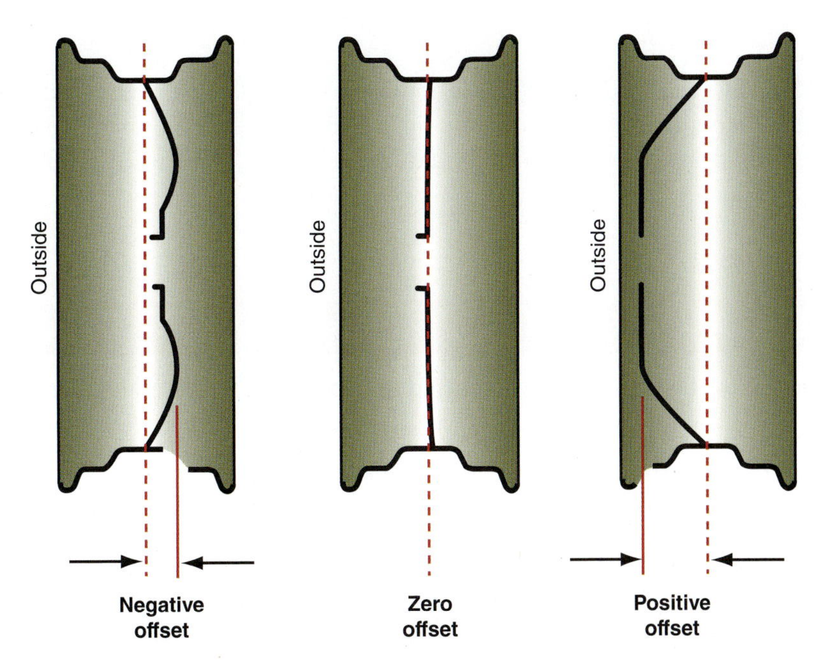

Figure 3-17 Wheel offset is the distance between the centerline of the rim and the mounting plane of the wheel.

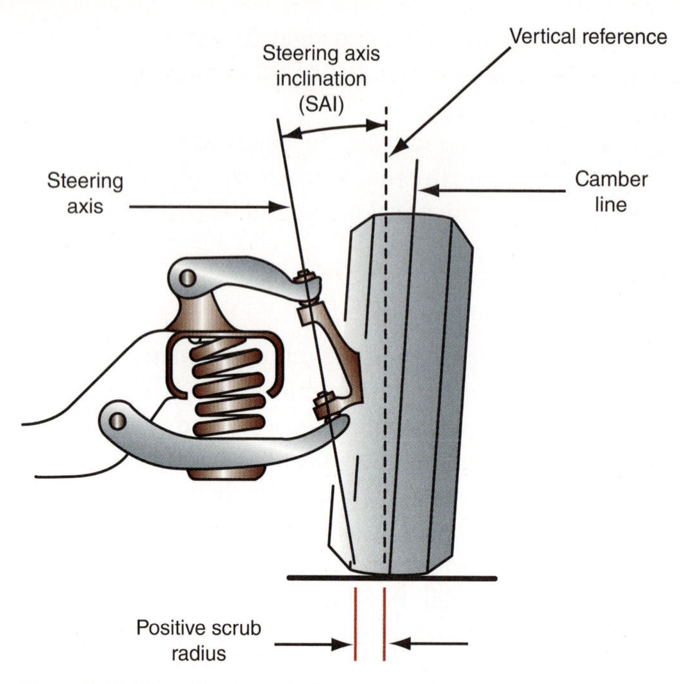

Figure 3-18 When the pivot point is inside the tire contact point, the scrub radius is positive.

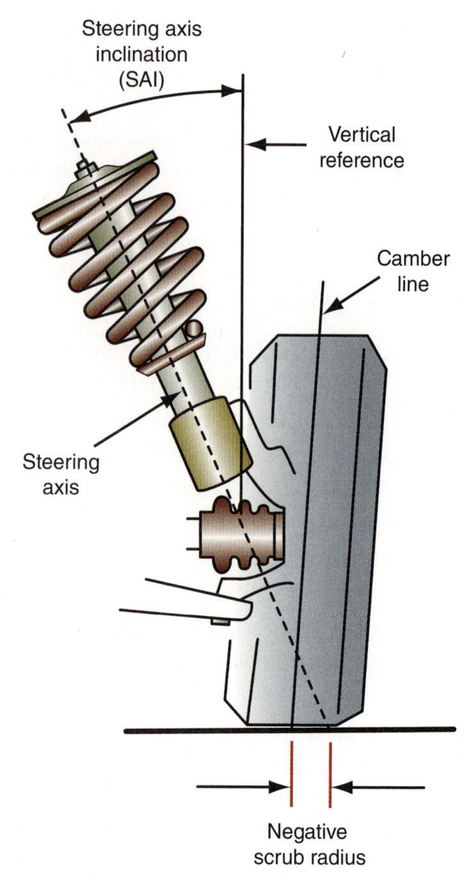

Figure 3-19 When the pivot point is outside the tire contact point, the scrub radius is negative.

radius helps to maintain vehicle stability. If scrub radius is changed from negative to positive by the installation of extremely offset wheels, the vehicle become unstable in case of tire or brake failure.

AUTHOR'S NOTE Scrub is not the amount of rubbing an oversize tire does on a vehicle component. *Scrub* is an engineering term involving the movement of the tire on the road and must be considered during steering, suspension, and other system designing.

WHEEL BEARINGS

Three basic kinds of wheel bearings are used on late-model cars and light trucks:

Shop Manual
page 106

1. Tapered roller bearings
2. Straight roller bearings
3. Ball bearings

Different types of bearings are designed for different types of loads. Radial loads are from the top down; the weight is on the radius of the bearing. An axial load is along the axis or side of the bearing. Ball bearings (**Figure 3-20**) have a small point of contact and therefore offer less resistance to turning and can handle radial as well as axial loads. For added strength, ball bearings can be installed in double rows. Straight roller bearings

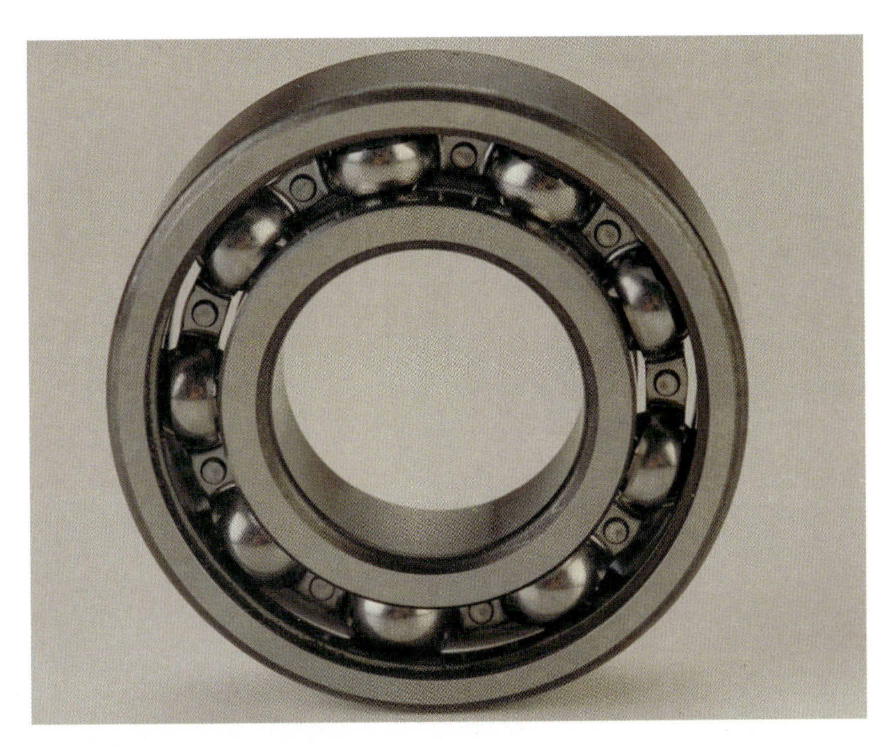

Figure 3-20 Ball bearing assembly.

(**Figure 3-21**) have a line contact that offers more resistance but can handle higher loads. Tapered roller bearings are designed for a high radial load as well as a high axial load, in one direction, and therefore are installed in pairs

Tapered roller bearings are commonly used on the front or rear wheels of a vehicle. A complete tapered roller bearing assembly consists of an outer race (bearing cup), an inner race (cone), tapered steel rollers, and a cage that holds the rollers in place (**Figure 3-22**). The inner race, the rollers, and the cage are a one-piece assembly and cannot be separated from one another. The outer race is normally pressed into the wheel hub. Tapered roller bearings are designed to take a radial as well as an axial load in one direction, so tapered roller bearings are installed in pairs with one large bearing set on the inboard side of the hub and a smaller bearing set on the outboard side. Tapered roller

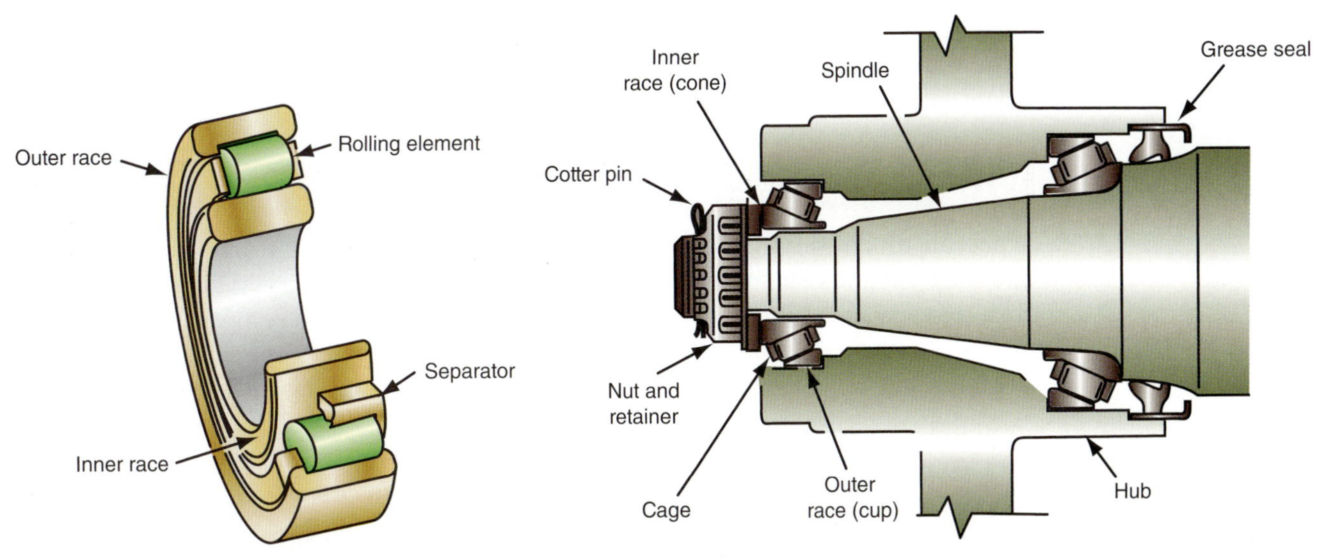

Figure 3-21 Roller bearing assembly.

Figure 3-22 Typical tapered roller bearing installation.

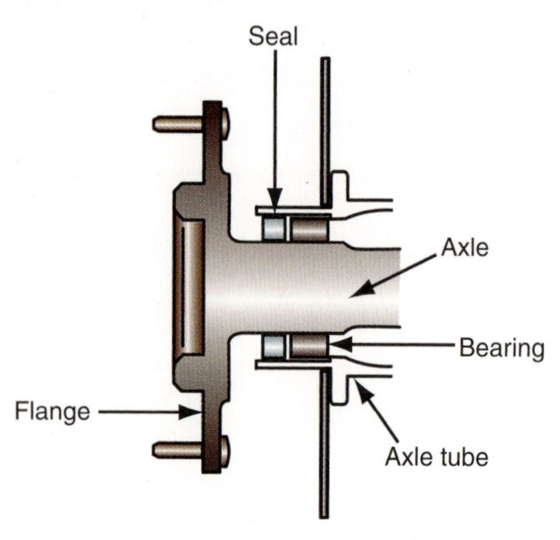

Figure 3-23 Straight roller bearings are used on most RWD axles.

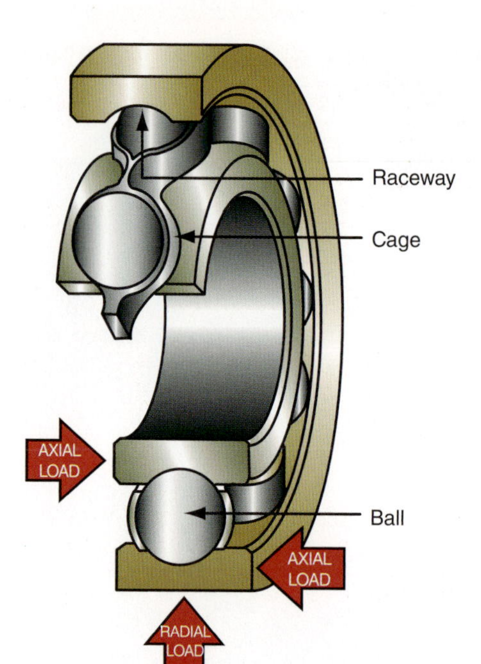

Figure 3-24 Typical ball bearing construction.

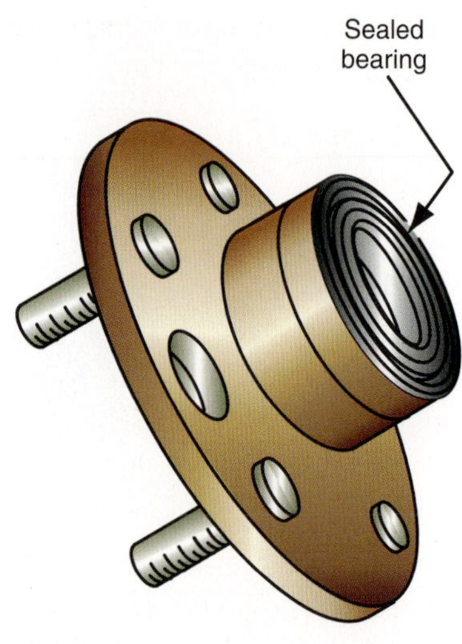

Figure 3-25 A typical sealed bearing may be pressed into the hub or may be an integrated part of the hub.

bearings must be cleaned and repacked with grease periodically and then have the end play adjusted. These bearing services are usually part of a complete brake job.

Straight roller bearings are used on the drive axles of most rear-wheel drive (RWD) vehicles (**Figure 3-23**). These bearing assemblies consist of an outer race pressed into the axle housing, straight steel rollers, and a cage to hold the rollers in place. The axle shaft acts as the inner bearing race. Because straight roller bearings are installed in the drive axle housing, they are lubricated by the final-drive lubricant. Periodic adjustment and repacking are unnecessary.

Ball bearings may be used on rear drive axles, and sealed double-row ball bearings are used on the front and rear wheels of many FWD vehicles. A typical ball bearing assembly consists of inner and outer races, steel balls, and a cage to hold the balls in place (**Figure 3-24**). Although ball bearings were used on the nondriven front wheels of many cars before 1960 and required periodic service, most modern ball bearing assemblies are sealed and permanently lubricated.

Most late-model cars have sealed double-row ball bearings or tapered roller bearings in an assembly that includes the wheel hub, the bearing races, and a mounting flange (**Figure 3-25**). These are nonserviceable, nonadjustable assemblies that must be replaced if damaged or defective.

WHEEL ALIGNMENT FUNDAMENTALS

Shop Manual
page 118

Wheel alignment service is based on maintaining the geometric relationships designed into the front and rear suspension and steering systems of a vehicle. These relationships are a series of angles in the suspension and steering. Some of these angles affect braking more than others. Improper alignment angles can cause a pull to one side during braking and improper handling. Traditionally, there are five basic steering and suspension angles:

1. Camber
2. Caster
3. Toe
4. Steering axis inclination
5. Toe-out on turns, or turning radius

Two other steering and suspension angles indicate the relationship among all four wheels. They are:

1. Thrust angle
2. Setback

Camber

Camber is the inward or outward tilt of the wheel measured from top to bottom and viewed from the front of the car. If the wheel tilts outward at the top, it has positive camber. If it tilts inward, camber is negative (**Figure 3-26**). Each front and rear wheel has its individual camber angle. Camber is measured in degrees.

Improper camber will wear the tire on one edge and cause steering problems based on the camber error. A vehicle will pull to the side with the most positive camber.

Caster

Caster is the backward or forward angle of the steering axis viewed from the side of the car. If the steering axis tilts backward at the top, the wheel has positive caster. If the steering axis tilts forward at the top, caster is negative (**Figure 3-27**). Each front

Improper caster usually will not cause abnormal tire wear but may cause steering problems.

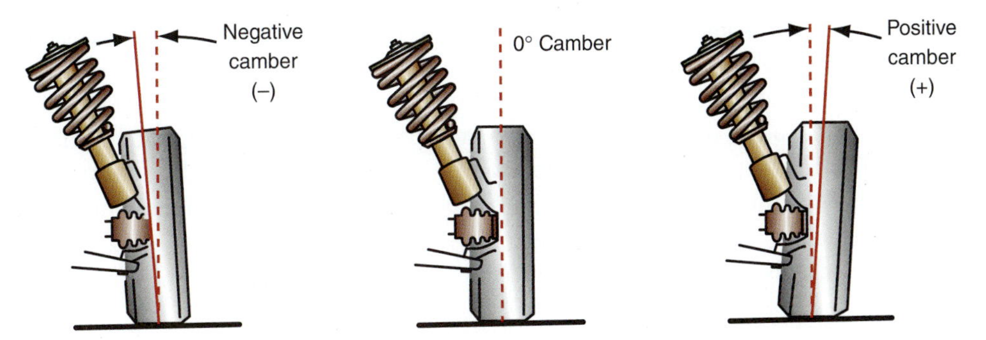

Figure 3-26 Camber is the inward or outward tilt of the wheel measured from top to bottom and viewed from the front.

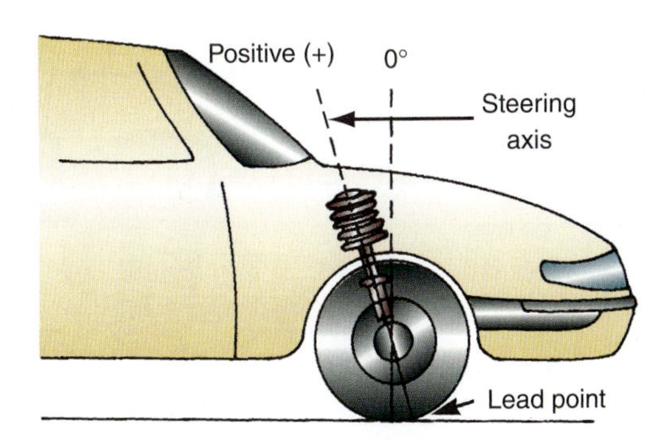

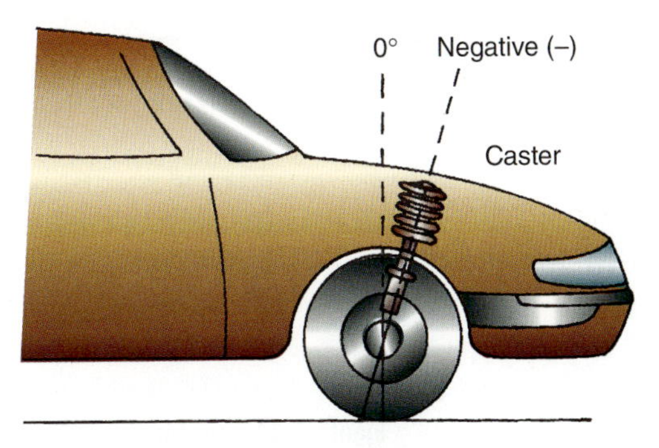

Figure 3-27 Caster is the backward or forward angle of the steering axis viewed from the side.

wheel has its individual caster angle. Most vehicles have a positive caster angle. Although front wheel drive vehicles are not as sensitive as rear wheel drive vehicles, a vehicle will pull toward the side with the most negative caster. Because the rear wheels are not steering wheels, rear caster is always zero. Caster is measured in degrees.

Toe

Improper toe will feather wear the tread and may cause steering problems.

Toe angle is the difference in the distance between the centerlines of the tires on either axle (front or rear) measured at the front and rear of the tires and at spindle height. If the centerlines of the tires are closer together at the front of the tires than at the rear, the wheels are toed in (**Figure 3-28**). If the centerlines of the tires are farther apart at the front of the tires than at the rear, the wheels are toed out. If the centerline distances are equal from front to rear on the tires, the toe angle is zero.

Steering Axis Inclination

Steering axis inclination (SAI) is the angle formed by the steering axis of a front wheel and a vertical line through the wheel when viewed from the front with the wheels straight ahead (**Figure 3-29**). SAI is measured in degrees independently for each wheel. SAI works with camber to maintain the vehicle load inboard on the larger front-wheel bearing. SAI is not adjustable, but it may be measured to identify suspension component damage. SAI, toe-out on turns, thrust angle, and setback are sometimes measured by alignment shops to determine why the wheel alignment cannot be brought to specifications. The technician uses measurements to determine if the vehicle has suffered accident damage or perhaps the steering wheels were bounced off a road curb.

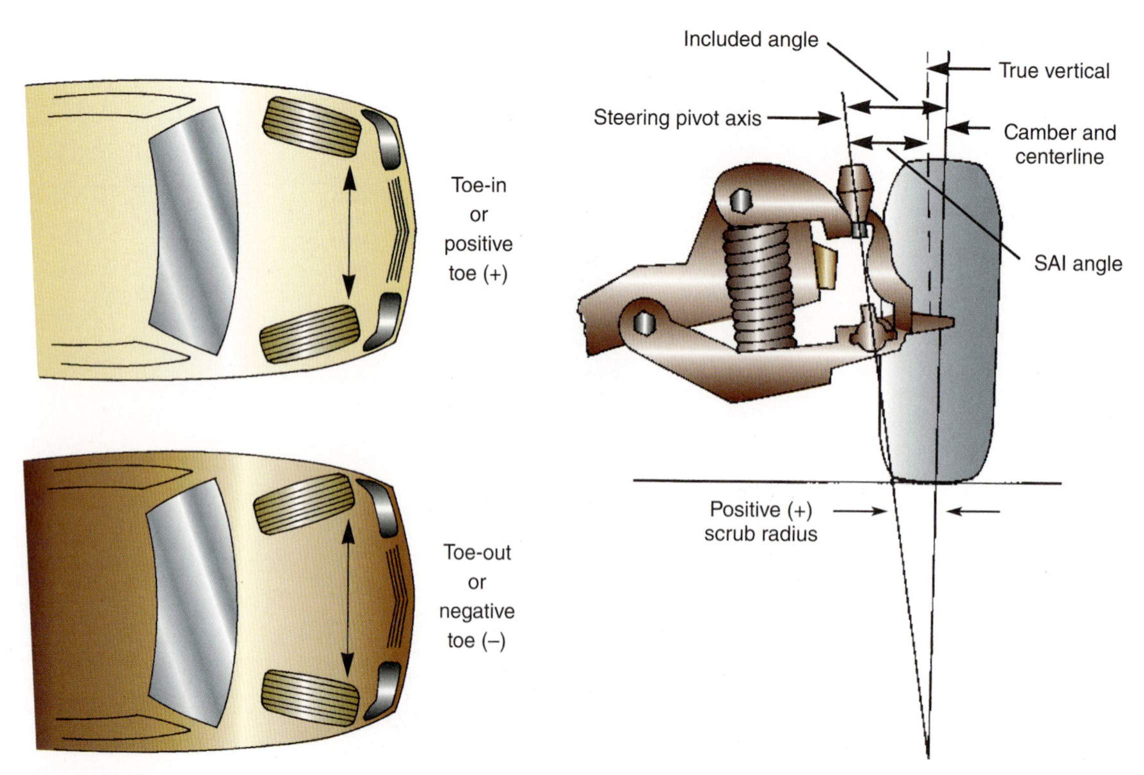

Figure 3-28 The toe angle is the difference in the distance between the centerlines of the tires measured at the front and rear of the tires.

Figure 3-29 Steering axis inclination is the angle formed by the steering axis of a front wheel and a vertical line through the wheel.

Toe-Out on Turns

Toe-out on turns (turning radius) is the difference between the angles of the front wheels in a turn. Because of suspension and steering design, the inside front wheel turns at a greater angle during a turn (**Figure 3-30**). Thus, it toes out.

Toe-out on turns is measured in degrees by turning one wheel a specified amount and then measuring the angle of the other wheel. The difference between the two is the turning angle or the toe-out on turns. Toe-out on turns prevents tire scuffing and tire squeal during cornering. Toe-out on turns is not adjustable, but it may be measured to identify suspension or steering component damage.

Thrust Angle

To understand the **thrust angle,** you must understand the geometric centerline of the vehicle and the thrust line of the vehicle. The thrust angle is the angle between the geometric centerline and the thrust line (**Figure 3-31**).

The **geometric centerline** is a static dimension represented by a line through the center of the vehicle from front to rear. It can be used as a reference point for individual front- or rear-wheel toe, but it does not represent the direction in which the rear wheels are steering the vehicle. The **thrust line** is the bisector of total toe on the rear wheels. Put more simply, the thrust line is the direction in which the rear wheels are pointing.

Setback

Setback refers to a difference in wheelbase from one side of the car to the other (**Figure 3-32**). Some vehicles are built with setback as a design feature. Ford trucks with twin I-beam front suspension are some of the best-known examples. Setback also can be

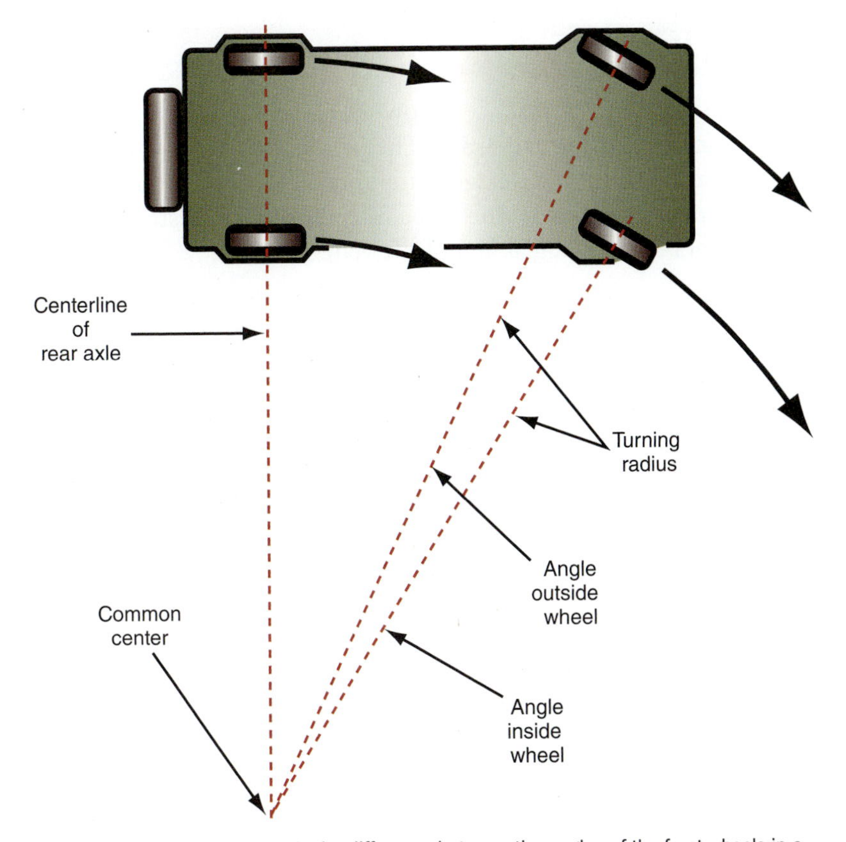

Figure 3-30 Toe-out on turns is the difference between the angles of the front wheels in a turn.

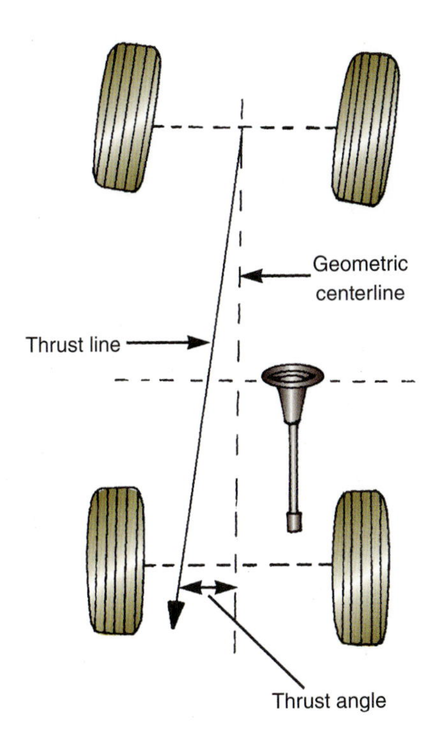

Figure 3-31 The thrust angle is the angle between the geometric centerline and the thrust line.

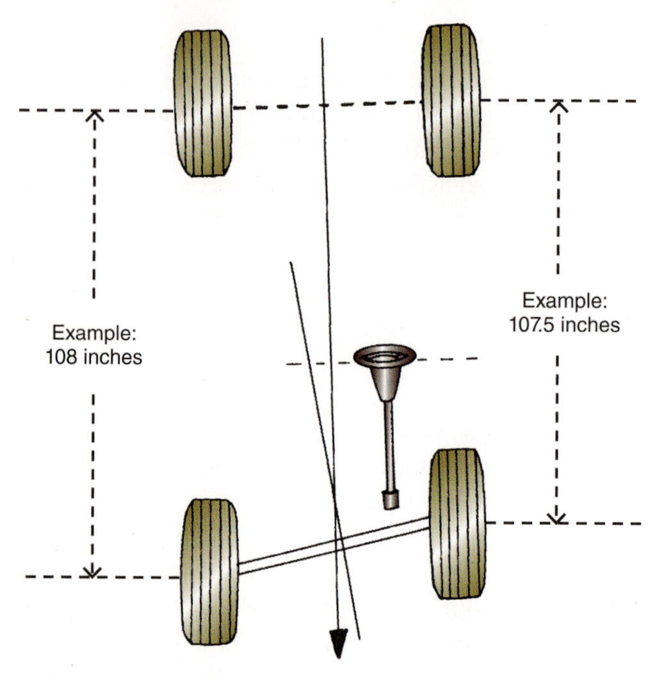

Figure 3-32 Setback is the difference in wheelbase from one side of the car to the other (exaggerated examples).

Shop Manual
page 121

present in a vehicle where it is not supposed to be. This is usually the result of a collision in which the frame is bent or suspension components are severely knocked out of position. Accidental setback also can change the vehicle thrust angle and cause handling problems.

EFFECTS ON BRAKING PERFORMANCE

The bottom line of the braking system performance is how the tire contacts and holds the pavement, or the **tread contact patch** (**Figure 3-33**). The previous sections outline the overall requirements for tires, suspension, and related components. Major deviations from these requirements not only can result in poor ride and control through the steering, but may and will in some cases affect the braking ability of the vehicle.

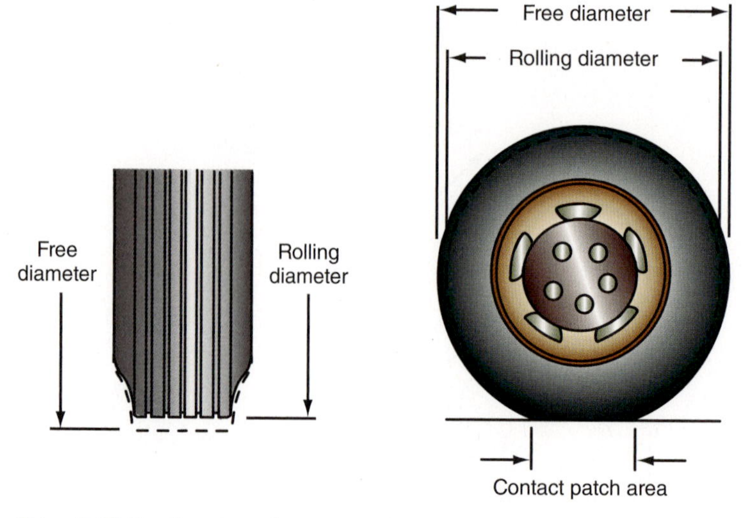

Figure 3-33 Tire diameter and contact patch area are important factors in trouble-shooting brake performance.

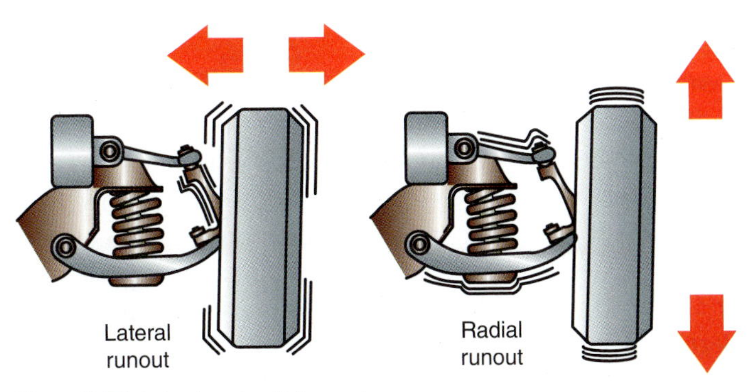

Figure 3-34 Lateral and radial runout.

The tread contact patch's friction ability is determined not only by the tire size, but also by how the tires contact the road. One primary failure is over- or underinflated tires. This causes a larger or smaller contact patch and a corresponding reduction or increase of braking friction. The angle at which the tire contacts the road may also affect braking.

Camber, caster, or toe angles that are excessively negative or positive will affect tire wear and the angle and, to some extent, the size of the contact patch. The "perfect" contact patch has all the treads on the bottom of the tire in contact with the road. Excessive alignment angles may cause the tire to push toward one side during driving. This push or pull can be greatly increased as the tire tries to push but the vehicle is trying to go straight ahead. Even if the alignment is correct, worn or broken steering and suspension components may cause the same problem.

Wheels and tires that do not meet the vehicle manufacturer's specifications can and do cause the steering and suspension angles to change. Lifting or lowering the vehicle will also change those angles. Worn or damaged wheel or axle bearings can cause the wheel and tire assembly to wobble during operation. Excessive wobble is called runout. Runout around the outside is called **radial runout,** so the tire moves up and down. Excessive side-to-side variation is called **lateral runout,** or **side-to-side** (**Figure 3-34**).

Technicians must pay attention to what their customers are saying when they complain of a brake problem. Chapter 3 of the *Shop Manual* discusses ways and means for the technicians to isolate apparent braking problems from reduced performance of related systems and components. Remember that the engineers designed each system to work together. If one system fails to perform properly, then the other systems will not perform properly. This is especially true of brake, suspension, and steering systems because of their shared components and overlapping operation.

PERFORMANCE TIRES, WHEELS, AND ALIGNMENT

AUTHOR'S NOTE This section should give the reader an idea of differences between production vehicles and race vehicles. There are two classifications of "performance" vehicles. One classification consists of production lines like the Viper, Corvette, and Mustang, whereas the other consists of actual racing vehicles. The theory of operation for all performance vehicles is the same as that of a straight production product. However, race vehicles are usually "hand- engineered" and "hand-built." In other words, they work well on a controlled track, but not very well in downtown Atlanta traffic. A short section on the major differences between production and race products is included at the end of appropriate chapters.

Tires

Racing tires are engineered and manufactured for one reason: to provide traction for a particular track with a specific surface for speed and control. Most oval track (NASCAR) tires have no tread, but they provide much better traction than a typical treaded tire. However, there is one very distinct drawback: They do not work well on a damp track. Have you ever seen a NASCAR race conducted in the rain? Road course racers (Formula 1, Grand Prix) do use treaded tires because they run the race regardless of rain, but many times the rubber and chemical compound that these tires are made of is very similar to that of oval track tires.

Race tires for NASCAR are actually leased from the respective tire suppliers. A set of racing tires could cost more than some cars on the road. Goodyear is the primary tire supplier for NASCAR and develops a tire specifically for a certain track. Other tire manufacturers like Bridgestone and Michelin supply tires for road course competitors like Formula 1 and Grand Prix. The tire may be of a softer compound with a short life and better grip or a hard compound with long life but not quite as much traction. The tire used on the short track at Martinsville, Virginia, with top speeds of about 130 mph is very different from the tire for Atlanta with top speeds in the 200-mph range. Race vehicles are usually equipped with an "inner" tire to help prevent blowouts. This is basically a little smaller tire mounted inside the visible tire.

Racing tires provide an excellent proving ground for tire engineers to test rubber compounds for wear and traction. Most of today's production tire designs and chemical compounds originated from research data collected during races. In fact, some of the ideas behind run-flat tires came from the oval track.

Wheels

Like the tires, the basic design of a racing wheel is very similar to that of a production wheel. Depending on the track and expected speeds, it may have more positive or more negative offset. Also, the bead-holding area of the wheel is usually a little deeper for a better grip on the tire bead. They are almost never made of anything other than high-quality steel, although some manufacturers are working on designs from carbon-fiber. The side forces occurring during any race will destroy a production wheel in a fairly short period of driving. Typically racing wheels are a little wider than production wheels with the same size diameter mounted. Other than the extremely high cost to purchase wheels of this type, they could be used on production vehicles. Generally speaking, racing wheels are bland in appearance so few people want to pay an extra $5,000 to get wheels that would never wear out but do not look any better or racier.

Wheel Alignment

This area contains probably the biggest and most noticeable difference between production vehicles and racers. The theory behind wheel alignment is the same, but the application of that basic theory to racers is extreme. Road course racers use an alignment setup close to that of production, but oval track vehicles definitely appear to be completely abnormal to a nonexperienced passerby.

If a vehicle is set up for Atlanta, Daytona, or Talladega, it would appear that all four wheels are laid in (negative camber) (**Figure 3-35**). The camber is set very negative compared to production vehicles. This odd setup also has a very positive effect on vehicle handling. Because of the downforce on the vehicle racing on high-speed tracks, the wheels are forced out at the top, creating zero or positive camber. This is true of all vehicles when they are loaded or in some other condition in which the body is forced downward. However, routine operation on the public roads does not create the speed and resulting downforce required to change the camber setting. Also most oval track race vehicles are

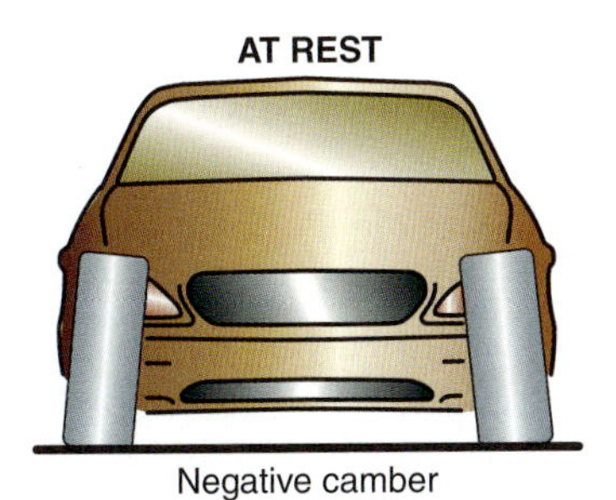

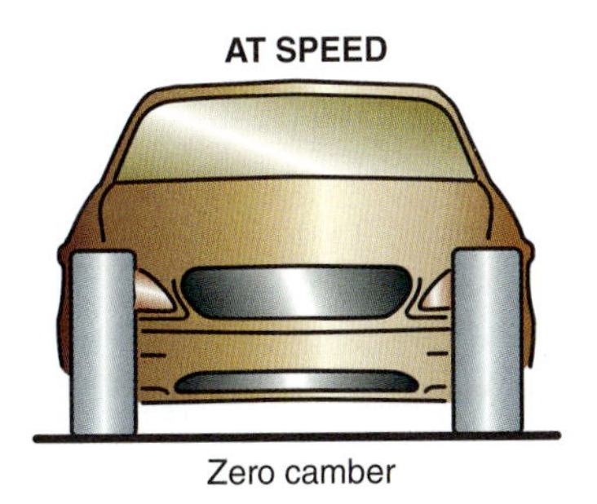

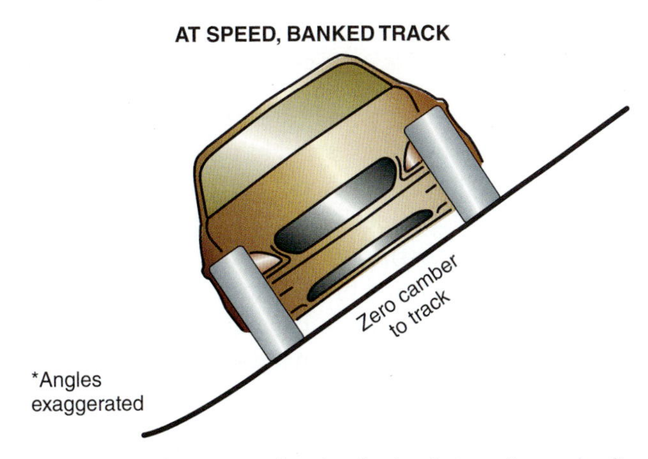

Figure 3-35 A simplistic race setup has the wheels set at negative camber (top left), and the downward force of the airflow and vehicle mass cause the wheels to move to zero or positive camber (bottom).

set up with a more positive camber on the left wheels than on the right. During that 200-mph left turn, the vehicle body is forced downward and leans or tilts hard right. This changes the camber of the right-side tires to be more positive than the left. If done correctly, all four tires try to stand straight up, and the tread becomes flat on the pavement during those high-speed turns; hence, more and better traction is present when it is needed the most. Short-track ovals require a different camber setup because the speeds are much slower and the track surface is flatter.

Toe on a racer is not quite as important as camber, but it is set up with some differences from production vehicles. Because of body lean, excessive camber, and the construction of the suspension system, toe measurements are set up not only to control the vehicle but also to help keep the tread straight on the track surface. Of more importance to a race vehicle is toe-out on turn. Like production vehicles, this angle is determined by the construction and initial service on the suspension and steering systems and is not adjustable.

Another factor influencing racing setup is the banking angle of the track surface when compared to horizon. The high-speed tracks have up to 36 degrees of banking in the turns. This banking is one of the main factors for selecting the camber and toe settings. A high-bank track does not require as much offset camber as a short track (1 mile or less in length) with little banking. This is because the centrifugal forces at a low-bank track affect the sideways movement of the vehicle much more. The vehicle tends to slide sideways because the centrifugal force is about 90 degrees to the surface of the track or horizon. The camber is set to account for this force and body lean. A vehicle on a high-bank track tends to be pushed into the pavement because the pavement is at an angle to the horizon. Although the centrifugal force may be much greater, the track surface is flatter with regard to the vehicle chassis, forcing the vehicle into the track instead of across it. All of this means nothing, however, unless the suspension and steering systems are also matched to the track and vehicle.

It would not be a good idea to set up a production vehicle like a racer with regard to tires, wheels, suspension, and steering. Setting the high costs aside, the tires would wear out very quickly and the vehicle would ride like a horse-drawn wagon. That is assuming the driver has the strength to fight for control during straight-ahead and right-turn driving because most racing vehicles do not steer easily until operating at high speeds. It is suspected that the driver would wear out before the tires.

SUMMARY

- If a vehicle has identical tires at all four wheels of the size and type recommended by the car manufacturer, brake performance should be at its best.
- Radial ply tires are used almost universally on late-model cars and light trucks.
- Tire size identification, maximum inflation pressure, and other specifications are molded into tire sidewalls.
- Tire and wheel runout, as well as wheel rim width and offset, can contribute to brake problems if they deviate from the car manufacturer's specifications.
- Late-model cars and light trucks use tapered roller bearings, straight roller bearings, and ball bearings at their wheels and axles.
- If wheel bearings have too much end play (looseness), wheel runout may be excessive and cause uneven braking and a pull to one side. Excessive runout also can contribute to pedal pulsation.
- The traditional wheel alignment angles of caster, camber, and toe can contribute to brake pull problems if they are out of specification or vary greatly from side to side.
- Loose or damaged steering and suspension parts also can cause brake pull problems.

REVIEW QUESTIONS

Essay

1. Explain the problem with having tires of different sizes on a vehicle as it relates to the ABS system.
2. Describe the information contained on the tire information placard.
3. Explain two characteristics of radial tires that won over drivers in the 70s.
4. Explain why radial tires have the best braking performance.
5. Describe the three most common tire measurements in the P-metric system.
6. Describe what is meant by a low-profile tire.
7. Describe the treadwear rating of the UTQG.
8. Explain how to decode the NHTSA date from the DOT number on the sidewall.
9. Explain why street tires all have a tread pattern, even though racing slicks have the best traction.
10. Describe how a straight roller bearing is suited to handling a radial load.

Fill in the Blanks

1. Brake systems are engineered in relation to many vehicle factors of ______________, ______________, and______________.
2. The cords in a ______________ply tire run at 90 degrees to the tread, creating a more flexible sidewall
3. The inside ______________is the same as the wheel size, or rim ______________on which the tire fits.
4. The tire ______________ ______________ is indicated by the letters "A" through "L," which often appear after the size specification on the tire sidewall.
5. Runout around the outside of the tire is called ______________ runout, so the tire moves up and down.

6. The bottom line of the braking system performance is how the tire contacts and holds the pavement, or the ______________ ______________ ______________.

7. ______________ refers to a difference in wheelbase from one side of the car to the other.

8. The ______________ ______________ is a static dimension represented by a line through the center of the vehicle from front to rear.

9. Because of suspension and steering design, the inside front wheel turns at a ______________ angle during a turn.

10. ______________ angle is the difference in the distance between the centerlines of the tires on either axle (front or rear) measured at the front and rear of the tires and at spindle height.

Multiple Choice

1. Technician A says brake shoes or pads apply friction to the wheels. Technician B says it is the friction between the tires and the road that actually stops the car. Who is correct?

   A. A only
   B. B only
   C. Both A and B
   D. Neither A nor B

2. Technician A says manufacturers may recommend one or two optional tire sizes at the rear that are larger than the front original equipment size. Technician B says a large variation from the carmaker's recommendation will not cause any problems. Who is correct?

   A. A only
   B. B only
   C. Both A and B
   D. Neither A nor B

3. Technician A says the tire casing is steel beads around the rim and layers of cords or plies that are bonded together to give the tire its shape and strength. Technician B says the tire tread is additional layers of sidewall and under-tread rubber added to the carcass. Who is correct?

   A. A only
   B. B only
   C. Both A and B
   D. Neither A nor B

4. Technician A says section width is the width of the tire across the widest point of its cross section, usually measured in inches. Technician B says aspect ratio is the ratio of the cross-sectional height to the cross-sectional width expressed as a percentage. Who is correct?

   A. A only
   B. B only
   C. Both A and B
   D. Neither A nor B

5. Technician A says a high aspect ratio indicates a tall tire. Technician B says most modern tires have aspect ratios from 45 to 80. Who is correct?

   A. A only
   B. B only
   C. Both A and B
   D. Neither A nor B

6. Technician A says the advantage of a low-profile tire is handling. Technician B says an advantage of a low-profile tire is ease of steering.

   A. A only
   B. B only
   C. Both A and B
   D. Neither A nor B

7. Technician A says an advantage of low-profile tires is the fact that there is more tire sidewall to absorb shocks from hitting irregularities in the road. Technician B says more sidewall of the low-profile tire can lead to a smoother ride and can also prevent wheel damage. Who is correct?

   A. A only
   B. B only
   C. Both A and B
   D. Neither A nor B

8. Technician A says the maximum cold inflation pressure for any tire is molded into the tire sidewall. Technician B says this is the inflation pressure after the tire has been standing for 1 hour. Who is correct?

   A. A only
   B. B only
   C. Both A and B
   D. Neither A nor B

9. Which is true of the Uniform Tire Quality Grading (UTQG) indicators?

   A. They have been used since 1985.
   B. They use a number to indicate relative tread life.
   C. They use a letter to indicate wet weather traction and heat resistance.
   D. All the above

10. All passenger car and light truck tires have tread wear indicators that appear as continuous bars across the tread when the tread wears down to the last 2/32 (1/16) inch. This practice started in:

    A. 2002
    B. 1968
    C. 1955
    D. 1974

CHAPTER 4
MASTER CYLINDERS AND BRAKE FLUID

Upon completion and review of this chapter, you should be able to:

- Explain the differences between different DOT brake fluid specifications.
- Explain proper brake fluid handling procedures.
- Identify the parts and explain the operation of a brake pedal and pushrod.
- Describe the mechanical advantage provided by the brake pedal and linkage.
- Explain the purpose and operation of the front-to-rear and diagonally split hydraulic systems.
- Describe the purpose of the master cylinder.
- Identify the main parts of a master cylinder.
- Explain the operation of a basic master cylinder.
- Describe the parts and operation of a dual-piston master cylinder.
- Describe the parts and explain the operation of a quick take-up master cylinder with a fast-fill valve.

Terms To Know

Cup seal
Diaphragm
Free play
Hydraulic system mineral oil (HSMO)
O-ring
Polyglycol
Quick take-up master cylinder
Quick take-up valve
Replenishing port
Reservoir
Residual pressure check valve
Vent port

INTRODUCTION

This chapter covers hydraulic fluid, which makes the brake system work, as well as the master cylinder construction and operation. The master cylinder converts the force on the brake pedal to each of the four wheel brakes to stop the car. The master cylinder changes the driver's mechanical force on the pedal to hydraulic pressure, which is changed back to mechanical force at the wheel brakes. The master cylinder uses the fact that fluids are not compressible to transmit the pedal movement to the wheel brake units. The master cylinder also uses the principles of hydraulics to increase the pedal force applied by the driver. The brake pedal's construction and operation are also discussed in this chapter as part of the master cylinder operation.

HYDRAULIC BRAKE FLUID

The specifications for all automotive brake fluids are defined by Society of Automotive Engineers (SAE) Standard J1703 and Federal Motor Vehicle Safety Standard (FMVSS) 116. Fluids classified according to FMVSS 116 are assigned DOT numbers: DOT 3, DOT 4,

DOT 3/4, DOT 5, and DOT 5.1. Basically, the higher the DOT number, the more rigorous the specifications for the fluid.

Shop Manual
page 142

These specifications list the qualities that brake fluid must have, such as:

- Free flowing at low and high temperatures
- A high boiling point (over 400°F or 204°C)
- A low freezing point
- Noncorrosive to metal or rubber brake parts
- Ability to lubricate metal and rubber parts
- Ability to absorb moisture that enters the hydraulic system

Choosing the right fluid for a specific vehicle is not based on the simple idea that if DOT 3 is good, DOT 4 must be better, and DOT 5 better still. Almost all carmakers specify DOT 3 fluid for their vehicles; but Ford calls for a heavy-duty variation, which meets the basic specifications for DOT 3 but has the higher boiling point of DOT 4. Import manufacturers are about equally divided between DOT 3 and DOT 4.

DOT 3 and DOT 4 fluids are polyalkylene-glycol-ether mixtures, called **polyglycol** for short. The color of both DOT 3 and DOT 4 fluid ranges from clear to light amber. DOT 5 fluids are all silicone based because only silicone fluid—so far—can meet the DOT 5 specifications. No vehicle manufacturer, however, recommends DOT 5 fluid for use in its latest brake systems. Although all three fluid grades are compatible in certain aspects, they do not combine if mixed together in a system. DOT 5, a silicone-based fluid, should never be mixed with or used to replace other types of brake fluids. Therefore, the best general rule is to use the fluid type recommended by the carmaker and to not mix fluid types in a system.

Polyglycol stands for polyalkylene-glycol-ether brake fluids that meet specifications for DOT 3 and DOT 4 brake fluids.

Brake Fluid Boiling Point

The most apparent differences among the five fluid grades are the minimum boiling points as listed in **Table 4.1**.

The boiling point of brake fluid is important because heat generated by braking can enter the hydraulic system. If the temperature rises too high, the fluid can boil and form vapor in the brake lines. The stopping power of the system then will be reduced. The pedal must be pumped repeatedly to compress the vaporized fluid and build up pressure at the brakes to stop the vehicle. However, it would require a lot of frequent, very hard braking action to bring clean brake fluid to its boiling point. About the only people who would do this are race drivers during a race. But old contaminated brake fluid may contain water that could boil during braking.

The dry boiling point is the minimum boiling point of new, uncontaminated fluid. After brake fluid has been in service for some time, however, the boiling point drops because of water contamination. Polyglycol fluids are hygroscopic, which means that they readily absorb water vapor from the air. Brake systems are not completely sealed, and some exposure of the fluid to air is inevitable. SAE field evaluation program R11 showed that the average 1-year-old car has about 2 percent water in its brake fluid. Even this small

TABLE 4-1 MINIMUM BOILING POINTS FOR THE SIX BRAKE FLUID GRADES

Boiling Point	DOT 3	DOT 4	DOT 5	DOT 3/4	DOT 5.1
Dry	401°F(205°C)	446°F(230°C)	500°F(260°C)	500°F(260°C)	585°F(307°C)
Wet[1]	284°F(140°C)	311°F(155°C)	356°F(180°C)	347°F(175°C)	365°F(185°C)

[1]Wet means the fluid is saturated with water.
DOT 5.1 long-life has the same dry and wet boiling point of 424°F(218°C)

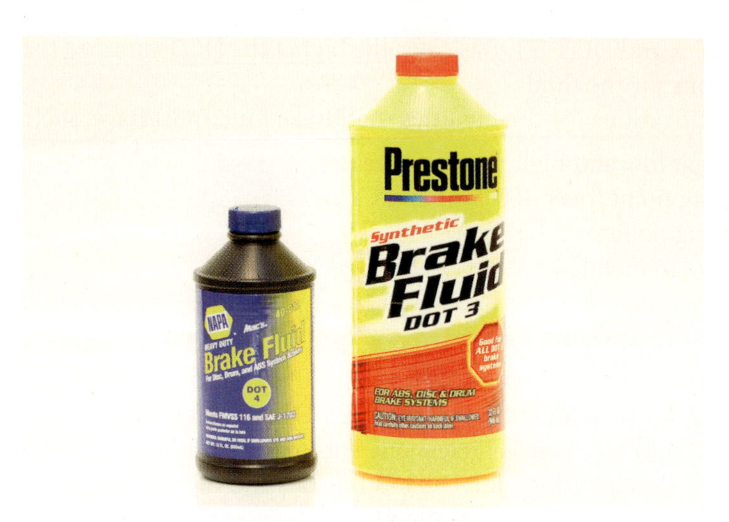

Figure 4-1 Brake fluid should be purchased only in the quantities needed for the job.

amount is enough to lower the boiling point of DOT 3 fluid from 401°F(205°C) to less than 320°F(160°C). DOT 4 fluid absorbs less moisture and maintains a higher boiling point than DOT 3 fluid.

Because both DOT 3 and DOT 4 fluids absorb moisture from the air, always keep containers tightly capped. Reseal brake fluid containers immediately after use. Although brake fluid is available in containers as small as 12 ounces to as large as 5 gallons (**Figure 4-1**), it is often wise to buy small containers and keep them sealed until needed. Taking these steps to minimize water in the brake fluid helps reduce corrosion of metal parts and deterioration of rubber components. It also keeps the brake fluid boiling point high to minimize the chance of vapor formation.

AUTHOR'S NOTE Although brake fluid may be purchased in quantities larger than 1 quart, it will be a special-order item. Very few public parts vendors or repair shops keep anything larger than a 1-quart container of brake fluid. Even a "brake-only" repair shop would seldom need brake fluid in quantities that would justify storing 5-gallon containers of it. Plus, when you consider that brake fluid must be placed in a smaller container for ease of pouring and labeled accordingly, it is not worth the trouble. It is recommended that brake fluid should always be purchased in 1-quart or smaller containers.

The DOT rating is found on the container of brake fluid (**Figure 4-2**). The vehicle service information and owner's manual specify what rating is correct for the car. Do not use a brake fluid with a lower DOT rating than specified by the vehicle manufacturer.

DOT 3 fluid will last longer than DOT 4 under ideal conditions; that is, the system is clean, sealed, and kept sealed or "perfect." DOT 3 is considered to be a lifetime fluid under those conditions, whereas DOT 4 still must be flushed out and replaced every 1–3 years under the same conditions. It is best to completely flush light vehicle brake systems about every 2–3 years because no typical brake system is "perfect."

The synthetic fluids have higher wet and dry boiling points. This makes them, at least in theory, better for braking and durability while in the braking system. The service life of DOT 3/4 is about 24 months (2 years), whereas DOT 5.1 is rated for about 60 months (5 years). Long-life DOT 5.1 is rated at a service life of up to 10 years.

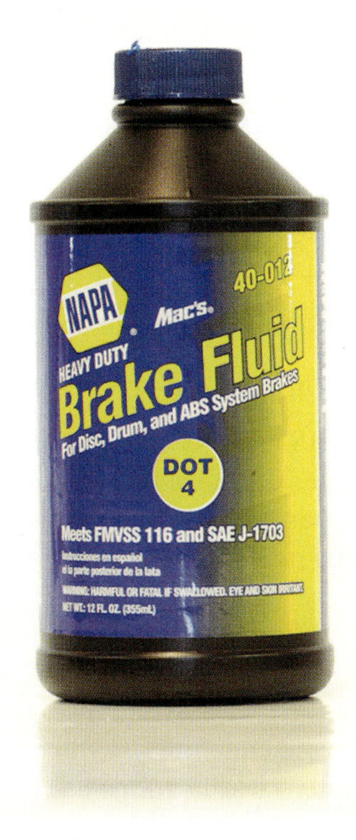

Figure 4-2 The DOT number is always on the label of a brake fluid container.

Other Brake Fluid Requirements

A high-temperature boiling point is not the only requirement that brake fluid must meet. Brake fluid must remain stable throughout a broad range of temperatures, and it must retain a high boiling point after repeated exposures to high temperatures. Brake fluid also must resist freezing and evaporation and must pass specific viscosity tests at low temperatures. If the fluid thickens and flows poorly when cold, brake operation would suffer at low temperatures.

In addition to temperature requirements, brake fluid must pass corrosion tests, it should not contribute to deterioration of rubber parts, and it must pass oxidation-resistance tests. Finally, brake fluid must lubricate cylinder pistons and bores and other moving parts of the hydraulic system.

DOT 5 Silicone Fluid

DOT 5 Silicone fluid does not absorb water. This purple fluid has a very high boiling point, is noncorrosive to hydraulic system components, and does not damage paint as does ordinary fluid. However, DOT 5 fluid has other characteristics that are not so beneficial.

Silicone fluid compresses slightly under pressure, which can cause a slightly spongy brake pedal feel. Silicone fluid also attracts and retains air more than polyglycol fluid does, which makes brake bleeding harder; it tends to out gas slightly just below its boiling point; and it tends to aerate from prolonged vibration. DOT 5 fluid has other problems with seal wear, water accumulation, and separation in the system. All of these factors mean that DOT 5 silicone fluid should never be used in an ABS.

Synthetic Fluids

As mentioned earlier, synthetic brake fluids are now available on the market. DOT 3/4 and DOT 5.1 are compatible with DOT 3 and DOT 4 fluids and can be mixed with them.

DOT 5.1 is based on polyglycol chemistry. Both can be used to flush and replace DOT 3 or DOT 4 fluid in a brake system and can be used in an ABS. They are distinguished from each other by color: DOT 3 and DOT 4 are clear and amber, DOT 3/4 is clear and pale blue, DOT 5.1 is crystal clear, and the long-life version of DOT 5.1 is clear and yellow in color. DOT 5.1 has extra solvency to put gum and sludge deposits into suspension with additional anticorrosion additives in the long-life version.

DOT 5.1 and DOT 5.1 long-life brake fluids are not silicone-based fluids. Do not mix or replace DOT 5 fluid with DOT 5.1 or DOT 5.1 long-life fluids. Damage to the brake system and possible injury could occur due to damage to brake system components.

Hydraulic System Mineral Oil Fluids

HSMO fluid has some good should only be used in Jaguar and Rolls Royce vehicles because it is mineral oil based.

Hydraulic system mineral oil (HSMO) is the rarest kind of brake fluid, used by only two carmakers: Jaguar and Rolls Royce. HSMO is not a polyglycol or silicone fluid; rather it is made from a mineral oil base. It has a very high boiling point, it is not hygroscopic, it is a very good lubricant, and it actively prevents rust and corrosion. HSMO fluid can be identified by its green color, and the reservoir cap will also be green in color.

Because HSMO is petroleum based, systems designed for its use also require seals made of special rubber. If polyglycol or silicone fluid is used in a system designed for HSMO, these fluids will destroy the HSMO system seals. Similarly, if HSMO is used in a system designed for polyglycol or silicone fluid, it will destroy the seals of those systems. HSMO is not covered by the DOT classifications of FMVSS 116 and is not compatible with DOT fluids.

Fluid Compatibility

Although the performance requirements of DOT 3, DOT 4, DOT 5, and synthetic fluids are different, FMVSS 116 requires that the four grades of fluid must be compatible with each other in a system. Mixing different types of fluid in a system is not recommended, but it can be done without damaging the system or creating a damaging reaction between two types of fluid.

It is important to remember that if DOT 3 and DOT 4 fluids are mixed in a system, for example, the boiling point of the DOT 4 fluid will be reduced by the same percentage as the percentage of DOT 3 fluid in the mixture. Thus, overall system performance may be compromised by mixing fluids.

Although FMVSS 116 requires DOT fluids to be compatible, it does not require them to be homogeneous. They are not required to blend into a single solution unless the fluids are of a single type, such as two DOT 3 fluids or two DOT 4 fluids.

Silicone DOT 5 fluid has a lower specific gravity than polyglycol fluid. If the two types are mixed, they do not blend; the silicone fluid separates and floats on top of the polyglycol fluid.

Although the synthetic fluids may be compatible with the DOT 3 and DOT 4 fluids, they also should not be mixed except in an emergency. They will blend, but the performance level of the synthetics will decrease in proportion to the amount of polyglycol-based fluid remaining in the system. If a synthetic fluid is requested by the customer, then a complete flushing and removal of all polyglycol fluids must be accomplished. This would also be a good time to check for any damage to the brake seals. If the fluid has been in the system for several years, it would benefit the customer to replace or rebuild those brake components housing flexible or fixed seals.

The best practice is to use a single, high-quality brand of brake fluid of the DOT type specified for a particular vehicle. Avoid mixing fluids whenever possible.

Brake Fluid Storage and Handling

Polyglycol fluids have a very short shelf life. As soon as a container of DOT 3 or DOT 4 polyglycol fluid is opened, it should be used completely because it immediately starts to absorb moisture from the air. DOT 5 silicone fluids and HSMO fluids are not hygroscopic and can be stored for long periods of time.

All brake fluids must be stored in clean containers in clean, dry locations. Preferably, brake fluid should be stored in its original container with all labeling intact. Containers must be kept tightly closed when not in use. Transfer containers must be properly labeled.

When working with brake fluid, do not contaminate it with petroleum-based fluids, water, or any other liquid. Also keep dirt, dust, or any other solid contaminant away from the fluid. When filling a system with polyglycol fluid, keep it off painted surfaces and off your skin.

If possible, buy brake fluid in small quantities to limit storage times. Plastic containers could allow air and/or moisture to seep through the container material, contaminating the fluid even when the container has not been opened. This is particularly true when the container is sitting in a shop or home garage where it is exposed to the environment to some extent. Contamination is not as great a concern at the vendor where the stock is stored and sold in a short period of time.

Other Brake Fluid Precautions

Brake fluid is considered a toxic and hazardous material. Used brake fluid must be disposed of in accordance with local regulations and EPA guidelines. Do not pour used brake fluid down a wastewater drain or mix it with other chemicals awaiting disposal. Observe these general precautions when working with brake fluid:

- Never mix polyglycol and silicone fluids because the mixture could cause a loss of brake efficiency and possible injury.
- Brake fluid can cause permanent eye damage. Always wear eye protection when handling brake fluid. If you get fluid in your eyes, rinse them thoroughly with water, then see a doctor immediately.
- Brake fluid may also irritate your skin. Wear chemical-resistant gloves. If fluid gets on your skin, wash the area thoroughly with soap and water.
- Always store brake fluid in clean, dry containers. Protect brake fluid from contamination by oil, grease, or other petroleum products, as well as contamination by water. Never reuse brake fluid.
- Do not spill polyglycol brake fluid on painted surfaces. Polyglycol fluid will damage a painted surface. Always flush any spilled fluid immediately with cold water.

Contaminated Fluid Problems

Always use the type of brake fluid recommended in the vehicle service manual or owner's manual. Using the wrong kind of fluid or using fluid that is contaminated with any other liquid can cause the brake fluid to boil or rubber parts in the hydraulic system to deteriorate.

Swollen master cylinder piston seals are the best indicator of contamination in the brake fluid. Deterioration also may be indicated by swollen wheel cylinder boots, swollen caliper boots, or a damaged master cylinder cover diaphragm.

If water or other contaminants is found in the brake system and the master cylinder piston seals have been damaged, replace all sealing parts in the system, including the brake hoses and any valves or switches with rubber seals. Also check for brake fluid on the brake linings. If you find any, replace the linings.

AUTHOR'S NOTE There is a quick, easy test to check the brake fluid for contamination. It is covered in the *Shop Manual.*

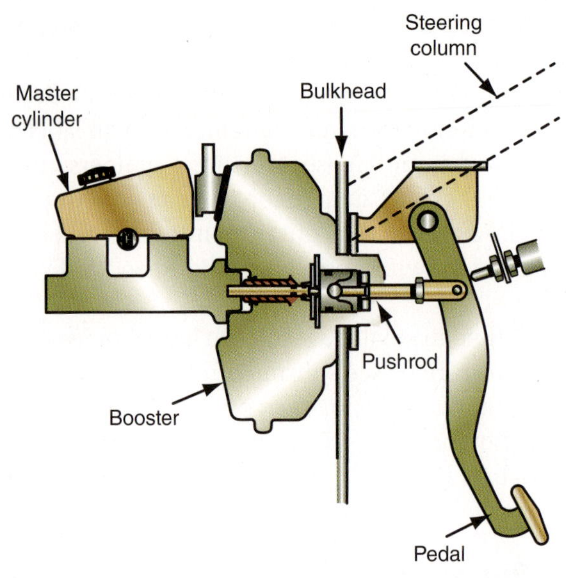

Figure 4-3 The brake pedal is suspended under the dash on modern vehicles. Its lever action will increase the driver input braking force depending on the length and curve of the pedal arm.

If water or contaminants are present in the system, but the master cylinder seals appear undamaged, check for leakage throughout the system or signs of heat damage to hoses or components. Replace all damaged components found. After repairs are made, or if no leaks or heat damage are found, drain the brake fluid from the system, flush the system with new brake fluid, refill, and bleed the system.

BRAKE PEDAL AND PUSHROD

Shop Manual
page 152

Braking action begins when the driver pushes on the brake pedal. The brake pedal (**Figure 4-3**) is a lever that is pivoted at one end, with the master cylinder pushrod attached to the pedal lever near the pivot point. This section explains how this lever arrangement multiplies the force applied at the brake pedal several times as it is applied to the master cylinder pushrod.

One end of the pushrod engages the master cylinder piston, and the other end is connected to the pedal linkage. The pushrod often is adjustable, with many made in two parts so they can be lengthened or shortened. Adjusting pushrod length is explained in the *Shop Manual.*

AUTHOR'S NOTE Vehicles with power brakes will have two pushrods, one between the pedal and power booster and another between the power booster and master cylinder. Both may be adjustable.

The brake pedal and master cylinder must be mounted close to each other to shorten the pushrod and so the pedal pushrod can operate the master cylinder pistons.

All late-model cars use a suspended pedal assembly. The pedal assembly is mounted to a support bracket that is attached to the inside of the engine compartment cowl or bulkhead under the dash. The pushrod that connects the pedal linkage to the master cylinder goes through a hole in the bulkhead. The pedal, pushrod, and master cylinder are

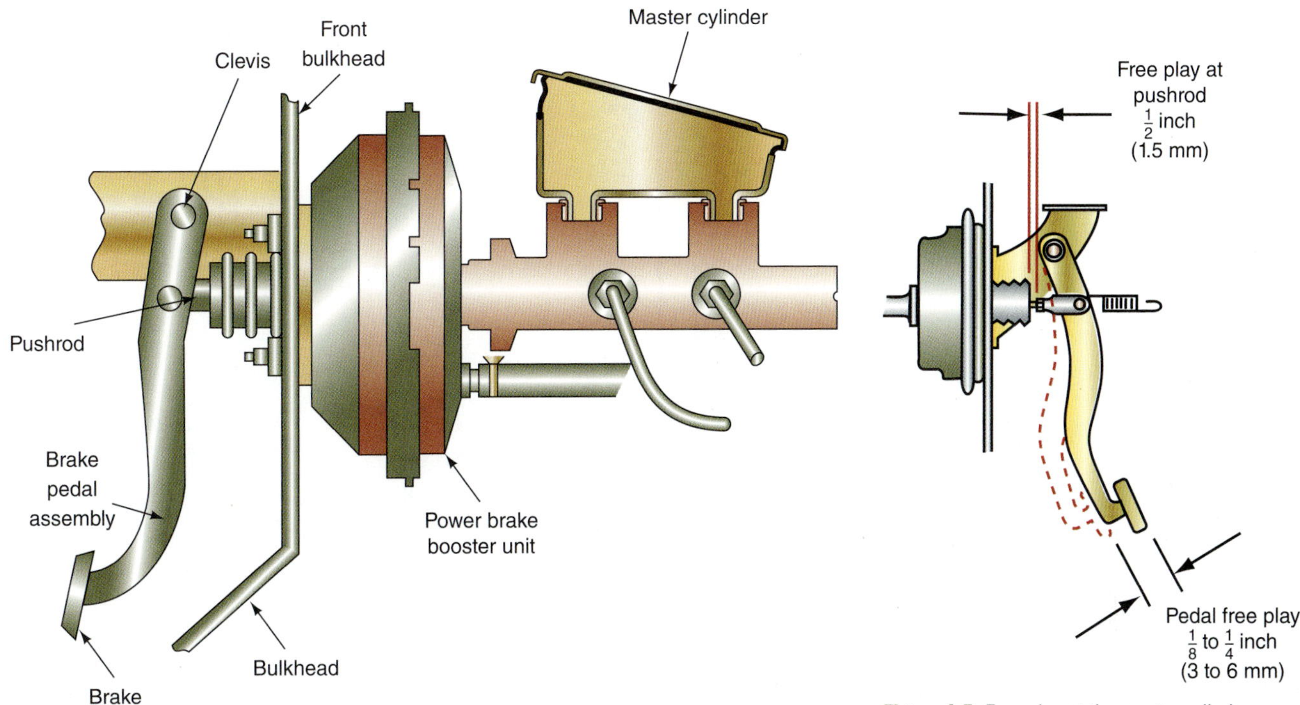

Figure 4-4 A suspended pedal installation with a vacuum power booster.

Figure 4-5 Free play at the master cylinder pushrod is multiplied by the pedal ratio, which increases free play when measured at the pedal.

mounted so, regardless of pedal position, the pushrod always pushes directly in line with the master cylinder pistons.

The master cylinder is mounted on the opposite side of the bulkhead in the engine compartment. The booster is mounted to the bulkhead, and the cylinder is mounted to the booster (**Figure 4-4**). With this arrangement, the master cylinder is easy to get to for checking or service.

Brake Linkage Free Play

All brake pedal linkage must provide some amount of **free play** between the master cylinder pistons and the pushrod. This free play is necessary to let the master cylinder pistons retract completely in their bores. Free play at the master cylinder is usually very slight: about 1/16 inch (1.5 mm to 20 mm). At the brake pedal, the free play is multiplied by the pedal ratio. Thus, if free play at the master cylinder is 1/16 inch (1.5 mm) and the pedal ratio is 4:1, free play at the pedal will be 1/4 inch (6 mm) (**Figure 4-5**).

Brake pedal **free-play** is needed to allow the master cylinder pistons retract fully in their bore.

Free play as measured at the pedal is usually specified by the vehicle manufacturer, at least as an inspection point. Free play is adjustable on some installations and not adjustable on others. Most adjustments are made by loosening a locknut at the pushrod clevis and turning the pushrod to lengthen or shorten it.

Vacuum power brake boosters have a second pushrod that transmits motion from the booster to the master cylinder. The booster pushrod may require adjustment separately from the brake pedal pushrod. The *Shop Manual* explains all necessary free play and pushrod adjustments.

SPLIT HYDRAULIC SYSTEMS

The dual master cylinder has two independent hydraulic systems. These two hydraulic systems can be connected to the wheel brake units in either a front-to-rear or diagonal arrangement.

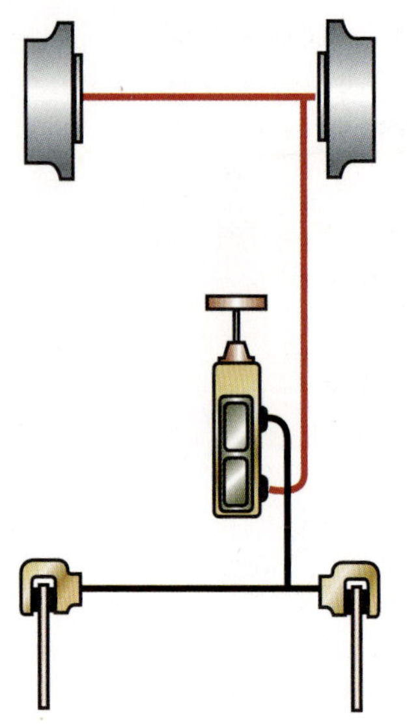

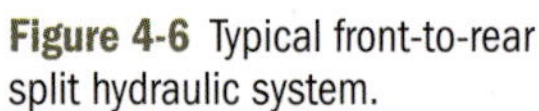

Figure 4-6 Typical front-to-rear split hydraulic system.

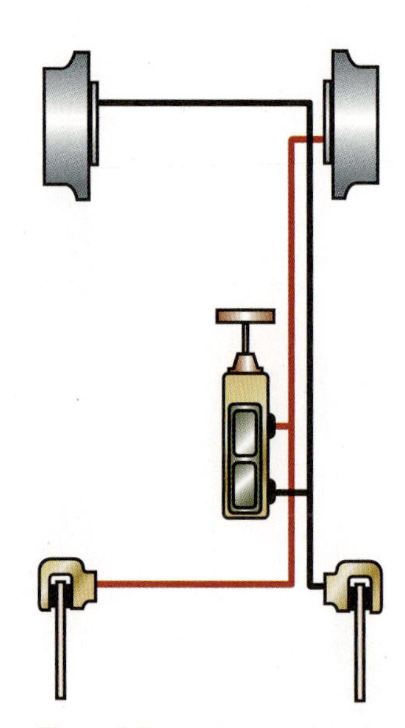

Figure 4-7 Typical diagonally split hydraulic system.

The front-to-rear hydraulic split system is the oldest split system. This system has two master cylinder outlets (**Figure 4-6**). One is connected to a line going to the two rear brakes and the other to a line going to the two front brakes.

Although this system provides two independent hydraulic systems, it has some disadvantages. The front brake system does 60 percent to 70 percent of the braking. If the front hydraulic system were to fail, all of the braking would have to be done by the rear wheels. This means that the car would have to be stopped with only 30 percent to 40 percent of its brake power.

Most late-model cars have a diagonally split hydraulic system. The diagonally split system (**Figure 4-7**) has one of the master cylinder circuits connected to the left front and right rear brakes. The other circuit is connected to the right front and left rear brakes. The connection can be made directly to the master cylinder outlets or externally with a valve. The advantage of this system is that if one system fails, the vehicle will have less than 50 percent of the braking action from one front and one rear brake. The driver may experience a left or right brake pull.

Split Hydraulic System Operation

In the released position, the cups of the primary and secondary pistons uncover the vent ports as is shown in Figure 4-13. Under normal conditions, when the brakes are applied, the primary piston moves forward, and a combination of hydraulic pressure and the force of the primary piston spring moves the secondary piston forward. When the pistons have moved forward so that their cups cover the vent ports, hydraulic pressure is built up and transmitted to the front and rear wheels. The replenishing ports work as described previously.

If there is a hydraulic failure in the brake lines served by the secondary piston (**Figure 4-8**), both pistons will move forward when the brakes are applied, as under normal conditions, but there will be nothing to resist piston travel except the secondary piston spring. This lets the primary piston build up only a small amount of pressure until the secondary piston bottoms in the cylinder bore. Then, the primary piston will build enough hydraulic pressure to operate the brakes served by this half of the system.

In case of a hydraulic failure in the brake lines served by the primary piston (**Figure 4-9**), the primary piston will move forward when the brakes are applied but will

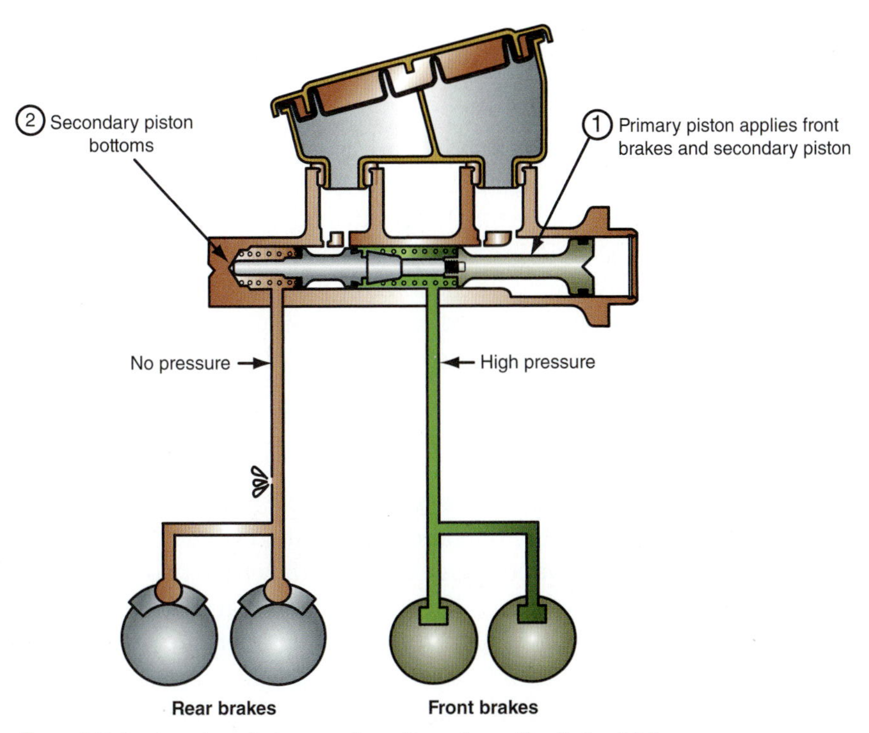

Figure 4-8 Dual master cylinder operation with a primary (front) circuit failure.

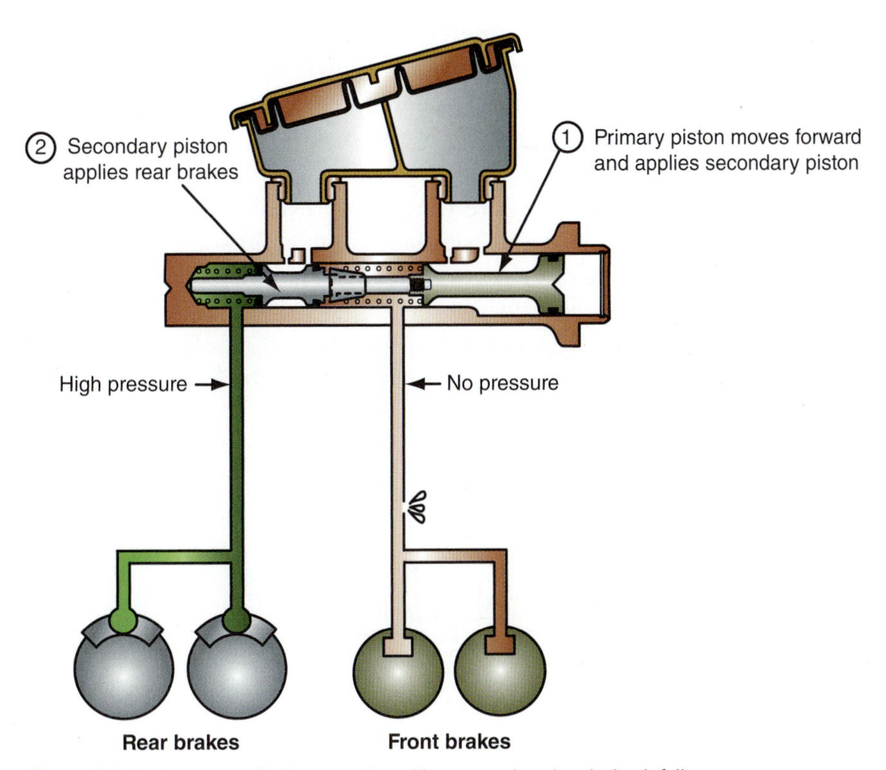

Figure 4-9 Dual master cylinder operation with a secondary (rear) circuit failure.

not build up hydraulic pressure. Very little force is transferred to the secondary piston through the primary piston spring until the piston stem comes in contact with the secondary piston. Then, pushrod force is transmitted directly to the secondary piston, and enough pressure is built up to operate its brakes.

Shop Manual
page 148

DUAL-PISTON MASTER CYLINDER CONSTRUCTION AND OPERATION

After 3 years brake fluid can contain at least 7 percent moisture. Some manufacturers call for periodic replacement of brake fluid; however, many do not.

Figure 4-10 is a simplified illustration of how the master cylinder and the brake hydraulic system work. The master cylinder pushrod is connected to a piston inside the cylinder, and hydraulic fluid is in front of the piston. When the pedal is pressed, the master cylinder piston is pushed forward. The fluid in the master cylinder and the entire system is non-compressible and immediately transmits the force of the master cylinder piston to all the inner surfaces of the system. Only the pistons in the drum brake wheel cylinders or disc brake calipers can move, and they move outward to force the brake shoes or pads against the brake drums or rotors.

The piston bore in a disc caliper is usually much larger than the one in a wheel cylinder, which is an additional reason for the disc side to have a larger reservoir chamber on a front/rear split system.

Master Cylinder Reservoir

The master cylinder has two main parts: a **reservoir** and a body (**Figure 4-11**). The reservoir supplies brake fluid for cylinder operation. All master cylinders and reservoirs built since 1967 are split or dual-chamber designs, meaning they have two separate chambers areas for two separate piston assemblies. The split design separates the hydraulic system into two independent sections providing reserve braking operation in case one section fails.

a master cylinder **reservoir** holds excess brake fluid for use in the braking system.

Old master cylinders were cast as a single unit, but the vast majority of master cylinders in use today are called composite master cylinders. The composite master cylinder has a separate aluminum body and a molded nylon or plastic reservoir. (**Figure 4-12**). All reservoirs have a removable cover so that brake fluid can be added to the system. The one-piece reservoirs typically have a single cover that is held on the reservoir with a retainer bail. Nylon or plastic reservoirs may have a single cover or two screw caps on top of the reservoir. Separate reservoirs may be clamped or bolted to the cylinder body, or they may be pressed into holes in the top of the body and sealed with grommets or **O-rings.**

O-rings are circular rubber seals shaped like the letter "O."

All master cylinder caps or covers are vented to prevent a vacuum lock as the fluid level drops in the reservoir. A flexible rubber **diaphragm** at the top of the master cylinder reservoir is incorporated in the screw caps or the cover. The reservoir diaphragm separates the brake fluid from the air above it while remaining free to move up and down with changes in fluid level. The diaphragm keeps the moisture in the air from entering the brake fluid in the reservoir. Moisture in the brake fluid will lower the fluid boiling point.

The master cylinder **diaphragm** allows the level of the brake fluid to change without direct contact with the air preventing moisture contamination.

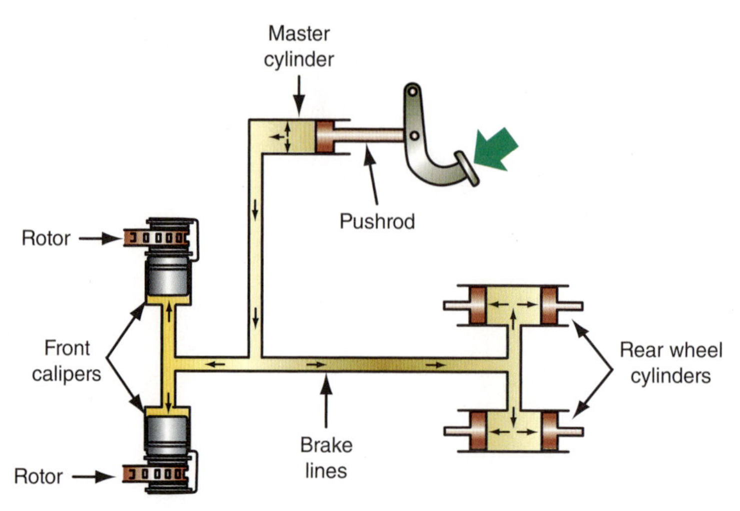

Figure 4-10 Simplified diagram of master cylinder and hydraulic system operation.

Figure 4-11 This master cylinder as viewed from the passenger side of the engine compartment. Note the power booster between the master cylinder and bulkhead. Also note that this master cylinder has a detachable reservoir.

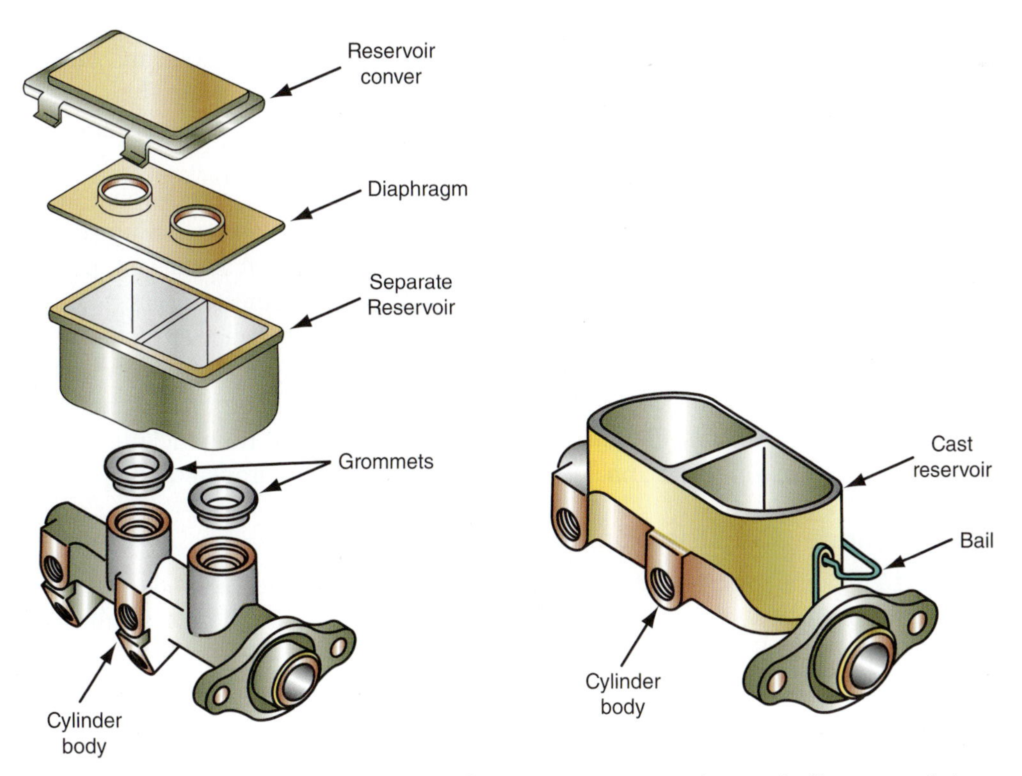

Figure 4-12 The master cylinder assembly to the left is the most common one in use today. The master cylinder on the right is an old style cast iron one-piece type.

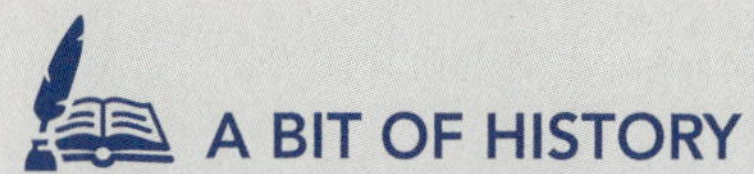

A BIT OF HISTORY

American Motors Corporation introduced the dual hydraulic system on its 1960 cars. In 1963, several manufacturers added dual tandem in-line master cylinders to their cars. By 1967, DOT established FMVSS 105 that requires that all cars sold in the United States have split brake hydraulic systems. In 1978, Chrysler Corporation introduced the first mass-produced cars with a diagonally split hydraulic system.

If a vehicle with front disc and rear drum brakes has the brake system split so that the front brakes are on one circuit and the rear brakes on the other circuit, the reservoir chamber for the disc brakes is larger than the chamber for the drum brakes. As disc pads wear, the caliper pistons move out farther in their bores. More fluid is then required to keep the system full from the master cylinder to the calipers. Drum brake wheel cylinder pistons always retract fully into the cylinders regardless of brake lining wear so the volume of fluid does not increase with lining wear. Vehicles with four-wheel disc brakes or diagonally split hydraulic systems usually have master cylinders with equal-size reservoirs because each circuit of the hydraulic system requires the same volume of fluid.

Plastic reservoirs often are translucent so that fluid level can be seen without removing the cover. Although this feature allows a quick check of fluid level without opening the system to the air, it should not be relied on for a thorough brake fluid inspection. Stains inside the reservoir can give a false indication of fluid level, and contamination cannot be seen without removing the reservoir caps or cover.

Master Cylinder Ports

Many different names have been used for the ports in the master cylinder, and often the same name has been applied to each of the ports. For the sake of uniformity, this text refers to the forward port as the "vent" port and the rearward port as the "replenishing" port, which are the names established by SAE Standard J1153.

The **vent port** is the forward port in the master cylinder bore.

The **replenishing port** is the rearward port in the master cylinder bore.

The **vent port** has been called the compensating port and the replenishing port by some manufacturers. To further confuse the nomenclature, the **replenishing port** has been called the compensating port by some manufacturers, as well as the vent port, the bypass port or hole, the filler port, or the intake port by many manufacturers.

However, it is not so important what these ports (or holes) are called as there is a good understanding of their purposes and operations. **Figure 4-13** is a cross section view of a dual master cylinder including the location and names for the two ports. Although the vehicle and parts manufacturers have not always observed the nomenclature of SAE Standard J1153, its use in this manual maintains consistency and understanding.

Master Cylinder Construction

The **cup seal** is the circular rubber seal with a depressed center section that is surrounded by a raised sealing lip to form a cup; cup seals are often used on the front ends of hydraulic cylinder pistons because they seal high pressure in the forward direction of travel but not in the reverse.

Figure 4-14 is a disassembled view of a dual piston master cylinder. A single-piston bore contains two piston assemblies with two pressure chambers. The piston assembly at the rear of the cylinder is the primary piston, and the one at the front of the cylinder is the secondary piston. Each piston has a return spring in front of it. Both pistons have a long extension at the front that is used in case of a hydraulic failure. There is a **cup seal** at the front of each piston and a cup or seal at the rear of each piston. The seals retain fluid in the cylinder chambers and prevent seepage between the cylinders. The secondary piston has a cup at each end. The front cup is to generate pressure in the secondary hydraulic system. The rear cup seals the primary chamber, and the hydraulic pressure generated by the primary piston is used to apply the secondary piston. This also causes the pressure generated by the secondary piston to be equal to that generated by the primary piston.

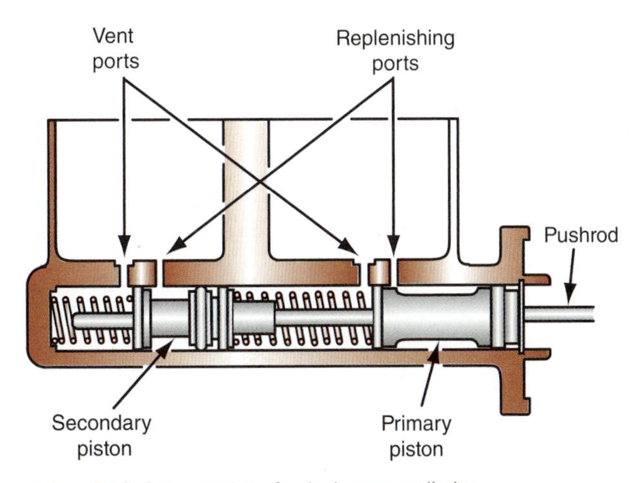

Figure 4-13 Cross section of a dual master cylinder.

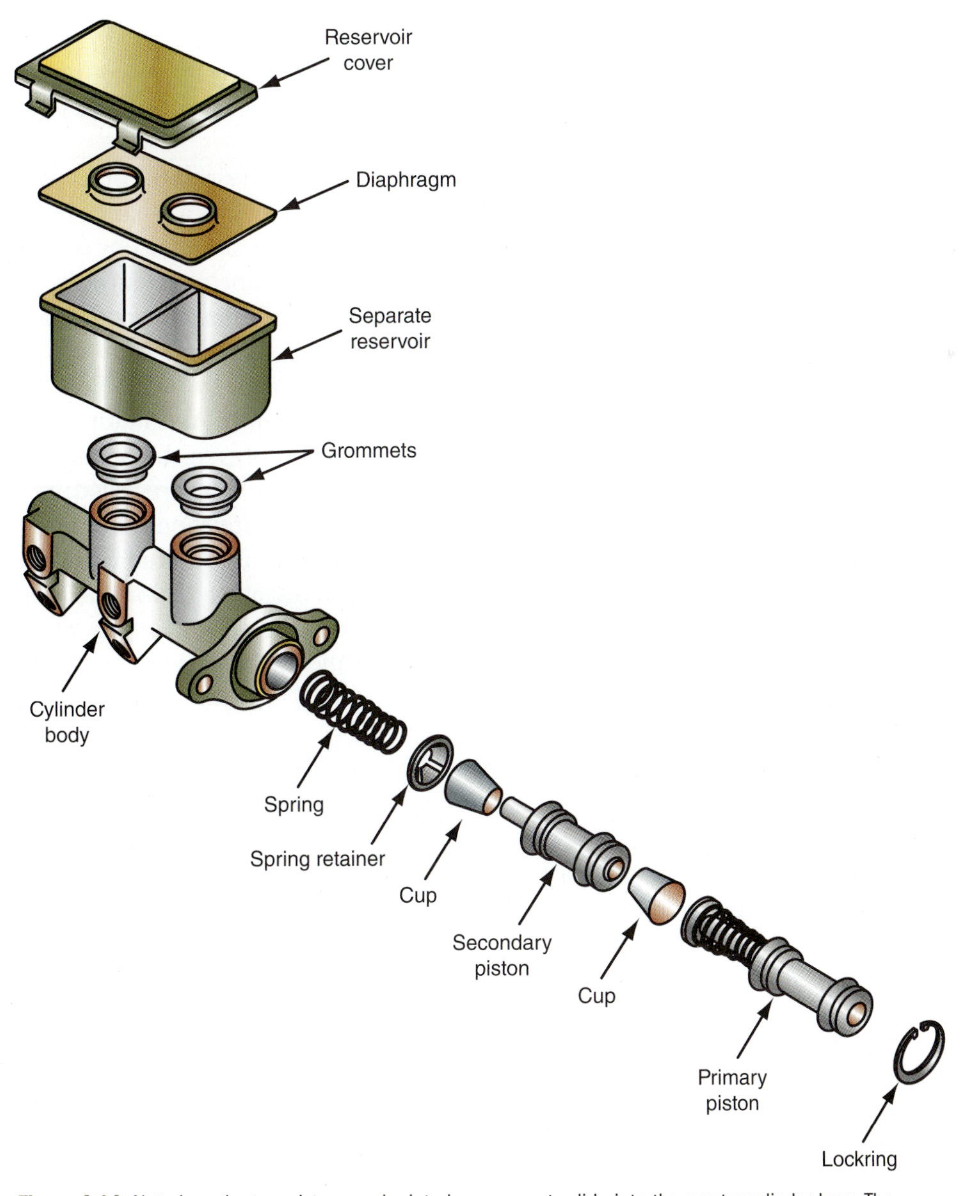

Figure 4-14 Note how the two pistons and related components slide into the master cylinder bore. The primary piston will be the one closest to the driver and will serve the front wheel brakes in most cases.

Two ports are at the bottom of each reservoir section. One set of ports is called the vent ports; the others are replenishing ports. The vent ports and replenishing ports let fluid pass between each pressure chamber and its fluid reservoir during operation.

Two spool-shaped pistons are inside the cylinder. **Figure 4-15** is a simplified illustration of a spool-shaped piston. The piston has a head on one end and a groove for an O-ring seal on the other end. The seal seats against the cylinder wall and keeps fluid from leaking past the piston. The smaller diameter center of the piston is the valley area, which lets fluid get behind the head of the piston. This prevents low pressures developing here as the piston moves forward.

Each master cylinder piston works with a rubber cup seal that fits in front of the piston head (**Figure 4-16**). The cup has flexible lips that fit against the cylinder walls to seal fluid pressure ahead of the piston head. The cup lip also can bend to let fluid get around the cup from behind. When the brakes are applied, pressure in front of the cup forces the lip tightly against the cylinder wall and lets the seal hold very high hydraulic pressure. The lip of a cup seal is always installed toward the pressure to be contained or away from the body of the piston. The cup seals only in one direction. If pressure behind the lip exceeds pressure in front of it, the higher pressure will force the lip away from the cylinder wall and let fluid bypass the cup.

Pistons have small coil springs (**Figure 4-17**) that return the pistons to the proper position when the brake pedal is released. Sometimes the springs are attached to the pistons; sometimes they are separate parts.

A dual master cylinder has two piston assemblies. The piston nearest the pushrod is the primary piston; the other is the secondary piston. Each piston provides a separate hydraulic system for the front and rear brakes or, on a diagonally split system, between one set of front brakes and one set of rear brakes. Two outlet holes provide the connection for the hydraulic lines. A snap-ring holds the components inside the cylinder, and a flexible boot fits around the rear of the cylinder and pushrod to keep dirt from entering the cylinder.

Figure 4-18 shows the major parts of the two-piece master cylinder. This master cylinder has an aluminum body to lower the weight of the assembly. Aluminum offers the advantage of being much lighter than cast iron but has large pores in the metal that can damage the lip seals of the bore pistons. The interior walls of the bore are rolled, or bearingized, to close these pores. The aluminum master cylinder bore should not be honed during service because the pores will be reopened. Replacement is required if the interior bore is worn or damaged. The removable nylon or plastic reservoir is also much lighter than a cast-iron unit. Because the master cylinder is made from two materials, it is often called a composite master cylinder. The pistons, cups, and springs used in a composite

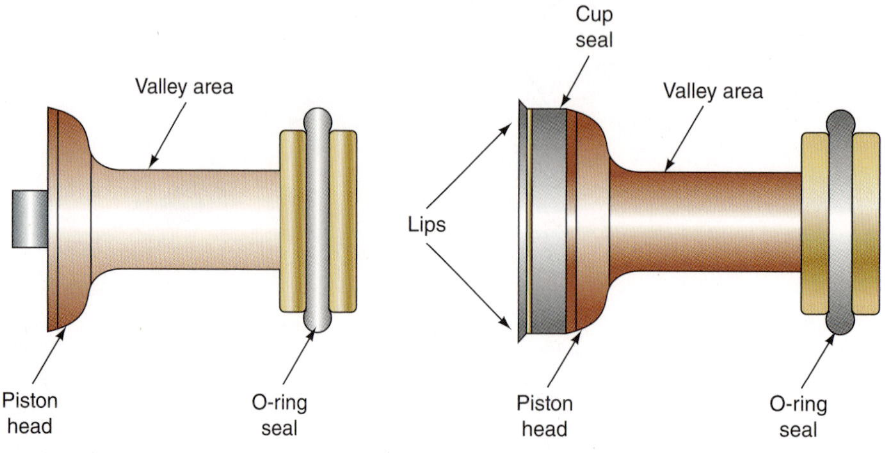

Figure 4-15 Simplified spool-shaped piston for master cylinder.

Figure 4-16 A cup seal and O-ring seal the piston in the cylinder bore.

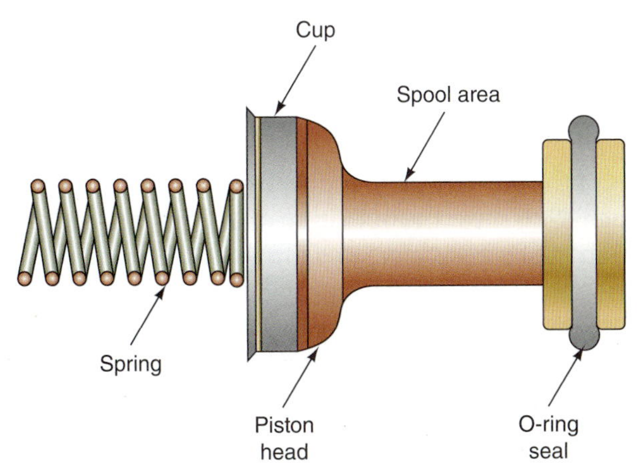

Figure 4-17 Note how the two pistons and related components slide into the master cylinder bore. The primary piston will be the one closest to the driver and will serve the front wheel brakes in most cases.

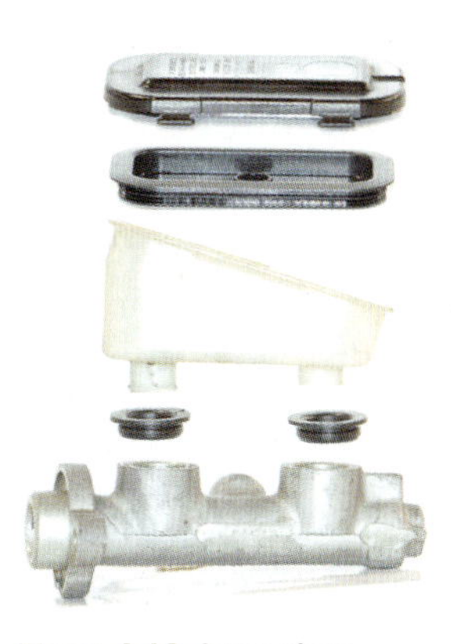

Figure 4-18 A two-piece composite master cylinder has a plastic reservoir mounted on an aluminum body.

master cylinder are essentially the same and work the same way as those in a one-piece master cylinder.

Portless Master Cylinder

Some Toyota master cylinders do not use replenishing or compensating ports. Instead they use a valve that is machined into the master cylinder pistons (**Figure 4-19**). When the master cylinder is at rest, fluid can flow into the area in front of the pistons. When the master cylinder is applied, the valves close under spring pressure

Master Cylinder with Hydraulic Brake Assist

A special master cylinder is required for vehicles with hydraulic brake assist. Brake fluid can be applied directly to the master cylinder pistons when necessary, as determined by the brake control module, via the brake stroke sensor (**Figure 4-20**). The brake control

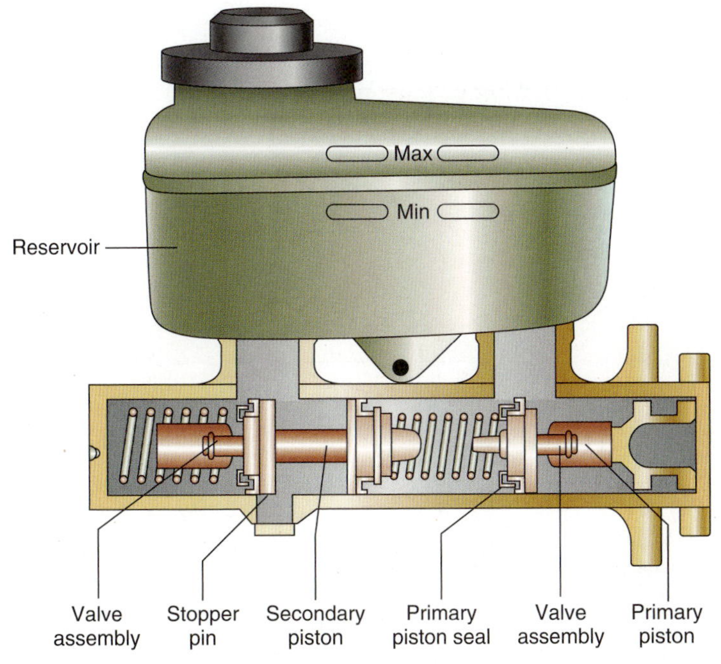

Figure 4-19 The portless master cylinder does not use the conventional vent and replenishing ports, but instead uses special valves in the primary and secondary pistons to refill the master cylinder.

Figure 4-20 Brake pedal stroke sensor.

module on a hybrid vehicle can also determine how much braking can be taken care of by regenerative braking versus the hydraulic braking.

Master Cylinder Operation

The primary and secondary pistons of the master cylinder operate in the same way during normal braking. **Figure 4-21** through **4-26** show the operation of the primary piston. The secondary piston is moved forward by the pressure created ahead of the primary piston. Each of the two pistons has a stop at its front end. This provides a means to allow one piston to build pressure if there is an external leak in the brake system. As the noted figures are examined, the secondary piston is performing the same action as the primary piston.

The low-pressure area is created by the fast-rearward movement of the piston and the slow-moving brake fluid. This sudden increase of empty volume causes the pressure to drop.

Figure 4-21 shows a piston assembly and reservoir with the piston in the released position. The reservoir is full of brake fluid. The vent port in the bottom of the reservoir is located just ahead of the piston cup. Fluid flows from the reservoir through the vent port into the pressure chamber in front of the piston cup. The replenishing port is located above the valley area of the piston, behind the piston head. The O-ring seal on the primary piston keeps the fluid from leaking out the rear of the cylinder. The return spring in front of the piston and cup returns the piston when the brakes are released.

When the driver presses on the brake pedal, the pushrod pushes the master cylinder piston forward (**Figure 4-22**). As the piston moves forward, the piston pushes the cup past the vent port. As soon as the vent port is covered, fluid is trapped ahead of the cup. The fluid, under pressure, goes through the outlet lines to the wheel brake units to apply the brakes and applies the secondary piston.

When the driver releases the brake pedal, the return spring forces the piston back to its released position. As the piston moves back, it pulls away from the fluid faster than the fluid can flow back from the brake lines to the pressure chamber. When this happens, low pressure is created ahead of the piston.

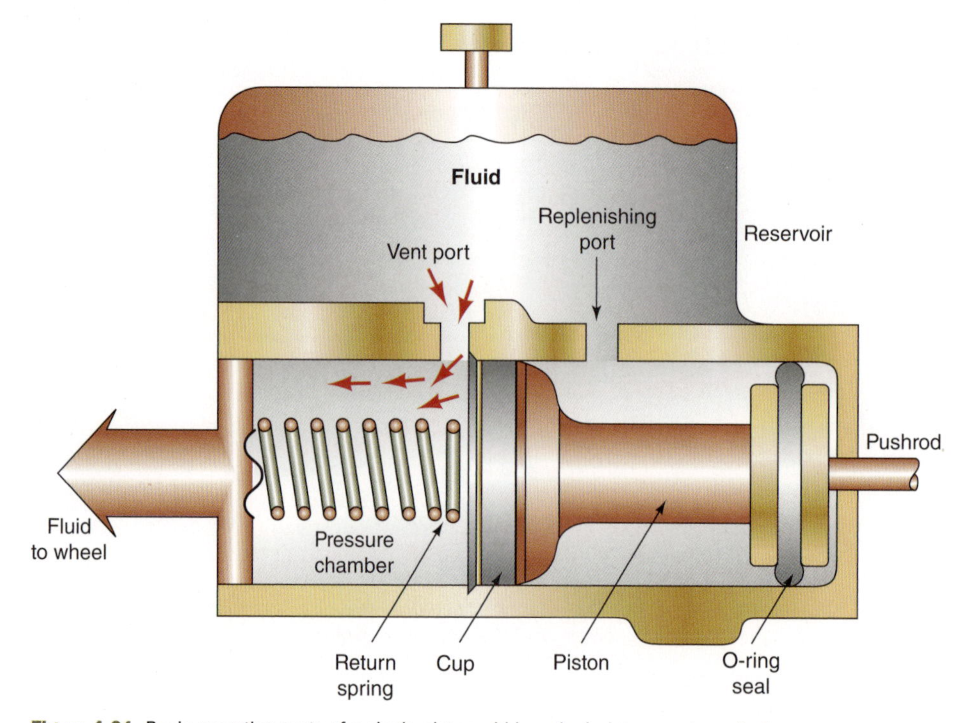

Figure 4-21 Basic operating parts of a single piston within a dual-piston master cylinder.

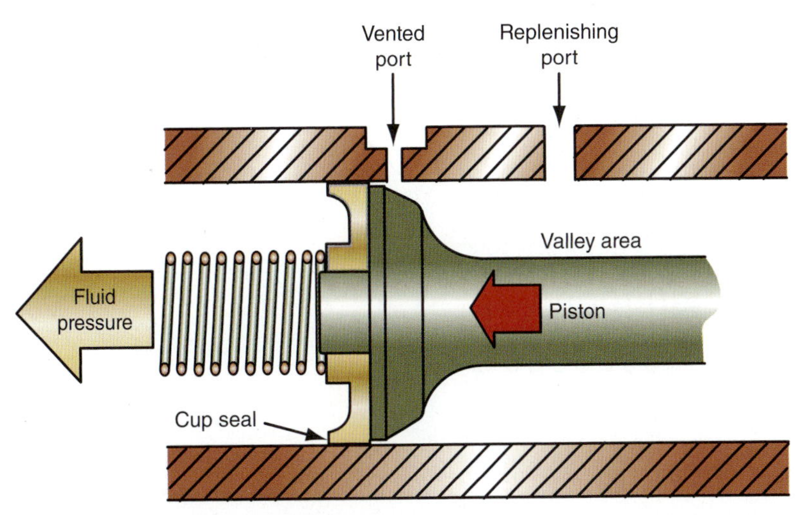

Figure 4-22 As the piston moves forward, the cup seal closes the vent port and lets the pressure develop ahead of the piston.

The piston must move back to the released position rapidly so it can be ready for another forward stroke, if necessary. The low-pressure area must be filled with fluid as the piston moves back. A path for fluid flow is provided from the valley area (**Figure 4-23**), past the primary cup protector washer and through several small holes in the head of the piston, or by having enough clearance between the piston head and the cylinder bore. Fluid flows through the piston or around the lip of the cup and into the chamber ahead of the piston. This flow quickly relieves the low-pressure condition.

The vent port allows pressure in the brake pressure chamber to be relieved back into the master cylinder when the brakes are fully released.

The fluid that flows from the valley area to the pressure chamber must be replaced. **Figure 4-24** shows how the replenishing port lets fluid from the reservoir flow into this area.

When the piston is fully returned to its released position, the space in front of it is full of fluid. The piston cup again seals off the head of the piston. In the meantime, the fluid from the rest of the system has begun to flow back to the high-pressure chamber. If this fluid pressure were not released, the brakes would not release. **Figure 4-25** shows how the returning fluid flows back to the reservoir through the vent port. The vent port is covered by the piston cup at all times, except when the piston is released.

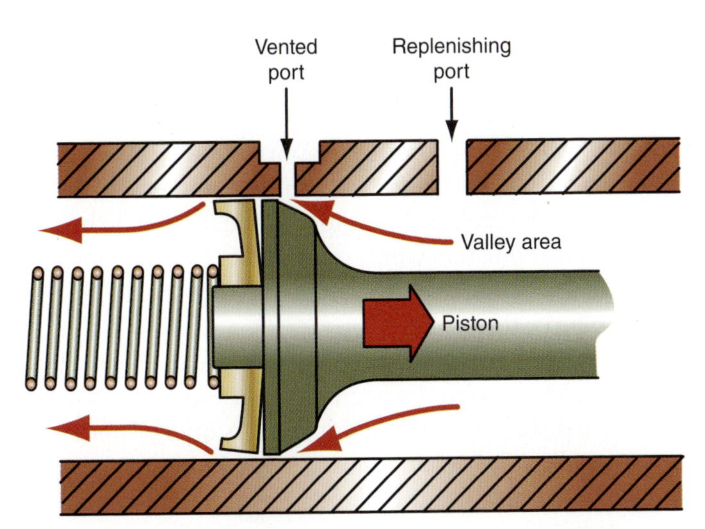

Figure 4-23 As the piston retracts, fluid in the valley area flows past the piston's cup seal.

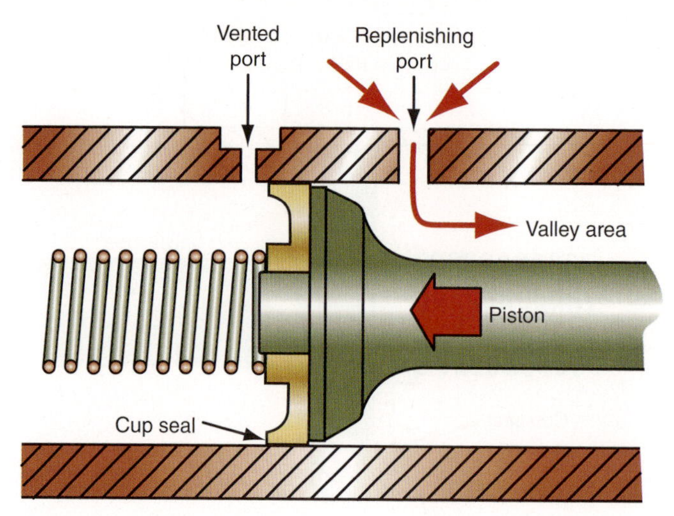

Figure 4-24 The low-pressure valley area is refilled through the replenishing port.

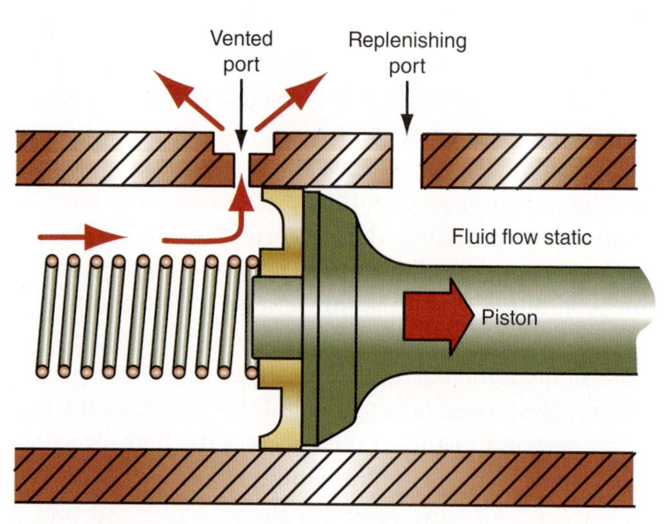

Figure 4-25 On the return stroke, fluid flows from the high-pressure chamber back to the reservoir through the vent port.

The replenishing port provides fluid to the spool valve valley from the master cylinder to prevent air pockets in front of the pistons on brake release. It also allows fluid from the master cylinder to compensate for brake wear.

The replenishing port also has another important job. There are times when the amount of fluid in the wheel brake units and lines must be increased. When disc brake pads wear, there is more space in the brake calipers for hydraulic fluid. When drum brake lining wears, before the automatic adjusters work, more fluid is needed in the wheel cylinders. When the brake system is serviced and air is in the system taking up space, more fluid is needed in the system to force the air out.

This is accomplished in the same way low pressure ahead of the piston is relieved. On the return stroke, fluid flows through the head of the piston and around the lip of the cup (**Figure 4-26**). When the piston returns, the vent port is open. There is not as much fluid returning from the wheel brake units and lines, so less fluid flows through the vent port and back into the reservoir. If the drum brakes are adjusted or new pads are installed on disc brakes, the system will automatically compensate for the amount of fluid needed on the next piston cycle.

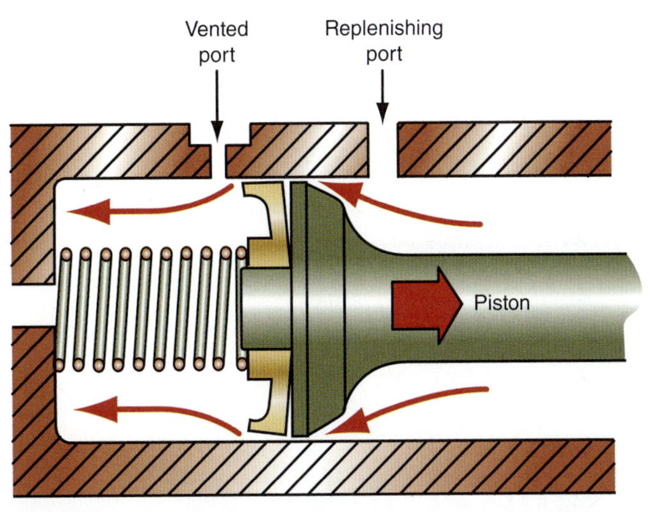

Figure 4-26 Fluid can flow around the cup seal when more fluid is required.

Master Cylinder Operation with an External Hydraulic Leak

The dual-piston master cylinder was designed to reduce total brake failure when an external hydraulic leak is present. Assume there is an external leak in the rear hydraulic circuit of a front/rear split hydraulic system. The application of the primary piston applies hydraulic pressure against the rear of the secondary piston (remember the rear cup). However, the rear system is open and secondary pressure cannot be generated, so both pistons move forward with little braking effect. The extension at the front of the secondary piston contacts the front of the bore and stops further movement of that piston. Now the secondary piston's rear cup becomes a seal, and the primary piston can build hydraulic pressure within its own hydraulic system applying that system's brakes.

A leak in the front (primary) hydraulic system works similarly except the extension at the front of the primary piston mechanically contacts and applies the secondary piston. Hydraulic pressure is now generated within the secondary circuit. The driver immediately senses a much lower brake pedal and poor braking in both conditions. The vehicle takes longer to stop, but at least there is no total brake system failure.

A BIT OF HISTORY

Single-piston master cylinders used before 1967 had one piston and one hydraulic circuit for all four-wheel brake units. The problem with this system was that a hydraulic failure in the master cylinder could cause a complete loss of brakes to all four wheels. They were used on four-wheel drum brakes without power boosters.

Residual Pressure Check Valve

Residual pressure check valves were once used on almost all drum brake systems, but their use has decreased since the late 1980s. Piston cup expanders were developed for wheel cylinders to hold the cups against the cylinder walls and to keep air from being drawn into the wheel cylinders (**Figure 4-27**). Cup expanders are simpler, cheaper, and more reliable than check valves.

Residual pressure check valves were once used to hold slight pressure on the drum brake pistons to maintain seal contact with the walls of the wheel cylinder- cup expanders are used now.

Figure 4-27 The cup expander, centered and attached to the spring, has eliminated the need for a residual check valve in the master cylinder.

FAST-FILL AND QUICK TAKE-UP MASTER CYLINDERS

Several carmakers use fast-fill or quick take-up master cylinders. These cylinders fill the hydraulic system quickly to take up the slack in the caliper pistons of low-drag disc brakes. Low-drag calipers retract the pistons and pads farther from the rotor than do traditional calipers. This reduces friction and brake drag and improves fuel mileage.

If a conventional single-bore dual master cylinder were used with low-drag calipers, excessive pedal travel would be needed on the first stroke to fill the lines and calipers with fluid and take up the slack in the pads. To overcome this problem, fast-fill and **quick take-up master cylinders** provide a large volume of fluid on the first stroke of the brake pedal.

The larger diameter primary piston in the **quick take-up master cylinder** moves a comparatively large amount of fluid with less pedal travel.

A fast-fill or quick take-up master cylinder is identified by the dual bore design that creates a bulge or stepped outside diameter of the casting (**Figure 4-28**). The cylinder has a larger diameter bore for the rear of the primary piston than for the front of the primary piston. Inside the cylinder, a fast-fill or quick take-up valve replaces the conventional vent and replenishing ports for the primary piston (**Figure 4-29**).

The quick-take-up valve allows fluid into the primary piston area from the high pressure area behind the primary piston on brake application.

The **quick take-up valve** contains a spring-loaded check ball that has a small bypass groove cut in the edge of its seat. The outer circumference of the quick take-up valve is sealed to the cylinder body with a lip seal. Several holes around the edge of the hole let fluid bypass the lip seal under certain conditions. Some valves (those more often called "fast-fill" valves) are pressed into the cylinder body and sealed tightly by an O-ring. A rubber flapper-type check valve under the fast-fill valve performs the same bypass functions as a lip seal of a quick take-up valve. The following sections explain the operation of a quick take-up or fast-fill valve.

Brakes Not Applied. When the brakes are off, both master cylinder pistons are retracted, and all vent and replenishing ports are open. Fluid to both ports of the primary piston must flow through the groove in the check ball seat, however.

Brakes Applied. As the brakes are applied, the primary piston moves forward in its bore. Remember that the diameter of the primary valley area is larger than the diameter of the rest of the cylinder. As the primary piston moves forward into the smaller diameter, the volume

Figure 4-28 A typical quick take-up master cylinder.

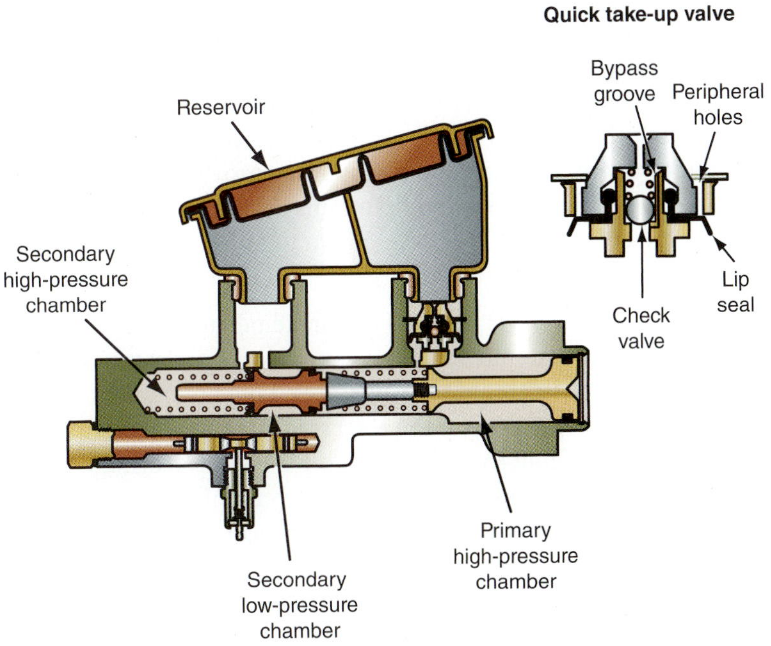

Figure 4-29 A quick take-up valve serves as the primary piston in this master cylinder designed for use with low-drag calipers.

of the valley area is reduced. This causes hydraulic pressure to rise instantly in the low-pressure chamber. The higher pressure forces the large volume of fluid in the valley area past the cup seal of the primary piston (**Figure 4-30**), providing the extra volume of fluid to take up the slack in the caliper pistons.

The lip seal of the quick take-up valve keeps fluid from flowing from the valley area back to the reservoir. Initially a small amount of fluid bypasses the check ball through the bypass groove, but this small amount is not enough to affect quick take-up operation.

As brake application continues, pressure in the valley area rises to about 70 to 100 psi. The check ball in the quick take-up valve then opens to let excess fluid return to the reservoir (**Figure 4-31**). Pressures in front and behind the primary piston equalize, and the

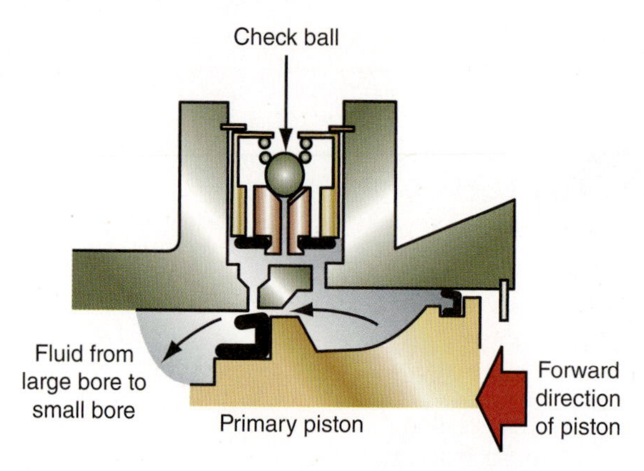

Figure 4-30 As the brakes are applied, fluid flows from the large bore to the small bore for the primary piston.

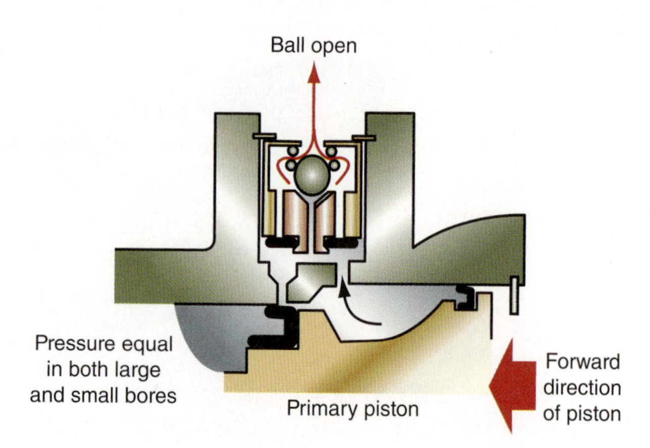

Figure 4-31 When pressures equalize in both chambers, fluid returns to the reservoir.

piston moves forward to actuate the secondary piston. These actions all occur in a fraction of a second.

All the actions described above apply to the primary piston if it is serving front disc brakes and the secondary piston is serving drum brakes. If the hydraulic system is split diagonally, or if the car has four-wheel, low-drag discs, the quick take-up fluid volume must be available to both pistons. Some master cylinders have a second quick take-up valve for the secondary piston. Others provide the needed fluid volume through the design of the cylinder itself. If the primary quick take-up valve stays closed, the fluid bypassing the primary piston cup causes the secondary piston to move farther. This provides equal fluid displacement from both pistons and maintains equal pressure in the system. When the quick take-up valve opens, both pistons move together just as in any other master cylinder.

Brakes Released. When the driver releases the brake pedal, the return springs force the primary and secondary pistons to move back. Pressure drops in the pressure chambers, and fluid bypasses the piston cup seals from the valley area. Low pressure is created in the valley area, which lets atmospheric pressure in the reservoir force fluid past the seal of the quick take-up valve (**Figure 4-32**). Fluid from the reservoir then flows through both the vent and replenishing ports to equalize pressure in the pressure chambers and valley areas.

On the return stroke, fluid flow to the secondary piston is through a normal replenishing port unless the secondary piston also has a quick take-up valve. If a secondary quick take-up valve is installed, it works as described for a primary quick take-up valve.

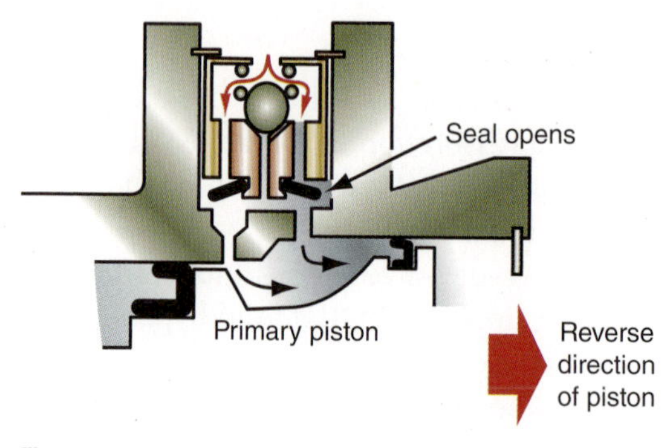

Figure 4-32 When the piston returns, low pressure draws fluid into the valley area.

CENTRAL-VALVE MASTER CYLINDERS

Some ABSs use master cylinders that have central check valves in the heads of the pistons. If the master cylinder provides pressure during antilock operation (a so-called open system) and the system also has a motor-driven pump, the master cylinder pistons may shift back and forth rapidly during antilock operation. This could cause excessive pedal vibration and, more important, wear on the piston cups where they pass over the vent ports.

To prevent seal damage and pedal vibration, spring-loaded check valves are installed in the piston heads (**Figure 4-33**). When the brakes are released, fluid flows from the replenishing ports to the valley areas, through the open central check valves, and into the pressure chambers. As the brakes are applied, the central valves close to hold fluid in the pressure chambers. When the brakes are released again, the central check valves open to let fluid flow back through the pistons to the valley areas and the reservoir.

The central check valves in this type of master cylinder provide supplementary fluid passages to let fluid move rapidly back and forth between the pressure chambers and the valley areas during antilock operations. This is not much different in principle from non-ABS fluid flow, but the extra passages reduce piston and pedal vibration and cup seal wear. Chapter 10 in this *Classroom Manual* provides more information on ABS master cylinders.

Performance Fluids and Master Cylinders

The pedal assembly, master cylinder, and brake fluid for racing vehicles are almost the same as those on a production vehicle. There are some differences in weight and increased performance, however.

Most race vehicles are equipped with low-drag disc brakes at all four wheels. To meet the initial braking requirement, a quick take-up master cylinder is needed. Racing master cylinders can be purchased with different bore sizes, resulting in more fluid pressure with reduced force by the driver.

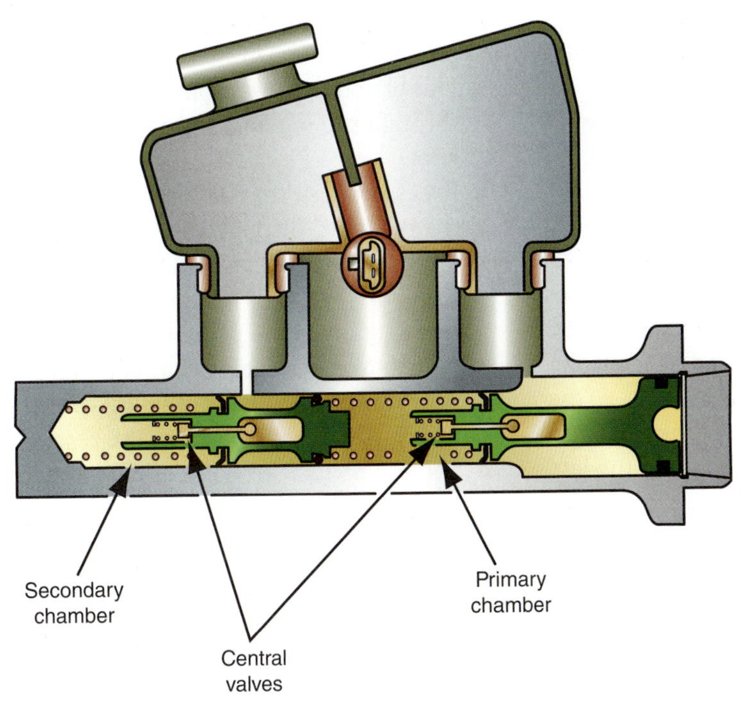

Figure 4-33 This ABS master cylinder has central valves in the piston heads in place of vent ports.

In most instances, only one dual-piston cylinder is used with some type of split system. However, some race crews opt for two identical single-piston master cylinders. The two master cylinders act like a split hydraulic system in that one master cylinder serves the front wheels, whereas the other serves the rear wheels. The master cylinders are applied by one brake pedal acting through a balance bar between the pedal lever and the two push-rods. Some race units are equipped with a brake power booster, and others are not. In this case, it is more an issue of weight than of driver endurance.

Of primary importance to race vehicle braking is the type of brake fluid used. On short tracks with a lot of braking, the boiling point of the fluid can be reached quickly and may be sustained for long periods. Brake fluids developed for racing purposes generally have the same chemical properties as conventional fluids, but they have much higher boiling points. Castrol offers a blend of polyglycol ester of dimethyl silane, ethylene polyglycols, and oxidation inhibitors. This blend has a dry boiling point of 450°F(232°C) and helps prevent fluid contamination during operation. Another brand, GS610, offers a fluid with a dry boiling point of 610°F(321°C). There are several manufacturers and suppliers of racing brake components. Brembo is one of the larger manufacturers of racing components, and some of its products are now being installed on some production performance vehicles.

SUMMARY

- Brake fluid specifications are defined by SAE Standard J1703 and FMVSS 116.
- Fluids are assigned DOT numbers: DOT 3, DOT 4, DOT 5, DOT 3/4, and DOT 5.1.
- Always use fluid with the DOT number recommended by the specific carmaker.
- Never use DOT 5 fluid in an ABS or mix with any other brake fluid.
- HSMO fluids are very rare and should never be used in brake systems designed for DOT fluids.
- The brake pedal assembly is a lever that increases pedal force to the master cylinder.
- The brake pedal lever is attached to a pushrod, which transmits force to the master cylinder pistons.
- A front-to-rear split hydraulic system has two master cylinder circuits. One is connected to the front brakes and the other to the rear brakes.
- A diagonally split hydraulic system is one in which one master cylinder circuit is connected to the left front and right rear brakes and the other circuit is connected to the right front and left rear brakes.
- The master cylinder has two main parts: a reservoir and a cylinder body.
- The reservoir can be a separate piece or cast as one piece with the cylinder.
- A dual-piston master cylinder has two separate pistons providing pressure for two independent hydraulic systems. Each of the two pistons in the master cylinder has a cup, a return spring, and a seal.
- During application, the piston and cup force fluid ahead of the piston to activate the brakes.
- During release, the return spring returns the piston.
- Fluid from the reservoir flows from the reservoir through the replenishing port around the piston cup.
- Excess fluid in front of the piston flows back into the reservoir through the vent ports.
- Quick take-up or fast-fill master cylinders have a step bore, which is a larger diameter bore for the rear section of the primary piston.
- Quick take-up master cylinders have a valve that provides rapid filling of the low-pressure spool area of the primary piston from the reservoir.
- Some ABS master cylinders have check valves in the heads of the pistons to reduce piston and pedal vibration and cup wear.
- Portless master cylinders do not use a replenishing or vent port. Fluid can flow between the reservoir and the area ahead of the master cylinder pistons by means of a valve machined into the master cylinder pistons when the master cylinder is at rest.

REVIEW QUESTIONS

Essay

1. Explain why DOT 5 brake fluid is not recommended by any manufacturer.
2. Explain why the boiling point of brake fluid is important.
3. Explain why it is not a good idea to mix DOT 5 fluids with DOT 3 and DOT 4.
4. Describe a sure sign of brake fluid contamination with mineral oil.
5. Explain why brake pedal linkage free-play is necessary.
6. Explain the split hydraulic system.
7. Describe a composite master cylinder.
8. Describe a master cylinder cup seal and how it is used.
9. What are the ports in the bottom of the master cylinder reservoir, and what do they do?
10. Explain the advantage of a quick take-up master cylinder.

Fill in the Blanks

1. A fast-fill or quick take-up master cylinder is identified by the dual bore design that creates a _______________ or _______________ _______________ of the casting.
2. DOT 3 and DOT 4 fluids are polyalkylene-glycol-ether mixtures, called _______________ for short.
3. Because both DOT 3 and DOT 4 fluids _______________ _______________ from the air, always keep containers tightly capped.
4. Silicone fluid _______________ slightly under pressure, which can cause a slightly spongy brake pedal feel.
5. Polyglycol fluids have a very _______________ shelf life.
6. The _______________ -to- _______________ hydraulic split system is the oldest split system.
7. Most late-model cars have a _______________ split hydraulic system.
8. The master cylinder has two main parts: a _______________ and a _______________.
9. All master cylinder caps or covers are vented to prevent a _______________ _______________ as the fluid level drops in the reservoir.
10. The piston assembly at the rear of the cylinder is the _______________ piston, and the one at the front of the cylinder is the _______________ piston.

Multiple Choice

1. Technician A says the master cylinder changes the driver's mechanical force on the pedal to hydraulic pressure. Technician B says this hydraulic pressure is changed back to mechanical force at the wheel brakes. Who is correct?
   A. A only
   B. B only
   C. Both A and B
   D. Neither A nor B
2. Technician A says choosing the right fluid for a specific vehicle is based on the simple idea that if DOT 3 is good, DOT 4 must be better, and DOT 5 better still. Technician B says most vehicle manufacturers recommend DOT 4. Who is correct?
   A. A only
   B. B only
   C. Both A and B
   D. Neither A nor B
3. Technician A says the dry boiling point of brake fluid is the minimum boiling point of new, uncontaminated fluid. Technician B says polyglycol fluids are hygroscopic, which means that they do not absorb water vapor from the air. Who is correct?
   A. A only
   B. B only
   C. Both A and B
   D. Neither A nor B
4. Technician A says a high-temperature boiling point is the only requirement that brake fluid must meet. Technician B says brake fluid also must resist freezing and evaporation and must pass specific viscosity tests at low temperatures. Who is correct?
   A. A only
   B. B only
   C. Both A and B
   D. Neither A nor B

5. Technician A says DOT 5 Silicone fluid does not absorb water. Technician B says DOT 5 silicone fluid is a purple fluid with a very high boiling point, is noncorrosive to hydraulic system components, and does not damage paint as does ordinary fluid. Who is correct?
   A. A only
   B. B only
   C. Both A and B
   D. Neither A nor B
6. Technician A says synthetic DOT 3/4 and DOT 5.1 are compatible with DOT 3 and DOT 4 fluids and can be mixed with them. Technician B says DOT 5.1 is based on polyglycol chemistry. Who is correct?
   A. A only
   B. B only
   C. Both A and B
   D. Neither A nor B
7. Technician A says if water or other contaminants are found in the brake system and master cylinder piston seals have been damaged, it is okay to flush the system with clean fluid and put it back in service. Technician B says replace all sealing parts in the system, including the brake hoses and any valves or switches with rubber seals. Who is correct?
   A. A only
   B. B only
   C. Both A and B
   D. Neither A nor B
8. Brake pedal linkage free play at the master cylinder is usually very slight generally about:
   A. 1/16 inch (1.5 mm to 20 mm)
   B. 1/4 inch (6.25mm)
   C. 1/8 inch (3.17 mm)
   D. 5/16 inch (7.9mm)
9. Technician A says most modern vehicles have a diagonally split hydraulic system. Technician B says the advantage of this system is that if one system fails, the vehicle will have less than 50 percent of the braking action from one front and one rear brake. Who is correct?
   A. A only
   B. B only
   C. Both A and B
   D. Neither A nor B
10. Technician A says each master cylinder piston works with a rubber cup seal that fits in front of the piston head. Technician B says the cup has rigid lips that fit against the cylinder walls to allow fluid leakage around the seal. Who is correct?
    A. A only
    B. B only
    C. Both A and B
    D. Neither A nor B

CHAPTER 5

HYDRAULIC LINES, VALVES, AND SWITCHES

Upon completion and review of this chapter, you should be able to:

- Describe the purpose and types of hydraulic brake lines.
- Identify the two types of flares used on brake line tubing.
- Explain the purpose and identify the types of brake line fittings.
- Describe the purpose and explain the mounting of flexible brake line hoses.
- Explain the reasons for strength requirements for tubing and hoses and identify the construction features that fulfill these requirements.
- List the general precautions for working with brake tubing and hoses.
- Explain the purpose, parts, and operation of a metering valve.
- Explain the purpose, parts, and operation of a proportioning valve.
- Define the split point and slope of a proportioning valve.
- Explain the purpose and describe the operation of a fluid level switch and a stop lamp switch.
- Explain the function of a brake pedal position switch.

Terms To Know

Banjo fitting
Brake pedal position (BPP) sensor
Center high-mounted stop lamp (CHMSL)
Double flare
Dynamic rear proportioning (DRP) Fittings
ISO fitting
Metering valve
Pressure differential valve
Proportioning valve
Split point

INTRODUCTION

Hydraulic lines made of tubes and hoses transmit fluid under pressure from the master cylinder to each of the wheel service brakes (**Figure 5-1**). Valves are used in the system to control hydraulic pressure and as safety devices. Electrical switches to operate the stop lamps, as well as sensors to indicate low brake fluid level, also are important parts of the brake system.

Brake lines are made up of solid brake lines and flexible brake hoses.

BRAKE LINES AND HOSES

Brake lines or tubing consists of steel tubes or pipes and flexible hoses connected with fittings. Rigid tubing is used everywhere except where the lines must flex. Flexing of the brake lines is necessary between the chassis and the front wheels and between the chassis and the rear axle or suspension. Brake tubing and hoses are manufactured to strict specifications developed by SAE and the International Standards Organization (ISO).

Brake lines are also called tubing, or tubes by some technicians.

Shop Manual
page 206

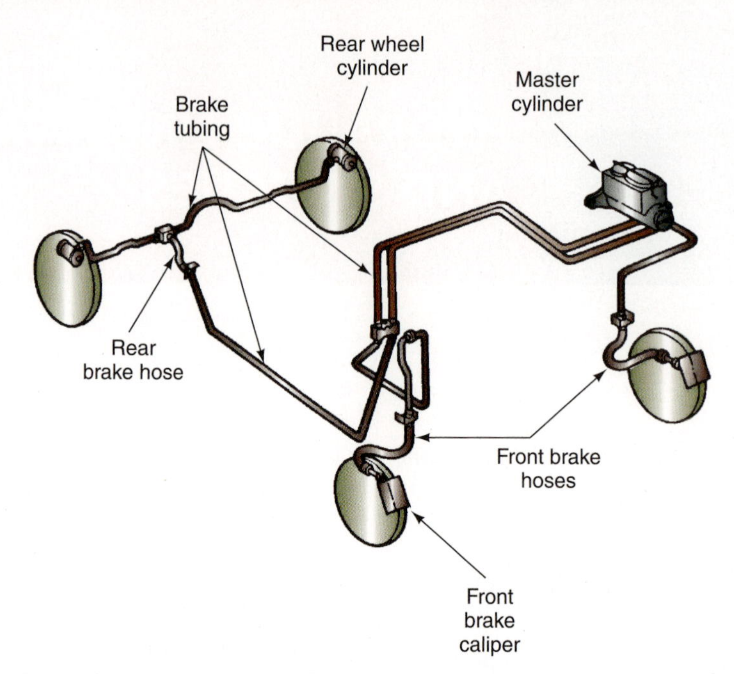

Figure 5-1 Brake lines consist of rigid tubing or pipes and flexible hoses that carry brake fluid from the master cylinder to the service brakes.

The general requirements for automotive brake fluid tubing and hoses are:

- Good corrosion resistance against chlorides (road salt)
- Strength, high burst pressure, and good fatigue or corrosion resistance
- Smooth bores allowing non-restricted flow
- Good resistance to surface fretting and stone pecking
- Ready availability at a realistic cost
- Easy to form

Brake Tubing

The hydraulic tubing used in the brake system is double-wall, welded steel tube that is coated to resist rust and other corrosion. Double-wall tubing is made in two ways: seamless and multiple ply. Each must meet the specifications of SAE Standard J1047 as amended.

WARNING **Double-wall steel brake tubing is the only type of tubing approved for brake lines. Never use 100 percent copper tubing as a replacement; it cannot withstand the high pressure or the vibrations to which a brake line is exposed. Fluid leakage and system failure will result.**

There is a copper-nickel alloy brake tubing that meets SAE Standard J1047 and ISO 4038. The alloy is 10 percent nickel, 1.7 percent iron, 0.8 percent manganese, and about 90 percent copper. This tubing meets all international and U.S. requirements for brake tubing, and it has the added advantage of being more corrosion resistant. Audi, Porsche, and Aston Martin vehicles use the copper-nickel brake tubing for their hydraulic brake systems. This chapter confines the discussion to steel tubing, however, because it is the most common, and the use, treatment, and fittings for the copper-nickel alloy tubing are exactly the same as those for steel. Seamless tubing is made by rolling a steel sheet twice around a mandrel so the edges *do not* adjoin each other to form a seam (**Figure 5-2**). The tubing is then run through a furnace where copper plating is applied and brazed to form the tubing into a single, seamless piece.

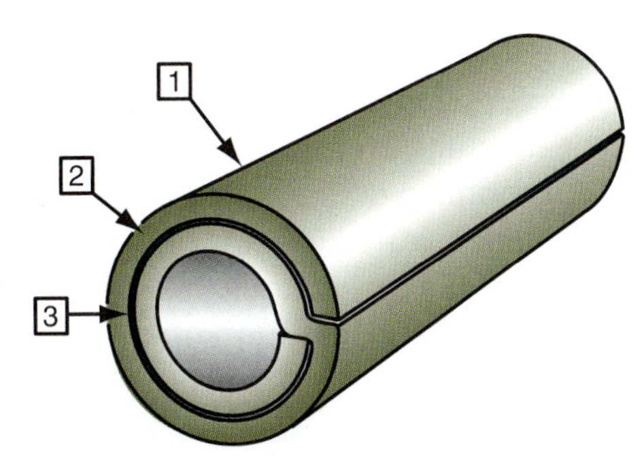

Figure 5-2 Cross-sectional view of a brake line.

Multiple-ply tubing is formed as two single-wall tubes, one inside the other. The seams of each section must be at least 120 degrees apart. Then the two-ply tubing is furnace brazed, just like seamless tubing, to form a single, seamless length.

All brake tubing is plated with zinc or tin for corrosion protection. In addition, all brake tubing must meet the burst specification of SAE Standard J1047, which requires that an 18-inch length of tubing must withstand an internal pressure of 8,000 psi. These requirements by themselves make it clear why copper tubing cannot be used for brake lines.

Tubing Sizes. Brake tubing is made in different diameters, lengths, and shapes (**Figure 5-3**). The most common diameters for steel tubing in the inch system are 3/16 inch, ¼ inch, and 6/16 inch. Other tubing diameters from ⅛ inch to ⅜ inch also are available.

Some vehicles use tubing sized in metric diameters. Metric diameters are specified in millimeters by SAE Standard J1290. Common metric diameters are 4.75 mm, 6 mm, 8 mm, and 10 mm.

AUTHOR'S NOTE It is almost impossible to get metric-sized brake tubing in a small town. Most parts vendors sell "metric" brake tubing, but in reality, it is an SAE-sized tubing (usually 3/16-inch inside diameter) with metric-sized and metric-threaded fasteners. A standard double flare tool will work on these lines, but the ISO flare tool will not work because the tubing collapses before flaring. The problem with double flaring an SAE-sized tube for connection to a vehicle's ISO fitting is the flare angles. They are a little different and will not seal properly.

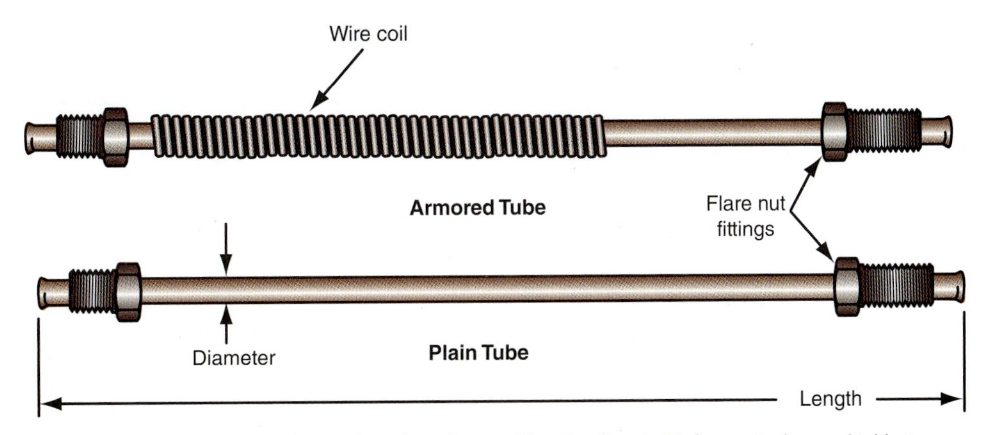

Figure 5-3 Brake tubing (line) comes in various sizes and lengths, flared with flare-nuts. Armored tubing uses a wire coil for protection.

It is important to note that ISO flare fittings may have a different pitch thread depending on the make and model of the vehicle. Make sure to carefully choose any replacement fittings.

Tubing Installation. Tubing installed on a car when it is manufactured is shaped properly to fit into the brake system. As are other parts made by the carmaker, tubing is referred to as an original equipment manufacturer (OEM) part. Aftermarket replacement tubes are most often available straight and in different lengths. If available, however, it is preferable to use an OEM-shaped, prefabricated tube as a replacement. Most often, the tubing is purchased as a coil of tubing, and the technician will fabricate a new line. Generally, one-piece tubing is used between the different components of the brake system. For instance, there is a brake tubing from the master cylinder to a control valve, then another tubing from the valve to either a wheel or the junction block where the tubing is branched to each wheel. If possible, do not cut the tubing to add a section. Replace the whole section of tubing instead of patching it.

Each end of the tubing has a fitting for connection into the system. The fitting, which is described later, fits over the tube and seats against a specially formed end of the tubing.

ISO flares tend to seal better while using less torque on the threaded nut.

The formed end of the tubing is called a flare. There are two common types of flares formed on the end of brake tubing, the double flare and the ISO flare (**Figure 5-4**). The inverted double flare has the end of the tubing flared out, then it is formed back onto itself. The ISO flare has a bubble-shaped end formed on the tubing. Each type of flare is used with a different type of fitting, and they are not interchangeable. Tools are available to form these flares when fabricating new tubing for a repair. Flares and their fittings are described in more detail later in this chapter.

The brake line tubes are routed from the master cylinder along the car frame or body toward the wheel service brakes (**Figure 5-5**). Clips hold the tubes in position. The clips

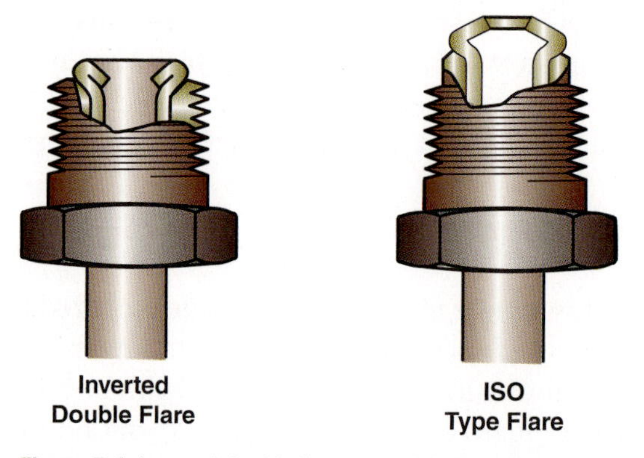

Figure 5-4 Inverted double flares and ISO flares are used on brake lines.

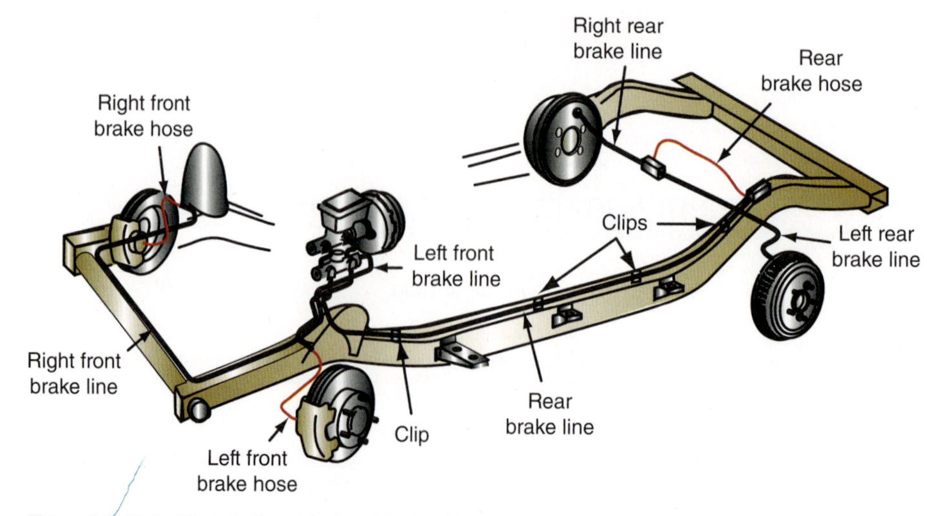

Figure 5-5 Typical installation of brake tubing and hoses. Brake tubing is held to the vehicle frame by clips and brackets.

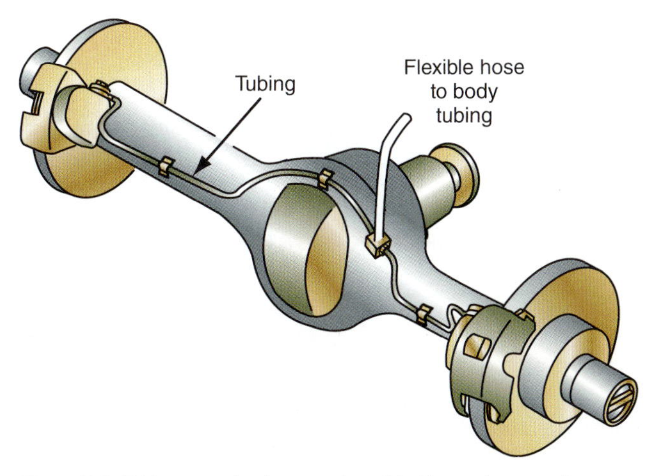

Figure 5-6 Tubing may also be routed on driveline and suspension parts, such as a solid rear axle housing.

usually have rubber isolators to cushion the tubes. Clips are important because they prevent the tubes from vibrating, which could cause metal fatigue and eventual rupture. Tubing also may be routed on suspension and driveline parts, such as a rigid rear axle and differential (**Figure 5-6**).

Brake Hoses

Brake hoses (**Figure 5-7**) are the flexible links between the wheels or axles and the frame or body. Hoses must withstand high fluid pressures without expansion and must be free to flex during steering and suspension movement.

Figure 5-8 shows the parts of a brake hose. The hose is made from materials that resist damage from both brake fluid and petroleum-based chemicals. Brake hoses are manufactured and tested to the specifications of SAE Standard J1401 and FMVSS 106. Among these specifications is the requirement that a brake hose must withstand 4,000 psi of pressure for 2 minutes without rupturing. The test pressure is then increased at a rate of 25,000 psi per minute until the hose bursts. The burst pressure is recorded as the final test measurement. These performance requirements make it clear why brake hoses are important safety devices.

FMVSS 106 further requires that brake hoses must be marked with raised longitudinal ribs or two $^{1}/_{16}$-inch colored stripes on opposite sides to indicate twisting of the hose during installation. Twisting creates stress that could lead to rupture. Twisting also may cause the hose to kink, cause ply separation, and block fluid pressure. To further prevent

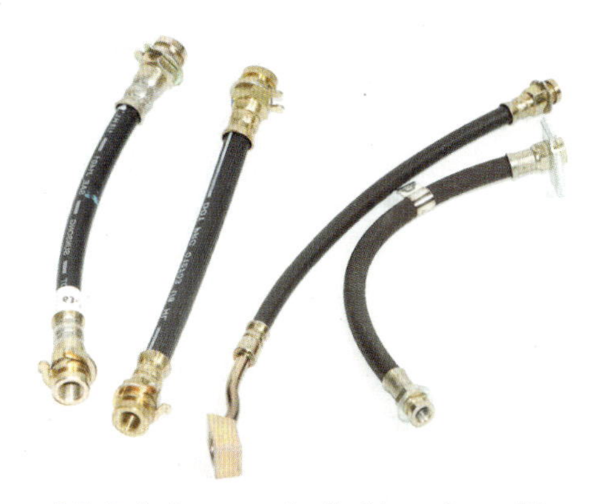

Figure 5-7 Brake hoses are the flexible sections of the brake line.

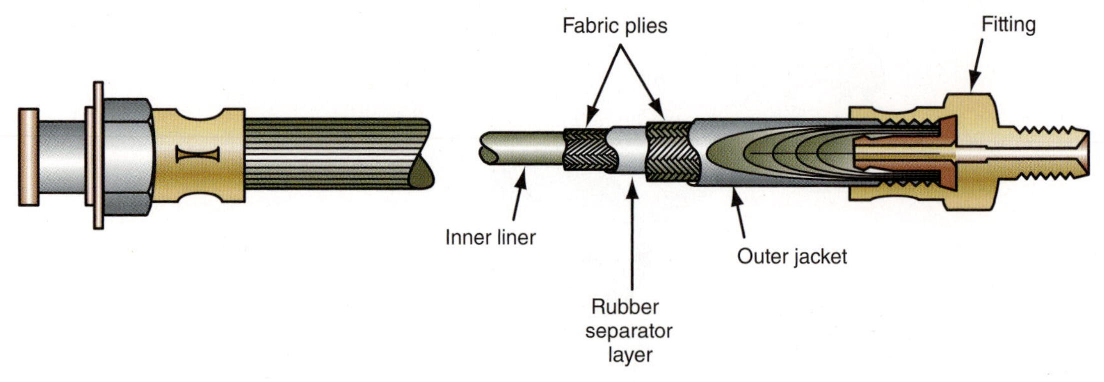

Figure 5-8 Brake hoses are made with two fabric layers alternating with two rubber layers. The outer jacket is ribbed to indicate if the hose is twisted during installation.

twisting, at least one end fitting on a hose usually can be rotated with the hose before fastening. This normally is the end at the body or frame.

Brake hoses are reinforced with metal or synthetic cords to withstand high pressures. Each end has a fitting so that it can be connected to other parts of the brake system. The brake hose length is specified from the end of one fitting to the end of the other fitting. Hose diameter is specified as the inside diameter of the hose. Hoses are available in both inch and metric diameters. Brake hoses come in different sizes and lengths with various end fittings for different vehicle requirements and are interchangeable between different makes and models.

AUTHOR'S NOTE Unlike the brake tubing, metric-sized brake hoses are, in fact, metric sized. This is primarily because the metric fittings are crimped or swagged to the hose. Using metric fittings on an SAE-sized hose could cause the same problems as a double flare fitting connected to a metric fitting.

The fittings are crimped, or swagged, to the brake hose ends at very high pressures. Clamp-on or crimped fittings used with low-pressure hydraulic hoses or oil hoses cannot be used for brake hoses.

When purchasing brake hoses, make sure to get a hose that is DOT approved to ensure the hose has been tested for quality.

For all of their strength and durability, brake hoses are the weakest links in the brake hydraulic system. Atmospheric ozone attacks the rubber material and, over a long period, causes the hoses to deteriorate. Slight porosity of the hose material also lets air enter the system, which contaminates the brake fluid—again, over a long period of time, however.

Hoses also are subject to wear, both externally and internally. Hoses must be installed so that they do not rub against vehicle parts. Some hoses have rubber ribs around their outer circumference to protect them from rubbing on suspension and chassis parts. A bulge in a hose usually indicates that the hose is failing because of internal wear or damage.

AUTHOR'S NOTE There are many technical service bulletins (TSBs) on brake diagnosis and service. When starting the repair order, consult the shop's service database to determine if any of these TSBs apply to the job at hand.

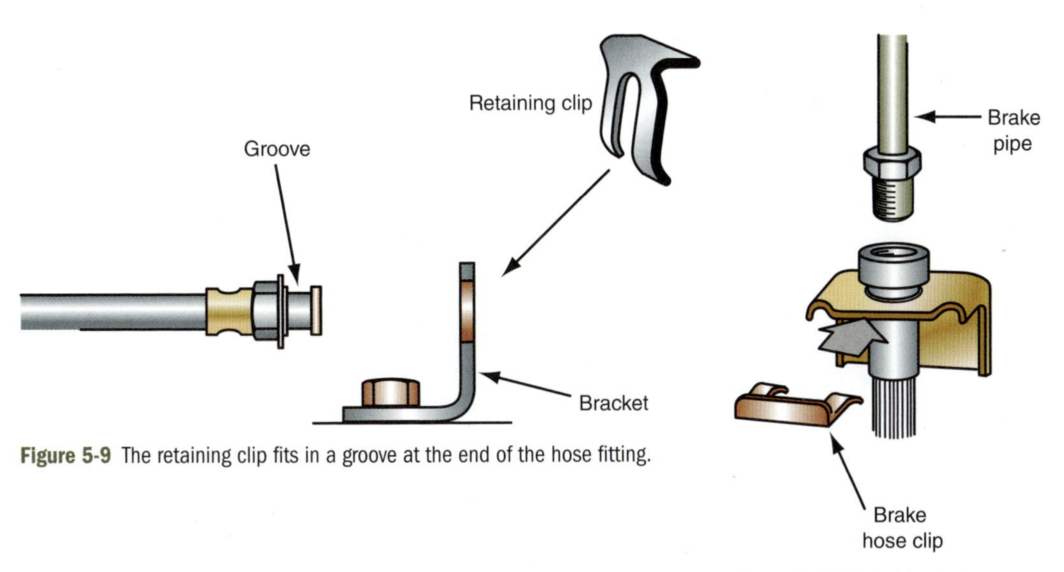

Figure 5-9 The retaining clip fits in a groove at the end of the hose fitting.

Figure 5-10 Typical brake hose mounting.

Another internal hose problem can occur when the inside of the hose wears to the point that a flap of rubber loosens from the hose wall. If the flap stays secured to the wall at one end and its loose end faces the master cylinder, it can delay fluid pressure to the wheel brake and cause uneven braking or pulling. If the loose end of the flap faces the wheel brake, it can delay pressure release from the wheel brake and cause brake drag. This type of hose defect is impossible to see from the outside and is difficult to pinpoint with any kind of test. If the symptoms described above exist, hoses usually are replaced to try to eliminate the problem.

A complete fitting usually consists of a nut fitted around a tube. The nut may be male (external threads) or female (internal threads) and will screw into or around its counterpart on a device or another tube. The flare is trapped and compressed by the two pieces of fitting.

The point at which a hose connects to a rigid tube usually is secured to a bracket on the frame or body. This is the end that can be moved so the hose is not twisted. A clip fits in a groove in the end of the hose fitting (**Figure 5-9**), and the end of the hose is inserted through a support bracket (**Figure 5-10**). The steel line is then threaded into the fitting on the end of the hose. The fitting on one side of the bracket and the clip on the other side hold the hose securely in position.

Brake Fittings

Fittings is a term applied to all plumbing connections used on the car. They may be referred to as flare nut, nut, or line nut. The term *fitting* is used in this text.

The threaded parts used to connect brake hoses and tubing together and to other brake components are called **fittings** (**Figure 5-11**). Fittings are made from steel to withstand brake system pressures. Fittings are threaded to allow connection to other brake parts. The ends of brake tubing are formed into either an inverted double flare or an ISO flare (described later) and fitted with a flare nut. Brake hoses can have either male or female

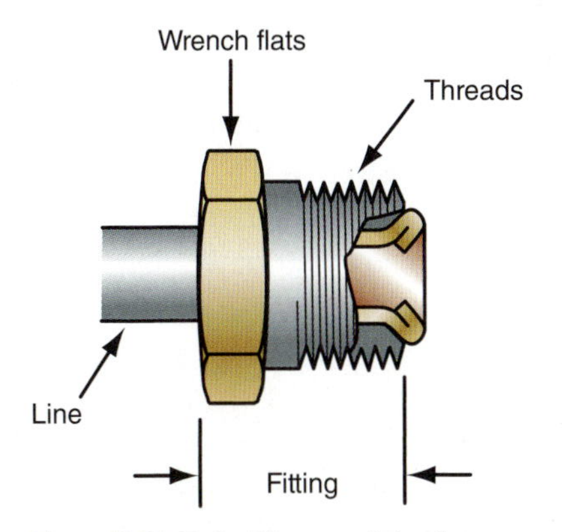

Figure 5-11 Typical flare-nut fitting for an inverted flare connection.

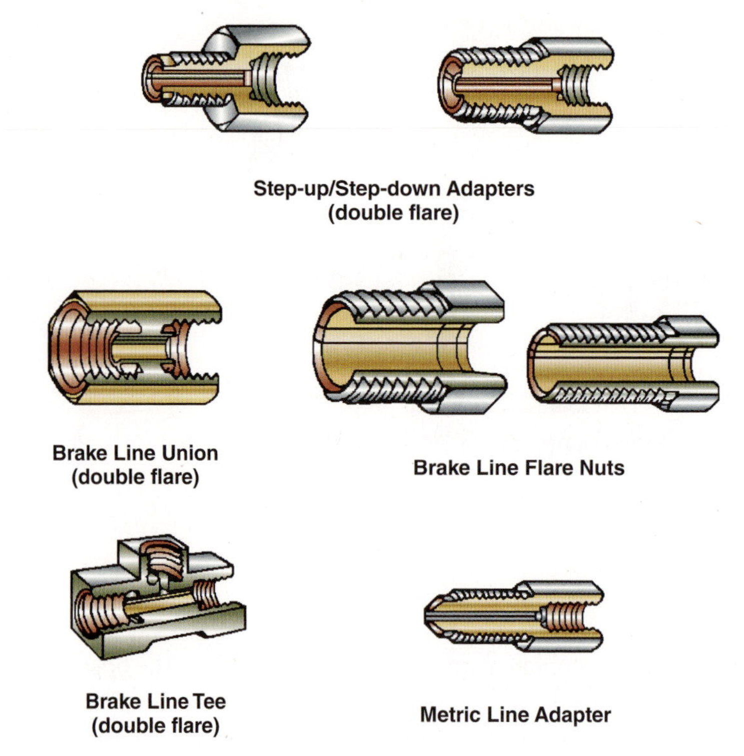

Figure 5-12 Assorted adapter and fittings are used for brake line connections.

fittings. The threaded connections of master cylinders, wheel cylinders, calipers, and most valves are female. Fittings are made with both SAE inch and ISO metric threads. Threads from one system do not fit threads of the other system.

AUTHOR'S NOTE Every time a fitting is added to a hydraulic circuit, a place is created for a leak. I would suggest that the number of fittings installed during a brake line repair be held to a minimum. Also, do not use "adapters" to join two sections of brake tubing or tube-to-hose connections. If the parts do not match, buy ones that do.

Figure 5-12 shows an assortment of different brake line fittings. Adapters can be used to connect two different sizes of fittings. Unions and tee fittings are used to connect two lines together. Again, the use of adapters should be held to a minimum; it is much more desirable to use lines that fit properly.

SAE Fittings. All SAE fittings used in brake systems have a 45-degree taper on the male nut and on the inside of the flare (**Figure 5-13**). The tubing seat in the female fitting has a 42-degree taper. The 3-degree mismatch forms an interference fit that creates a leak-free, high-pressure seal. SAE standards also exist for 37-degree fittings, but these are not used in brake systems.

A fitting with an external taper is simply called a standard flare (**Figure 5-13A**). When the fitting is tightened, the tapered surfaces of the male and female fittings create the seal. A fitting with an internal taper is called an inverted or LAP flare (**Figure 5-13B**). These are more common than standard flares on brake tubing. The male flare fitting compresses the bell-mouthed inverted flare against the seat in the female fitting, and the tubing is sandwiched between the two halves of the fitting to form the seal.

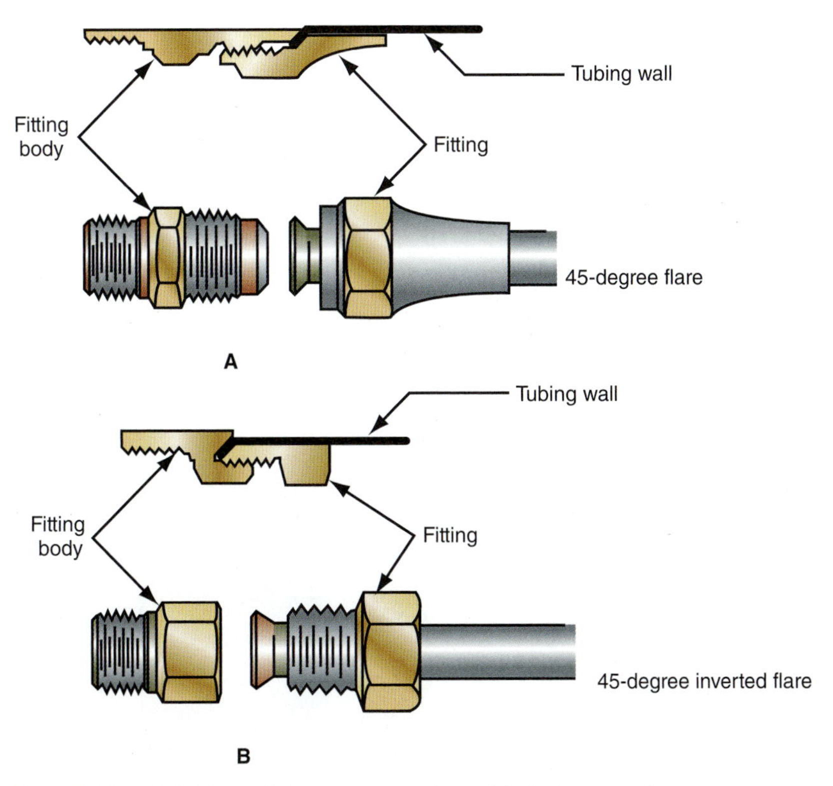

Figure 5-13 SAE 45-degree fittings are commonly used for brake connections.

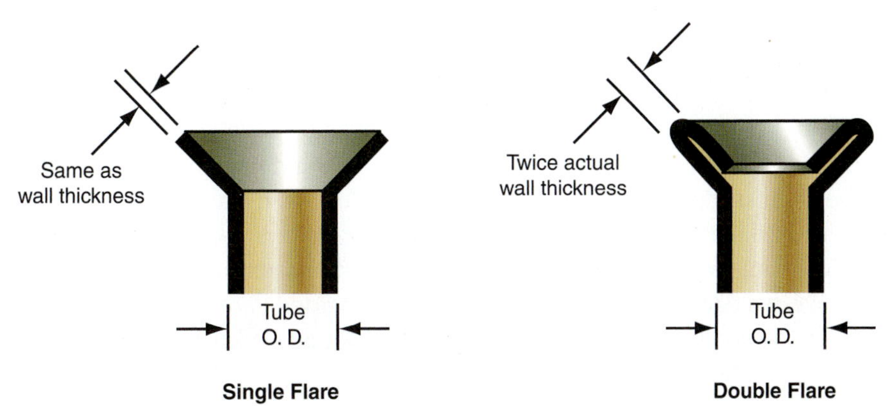

Figure 5-14 Brake line steel tubing must have a double flare. The copper-nickel alloy tubing requires the same flaring.

The flared tubing end can be formed as either a single flare or a **double flare** (**Figure 5-14**). A single flare does not have the sealing power of a double flare and is subject to cracking. Preformed replacement brake tubing is sold with a double flare on each end and the fittings in place on the tubing. If you cut and form a brake pipe from bulk tubing, you will have to form double flares on both ends. Flaring procedures are covered in the *Shop Manual.*

Double flare refers to a tubing connection in which the end of the tubing is flared out then formed back on itself.

ISO Fittings. The **ISO fitting** is a metric design, originally used on imported vehicles but now common on domestic brake systems. An ISO flare is not folded back on itself as is a double flare. The unique shape of an ISO fitting causes it to be called a "bubble" flare. **Figure 5-15** shows the differences between the shapes of an inverted double flare and an ISO fitting and the ways that the tubing ends are held in their fittings.

ISO fitting refers to a tubing flare in which a bubble-shaped end is formed on the tubing; also called a bubble flare. It does not need a separate tubing seat as an inverted flare does.

Like a standard or an inverted flare, an ISO fitting uses interference angles between the flare and its seat to form a leak-free seal. The angle of the outer surface of an ISO fitting flare is approximately 32.5 degrees. The flare seat is 30 degrees, and the angle at the end of the flare nut is 35 degrees.

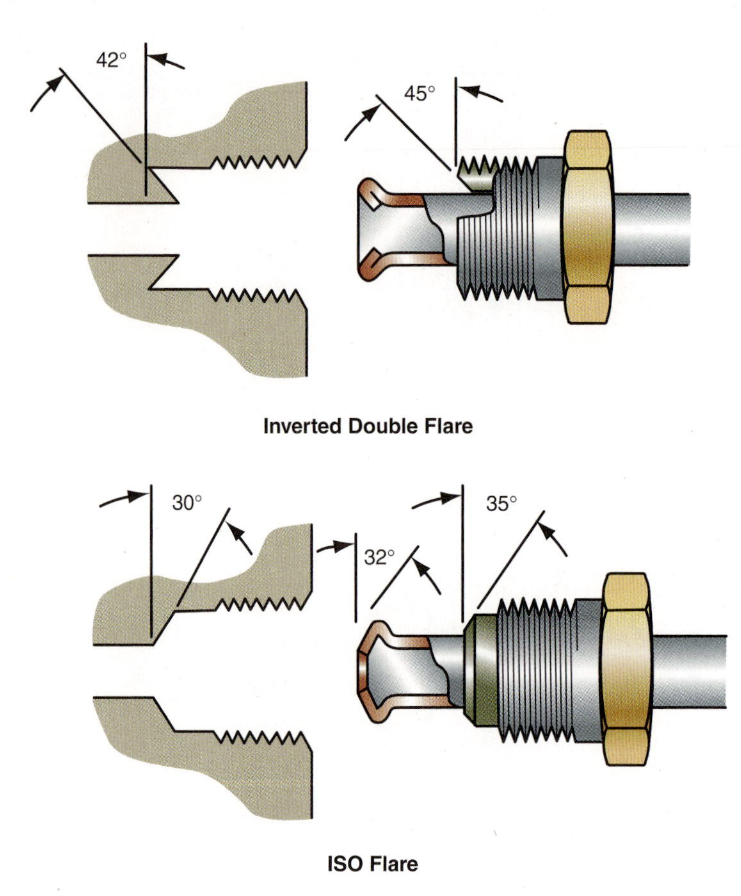

Figure 5-15 ISO and inverted flares are distinguished by the shape of the flare. An ISO flare does not need a separate tubing seat as does an inverted flare.

Compressed washer fittings are so named because the seal is made by compressing a washer. Change the washer if the fitting is loosened or removed.

ISO fittings have become popular with both domestic and foreign vehicle manufacturers because the outer surface of an ISO fitting will form a leak-free seal against a mating surface in a cylinder or caliper body that is simply drilled and countersunk with the right taper (see Figure 5-15). Manufacturing operations are simplified because an inverted, cone-shaped seat for an inverted flare is not required. A seat for an inverted flare requires extra machining operations or the addition of a steel insert. The seat for an ISO fitting is much simpler to machine and provides a seal that is equal to—or better than—a traditional inverted flare.

A **banjo fitting** is a round, banjo-shaped tubing connector with a hollow bolt through its center.

Compressed Washer Fittings. Straight compressed washer fittings are usually found on the ends of brake hoses that attach to calipers or wheel cylinders. As the rigid fitting on the end of the hose is tightened, it compresses a soft copper washer against a flat, machined surface on the cylinder or caliper (**Figure 5-16**).

Because the fitting is usually attached rigidly to the end of the hose and does not swivel, this end of the hose should be connected before the end with a swivel fitting. Compressed washer fittings also are sometimes used on hydraulic junction blocks and valve bodies.

Tightening and loosening of a compressed washer fitting will permanently compress the copper washer. Whenever a fitting is disconnected, replace the washer.

Always replace the washer(s) when the banjo fitting is loosened or removed.

Banjo Fittings. A **banjo fitting** is a circular fitting that looks like the instrument of the same name. A banjo fitting is used to attach a hose or tube to a port on a cylinder or caliper at a close right angle (**Figure 5-17**). Fluid passes from the brake line into the cutaway section inside the banjo fitting and then through the hollow bolt to the cylinder or caliper.

A banjo fitting is a kind of compressed washer fitting, and both flat surfaces of the fitting are sealed with soft copper washers. The washers should be replaced anytime the banjo bolt is loosened.

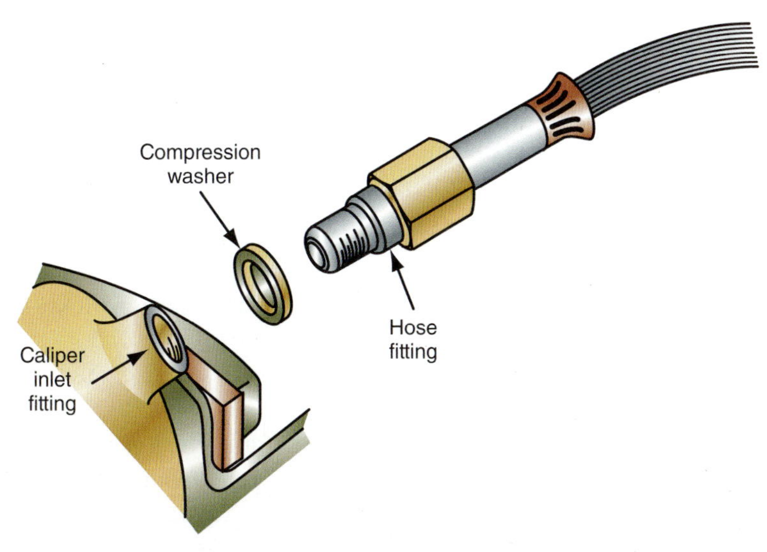

Figure 5-16 A copper washer is used with compressed fittings.

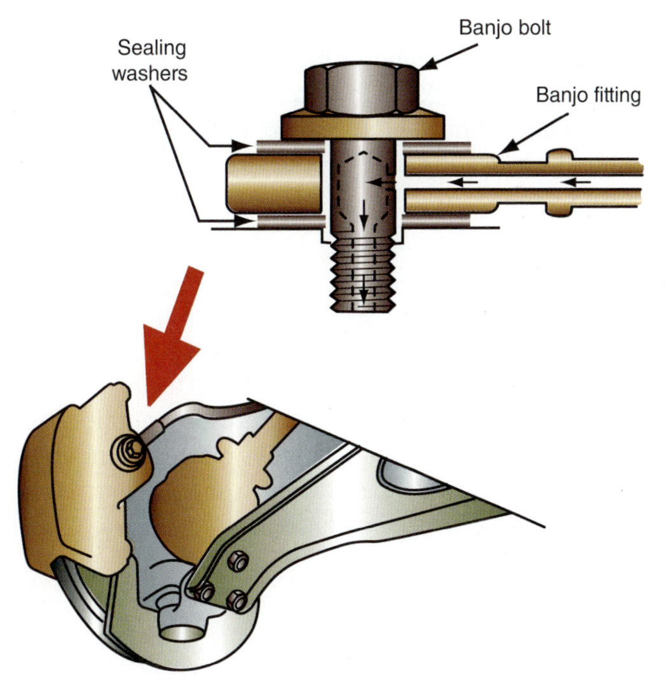

Figure 5-17 Banjo fittings are typically used at right-angle connections.

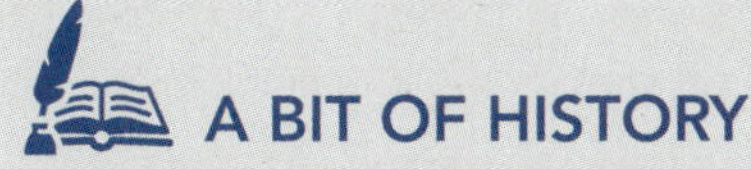

A BIT OF HISTORY

Dual-chamber master cylinders and split hydraulic systems became standard brake system hardware in 1967 by federal regulation. However, the first safety backup brake system was more than 30 years old by then. In 1936, Hudson introduced *Duo-Automatic Hydraulic Brakes,* which used four-wheel mechanical parking brake linkage superimposed on the hydraulic service brake system. Override linkage on the brake pedal applied the brake shoes mechanically if the pedal dropped low enough. Hudson's competitors apparently thought the system was more monkey business than safety business, and it remained a Hudson exclusive for more than 20 years. The last *Duo-Automatic Hydraulic Brakes* were used on some 1957 models, the last of the Hudsons, before the merger with Nash created American Motors.

Brake Line and Fitting Precautions

Brake lines and fittings have important safety requirements. Service instructions in the *Shop Manual* contain specific WARNINGS and CAUTIONS where needed. The following paragraphs summarize universally important precautions and guidelines for installing these parts:

- Never use straight copper tubing in place of double-wall steel brake tubing. Copper cannot withstand the high pressure or the vibrations to which a brake line is exposed. Fluid leakage and system failure can result. The approved copper-nickel alloy tubing is accepted.
- For similar reasons, never use copper fuel line fittings as a replacement for steel brake line fittings.
- Never use spherical-sleeve compression fittings in brake lines. Spherical compression fittings are low-pressure fittings for applications such as fuel lines. They will fail and leak under the high pressures and vibrations of a brake hydraulic system.
- Many states have outlawed the use of compression fittings on brake lines.Do not interchange metric-sized and SAE fittings. They have different threads and cannot be mixed. Do not interchange ISO flare nuts with SAE inch-sized flare nuts. Either of these conditions can cause fluid leakage and system failure.
- Do not use low-pressure fuel or oil hoses in place of brake hoses. Hoses not made for brake system use can fail under the high system pressures and may deteriorate when exposed to polyglycol brake fluids. Fluid leakage and system failure can result.
- When installing a brake hose, be careful not to twist or kink it. Fluid leakage and system failure can result.

Brake Hydraulic Valves

Before 1967, when most brake systems had drum brakes at all four wheels and master cylinders had only one chamber, the systems worked well enough with hydraulic pressure operating uniformly throughout the lines and cylinders. The development of combination disc and drum brakes required that the timing of pressure application had to be altered in certain ways. The requirement for dual-chamber master cylinders and split hydraulic systems also called for control valves and switches to operate warning lamps. The general kinds of control valves that may be used include:

Shop Manual
page 214

- Metering valve
- Proportioning valve
- Pressure differential valve (warning lamp switch)

With the advent of ABS and or four-wheel disc braking systems, many of these valves have been eliminated or incorporated into the ABS hydraulic modulator, although some manufacturers continue to use some of these valves or a variation of them. More on this in Chapter 10.

Metering Valve

Metering valves function only during the initial stage of braking.

Some cars with front disc and rear drum brakes have a **metering valve** in the hydraulic system to achieve balanced braking between the front and rear wheels. **Figure 5-18** is a simplified brake hydraulic system diagram that shows a metering valve in relation to the front disc brakes and a proportioning valve (described later) in relation to the rear drum brakes.

Metering valves were used primarily on rear-wheel drive (RWD) vehicles. A metering valve (**Figure 5-19**) is in the line to the front brakes and keeps the front disc pads from operating until the rear drum brakes have started to work. The valve delays pressure application to the front disc brakes because disc brakes are fast acting, whereas drum brakes have spring tension and linkage clearance to overcome.

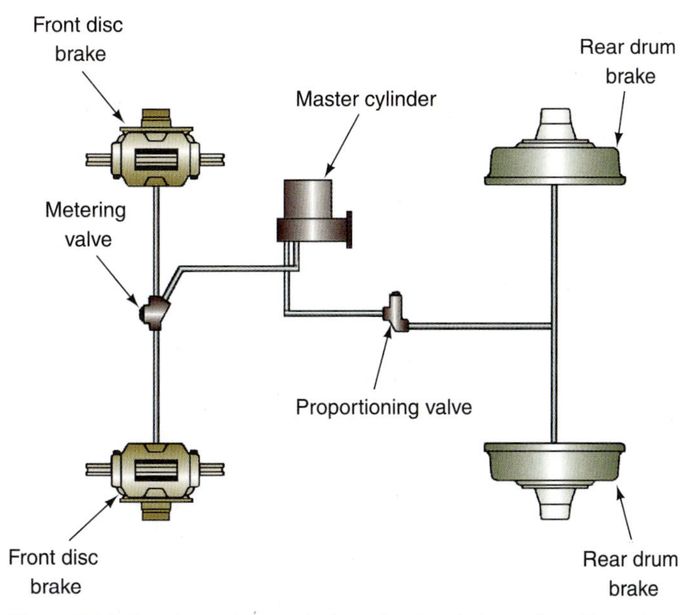

Figure 5-18 Metering and proportioning valves in relation to front disc and rear brake drums. Use of metering and proportioning valves has been reduced due to ABS braking systems.

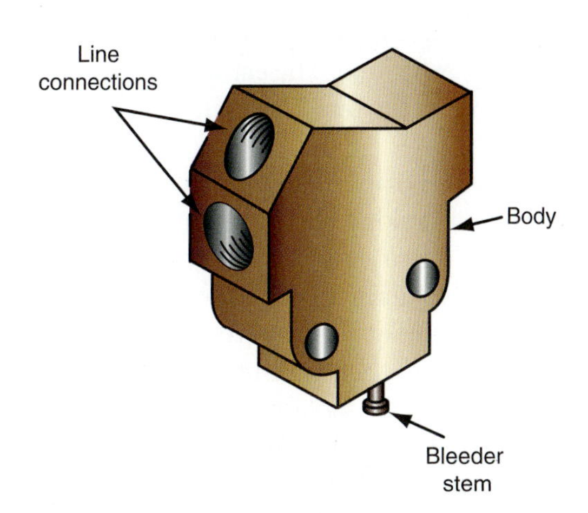

Figure 5-19 This traditional metering valve slightly delays application of front disc brakes while hydraulic pressure builds up in rear drum brake wheel cylinders.

Metering valves appeared with the first disc brake systems of the mid-1960s. Vehicles of that era were almost entirely RWD vehicles with front disc and rear drum brakes. Metering valves were required for all the reasons just described. With the increased use of front-wheel drive (FWD) and diagonally split hydraulic systems used since the early 1980s, however, metering valves have been eliminated from many vehicles.

On an FWD car, 80 percent of the braking is done by the front brakes, so it is desirable to apply them as quickly as possible. Until all clearance in the brake system is taken up, braking force is not great enough to overcome the torque of the front drive wheels. This driving torque and the forward weight bias of an FWD car eliminate the problem of front wheel lockup and any need for a metering valve. Furthermore, diagonally split brake systems used on most FWD cars would require two metering valves, one for the front brake on each side of the hydraulic system. Avoiding the complication of extra parts is another good reason to eliminate the metering valve.

On a vehicle with four-wheel disc brakes, the application times for the brakes at all wheels are about equal. A metering valve is therefore unnecessary.

Proportioning Valve

The **proportioning valve** restricts fluid to the rear wheels, thereby lowering the pressure to help prevent wheel lockup in the rear.

The amount of fluid restriction by a proportioning valve is changed proportionally to the vehicle load and amount of braking force being applied.

The **proportioning valve** was introduced in 1969 to help balance front and rear pressure on cars with disc and drum brakes. A metering valve controls the *timing* of pressure application to front disc brakes. A proportioning valve controls the actual *pressure* applied to rear brakes.

Inertia and momentum cause weight to shift forward during braking. The weight shift is proportional to the braking force and the rate of deceleration. During hard braking, the weight shift unloads the rear axle and reduces traction between the tires and the road. With reduced traction, the rear brakes may lock, and the vehicle may spin. Rear brake lockup can be avoided, however, by modulating the hydraulic pressure applied to the rear brakes. The goal for the best possible braking is to maintain an equal coefficient of friction between all tires and the road.

Disc brakes require higher hydraulic pressure than do drum brakes for equal braking force at the tires. Drum brakes use mechanical servo action (explained in Chapter 8 of this *Classroom Manual)* to increase force applied to the brake shoes. Because of this servo action, drum brakes require less hydraulic pressure to maintain braking force than to establish it. Disc brakes require a constant hydraulic pressure for a given amount of braking force. Overall, disc brakes always require higher hydraulic pressure than do drum brakes. A proportioning valve does exactly what its name indicates. It proportions hydraulic pressure between disc and drum brakes to maintain equal braking force at the tires.

Proportioning valves were originally designed for use with front disc, rear drum combinations for all the reasons just explained. The earliest proportioning valves were separate components, installed in the line to the rear brakes in front-to-rear split hydraulic systems (**Figure 5-20**).

Some vehicles with four-wheel disc brakes also have proportioning valves for the rear disc brakes. The goal of efficient braking always remains to maintain equal an coefficient of friction between the front and rear tires and the road. To reach this goal, the pressure to the rear disc brakes on some vehicles must be modulated to prevent lockup.

A proportioning valve has an inlet passage from the master cylinder at one end and an outlet to the rear brakes at the other end. Inside, a spring-loaded piston slides in a stepped bore. One end of the piston has a larger area than the other. The actual proportioning is done by a spring-loaded, check-valve-type stem that moves in a smaller bore through the center of the piston.

When the brakes are first applied and under light braking, the proportioning valve does nothing. Fluid enters the valve at the end with the smaller piston area (**Figure 5-21**), passes through the small bore around the stem, and exits to the rear brakes. The end of

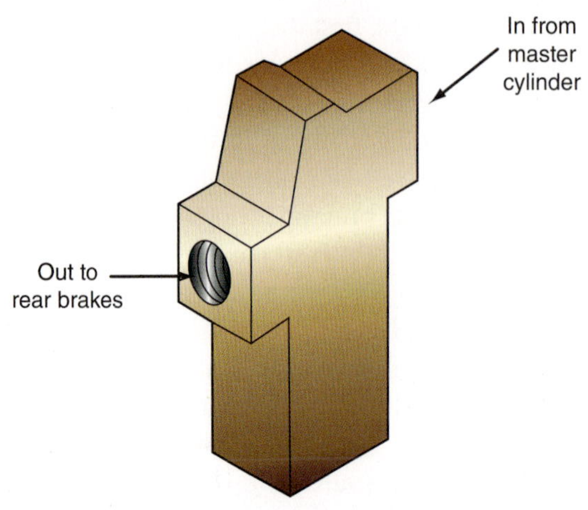

Figure 5-20 A simple proportioning valve is installed in the line to the rear in front-to-rear split hydraulic systems.

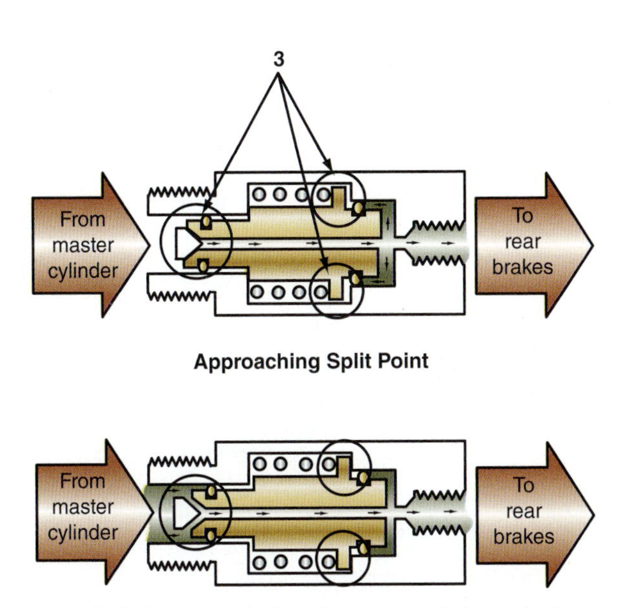

Figure 5-21 This proportioning valve serves a single rear brake circuit; typically found in a diagonally split hydraulic system.

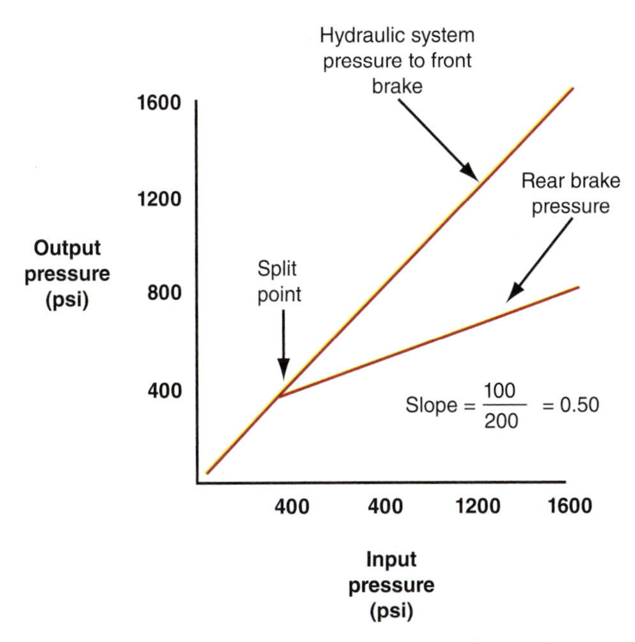

Figure 5-22 Split point and slope are the operating factors of a proportioning valve.

the valve piston at the outlet side of the valve is the end with the larger surface area. As outlet pressure rises in the valve, it exerts greater force on the piston than inlet pressure does and moves the piston toward the inlet, against spring pressure, thus closing the center valve stem and blocking additional pressure to the rear brakes (item 3 in Figure 5-21).

The pressure at which the proportioning valve closes is called the **split point** because the uniform system pressure splits at that point, with greater pressure applied to the front disc brakes and lower pressure applied to the drum brakes (**Figure 5-22**). As pressure continues to increase from the master cylinder, inlet pressure at the proportioning valve overcomes the pressure at the large end of the piston and reopens the valve. Fluid again flows through the center of the valve, pressure at the large end of the piston rises, and the valve closes again. The opening and closing action repeats several times per second.

The valve piston cycles back and forth and lets pressure to the rear drum brakes increase but at a slower rate than pressure to the front disc brakes. The pressure increase to the drum brakes above the split point is called the slope. The slope is the numerical ratio—or proportion—of rear drum brake pressure to full system pressure (refer to Figure 5-22). If half of the system pressure is applied to the rear brakes, the slope is 1:2 or 50 percent. When the brakes are released, pressure drops, and the spring moves the proportioning valve piston. This opens the valve for rapid fluid return to the master cylinder.

The first proportioning valves used since the early 1970s were installed in the single line to the rear brakes. One valve controlled pressure equally to both brakes. Diagonally split brake systems separated the rear brakes from each other, however. If a proportioning valve is required, two are needed.

Pressure Differential Valve (Failure Warning Lamp Switch)

The pressure differential switch was in use largely before the advent of ABS braking. Since there are so many still operating, we will include a small segment on their operation here. A **pressure differential valve** is a hydraulically operated switch that controls the brake failure warning lamp on the instrument panel. Each side of the pressure differential valve is connected to half of the hydraulic system (one chamber of the master cylinder). Each master cylinder piston provides pressure to a separate front/rear or diagonal hydraulic

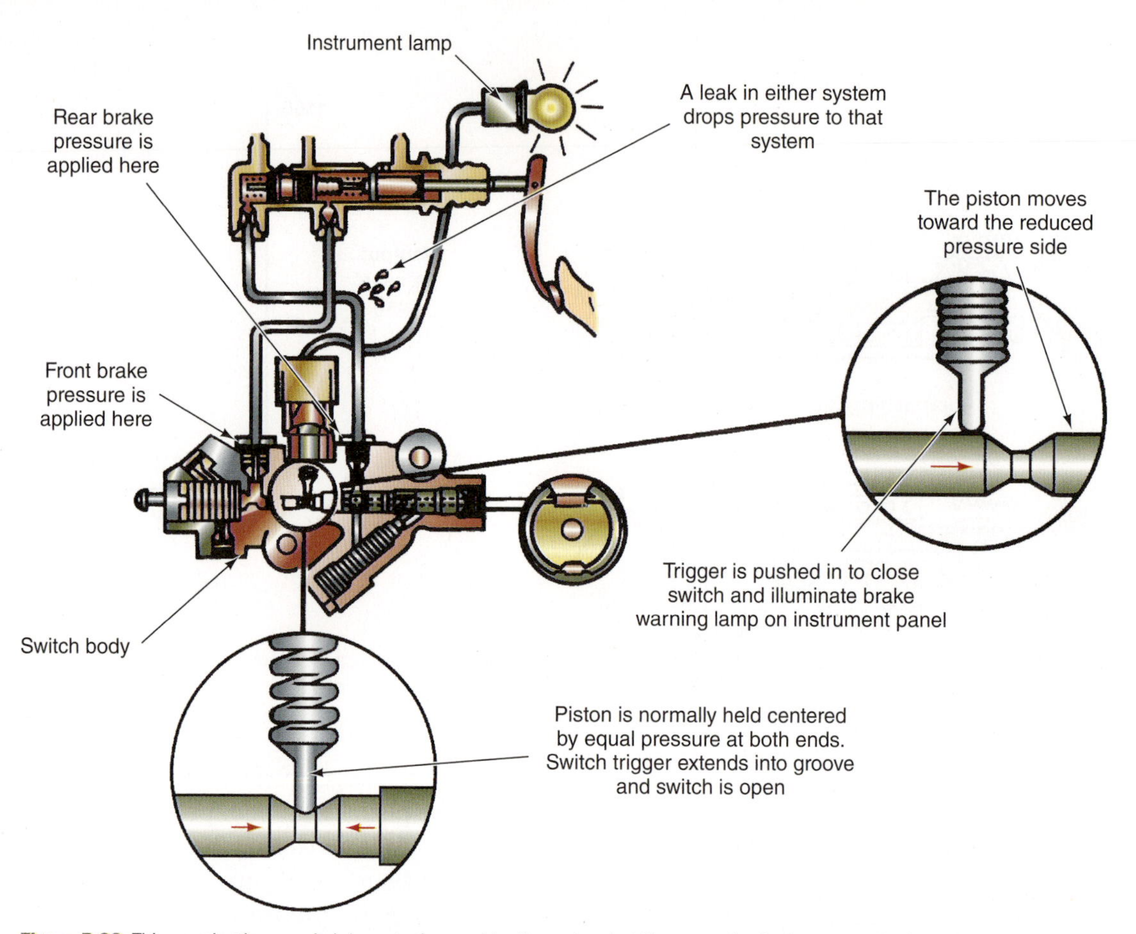

Figure 5-23 This warning lamp switch is part of a combination valve, but the operation is the same whether it is an individual component or combined with other valves.

system. If one of the systems fails, the brake pedal travel will increase and more brake pedal effort will be required to stop the car. The driver might not notice a problem, however, but the lamp on the instrument panel will provide a warning in case of hydraulic failure. **Figure 5-23** shows the operation of a pressure differential valve and warning lamp switch. The switch is a grounding switch, meaning it connects the light circuit to ground.

Combination Valve

Since the combination valve was actually a mix of metering and proportioning valves, combination valves have been replaced by ABS components. Combination valves were introduced in the 1970s to save money and to simplify the brake hydraulic system.

As the name indicates, a **combination valve** has two or three valve functions in one valve body. The most common three-function valve has three separate sections as shown in the sectional view (**Figure 5-24**)

Hydraulic Pressure Control with ABS

Antilock brakes are now standard equipment on new vehicles. Antilock brakes have a single, very simple operating principle: to prevent wheel lockup by modulating hydraulic pressure to the brake at any wheel that is decelerating faster than the others and is about to lock up. An ABS accomplishes this with wheel speed sensors and a control module that processes the wheel speed information and controls hydraulic pressure with electrically operated valves and a small high-speed pumps (**Figure 5-25**). ABS has prevented carmakers at having to guess at what the vehicle would do in a vast

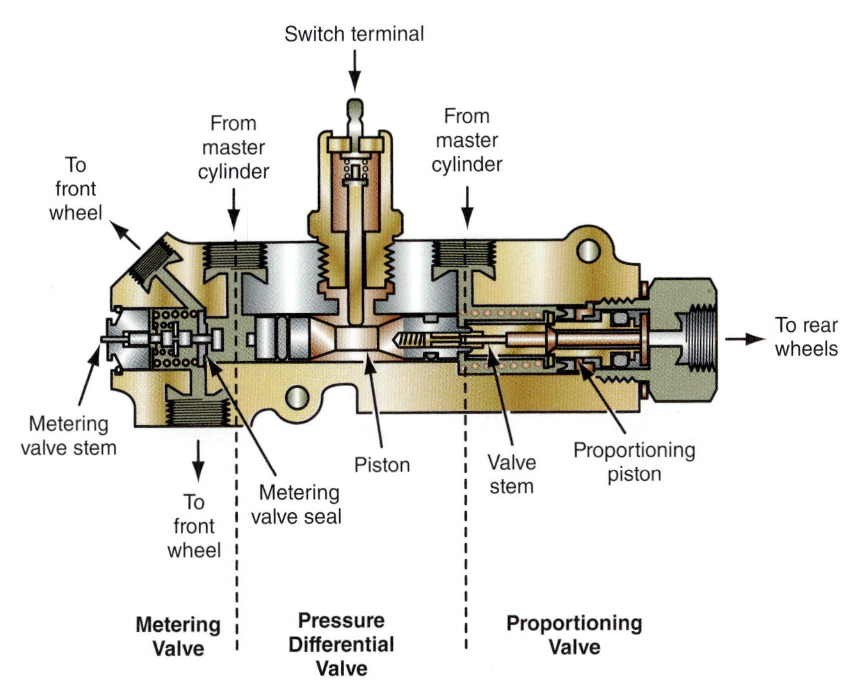

Figure 5-24 Cross-sectional view of a three-function combination valve.

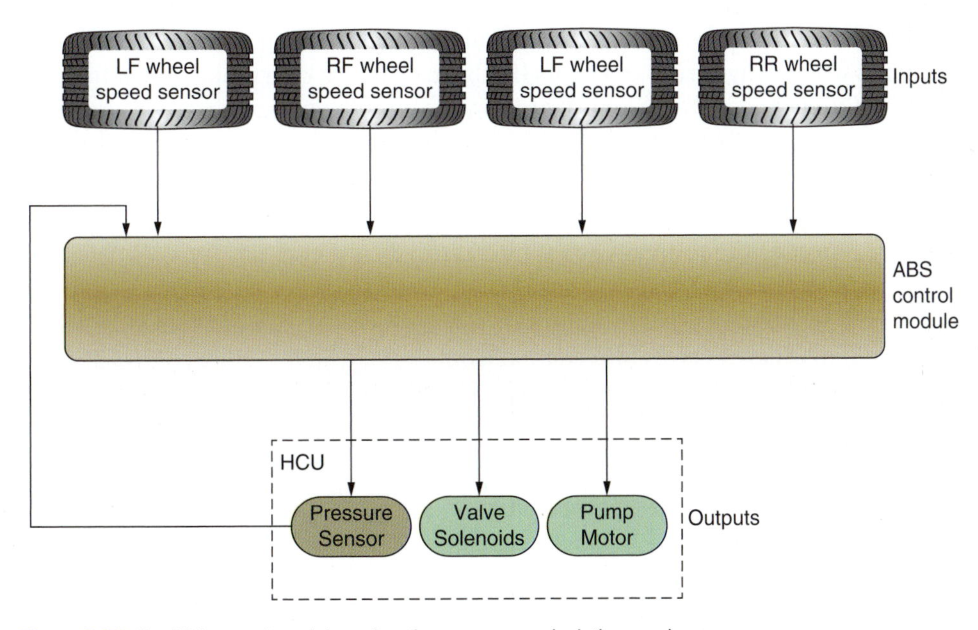

Figure 5-25 The ABS control module makes the necessary calculations and adjustments to hydraulic pressures to prevent wheel lock-up.

assortment of possible low-traction and panic braking circumstances. Instead, the vehicle itself could determine the conditions and apply corrections to all four wheels several times per second.

The pressure modulation or **dynamic rear proportioning (DRP)** provided by antilock brakes is no different than the modulation provided by metering valves and proportioning valves. In fact, ABS pressure control is more precise than a hydraulic or mechanical metering valve or a proportioning valve. Since antilock brakes became standard, engineers had the opportunity to eliminate the old valves and replace them with

Shop Manual
page 216

a computer-controlled system. Very few systems still use proportioning and metering valves in addition to the ABS function. This ensures that, in case of an ABS failure, the vehicle would still have safe braking. Chapter 10 in this *Classroom Manual* covers ABSs in detail.

AUTHOR'S NOTE ABS uses simple-function electrical solenoids that are switched on/off at a very fast rate. This switching allows instantaneous and continuous control of fluid pressures.

BRAKE ELECTRICAL WARNING SYSTEM

The following sections cover parking brake, fluid level, brake pad wear, stop lamp switch, and warning lamp operations.

Parking Brake Switch

The parking brake is used to hold a vehicle stationary. If the parking brake is even partially applied when driving, it will produce enough heat to glaze friction materials, expand drum dimensions, and increase pedal travel. On rear disc brake systems with integral actuators, it will distort rotors and reduce brake pad life.

A normally closed, single-pole, single-throw (SPST) switch is used to ground the circuit of the red warning lamp in the instrument cluster. This switch is located on the ratchet mechanism that locks the parking brake apply lever in place.

Vehicles with daylight running lights (DRLs) use the parking brake switch to complete a circuit that prevents the headlights from coming on if the parking brake is applied when the engine is started. When the parking brake is released, the DRLs function normally.

Electric Parking Brakes

A recent addition to the growing list of electronic devices on late-model vehicles is the electronic parking brake. These systems operate via electric actuators located on the rear brake calipers, or by electric cable actuators. These actuators are controlled by their own control module, which receives information from the vehicle's stability control system that the vehicle is moving, and automatically releases the parking brake. Of course, this system also has a "Parking Brake" indicator to alert the driver that the brake is applied and also to inform the driver that there could be a problem in the parking brake system itself. These systems have diagnostic trouble codes as well. More on these systems in Chapter 11.

Brake Pad Wear Indicators

Brake pads generally have a way to alert the driver when the brake pads are worn to the point of needing replacement. Most pads use a simple scraper built into the brake pad itself that touches the rotor when the pads are worn, producing an annoying squealing sound when the brakes are not applied; the sound goes away when the brakes are applied.

Both domestic and import car manufacturers have also built systems with an electronic wear indicator in the disc brake pads on some models. As the brake pads wear to a predetermined point, the red warning lamp notifies the driver they need attention.

In some systems, a small pellet is contained in the brake pad friction material. It is wired to ground the red warning lamp circuit whenever the brakes are applied. Each set of brakes contains a pellet. Each set offers a parallel leg to ground. In other systems, the pellets are wired into a series electrical circuit. As the pellet wears, it opens the electrical circuit and turns the red warning lamp on. This system self-tests the brake pad wear detection circuit all the time.

Master Cylinder Fluid Level Switch

Because brake fluid level is important to safe braking, many vehicles have a fluid level switch located in the reservoir to turn on the red brake warning lamp on the instrument panel when the fluid level drops too low. The fluid level switch has replaced the pressure differential valve warning switch on most vehicles. An added advantage of a fluid level switch is that it will alert the driver of a dangerous fluid level caused by inattention and poor maintenance practices.

Fluid level sensors are built into the reservoir body or cap. One kind of switch has a float with a pair of switch contacts on a rod that extends above the float (**Figure 5-26**). If the fluid level drops too low, the float will drop and the rod-mounted contacts touch a set of fixed contacts to close the lamp circuit; see item 4 in Figure 5-26. Another kind of switch uses a magnet in a movable float. If the float drops low enough, the magnet pulls a set of switch contacts together to close the lamp circuit. The contacts typically provide a ground path for the brake warning lamp (**Figure 5-27**).

AUTHOR'S NOTE If the lamp is on, then the fluid level is low. Before topping off the brake fluid, however, have the brakes checked. The fluid had to go somewhere, and unless there is an external leak, the only place for the fluid to go would be to the enlarged cavities in the wheel cylinders and calipers as the brake linings wore down.

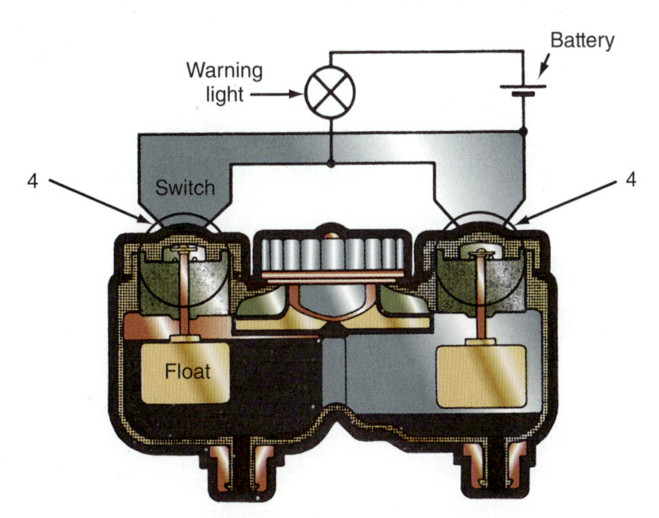

Figure 5-26 Float switches such as these are used in some master cylinders to warn of low fluid level.

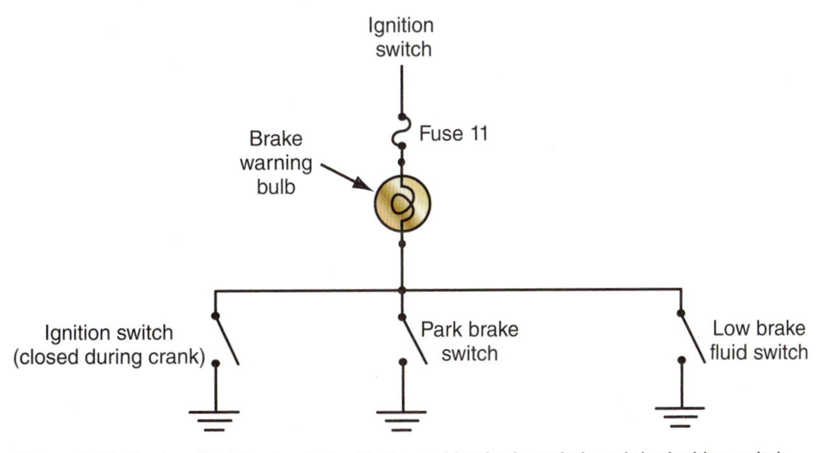

Figure 5-27 The low-fluid-level switch with the parking brake switch and the ignition switch on the ground side of the warning lamp.

Many European vehicles have a separate indicator lamp and circuit to indicate low brake fluid level. Manufacturers such as Mercedes-Benz, BMW, Jaguar, and Porsche use an electronic module to activate a special lamp on the dash.

Stop Lamp Switch and Circuit

Many older vehicles use a mechanical stoplight switch is mounted on the brake pedal bracket and activated by movement of the pedal lever (**Figure 5-28**).

Mechanical switches have been more common for many years because they can be adjusted to light the stop lamps with the slightest pedal movement.

Stop lamp switches also may be single-function or multifunction units. Single-function switches have only one set of switch contacts that control electric current to the stop lamps at the rear of the vehicle. Multifunction switches have one set of switch contacts for the stop lamps and at least one additional set of contacts for the torque converter clutch, the cruise control, ABS, body control module (BCM), and the powertrain control module (PCM). Some multifunction switches have contacts for all of these functions. Many switches require specific adjustment procedures. Consult the service information.

Stop lamp switches and lamps are more commonly referred to as brake light or stoplight switches.

The stop lamp switch is a normally open, momentary contact switch. The switch is attached to the brake pedal with a small clearance between the pedal arm and the switch lever or plunger (**Figure 5-29**). When the driver presses the pedal, the switch closes to complete the circuit and light the stop lamps. This same circuit is used to alert the ABS to monitor the wheel sensors during braking.

The conventional stop lamp switch receives direct battery voltage through a fuse. Therefore, the lamps operate even when the ignition is off. When the normally open switch is closed, voltage is applied to the stop lamps. The lamps on both sides of the vehicle and in the center high-mounted stop lamp (CHMSL) are wired in parallel. The bulbs are grounded through their mountings or by the use of remote grounds to complete the circuit.

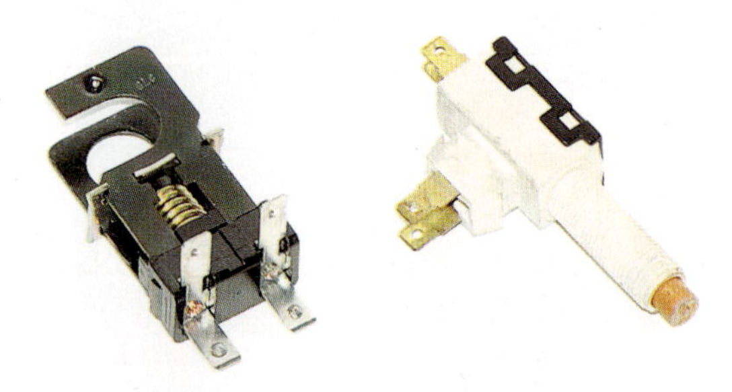

Figure 5-28 Mechanical stop lamp switches.

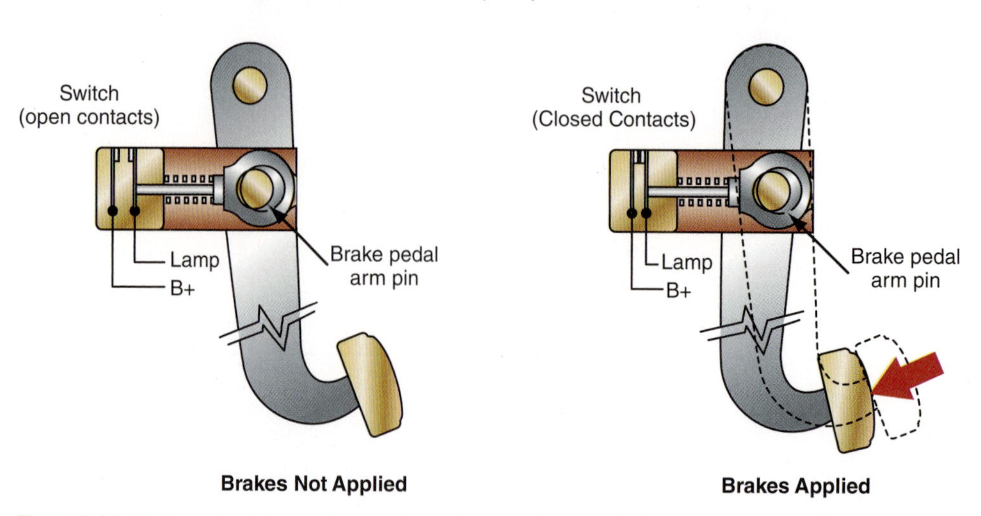

Figure 5-29 Most mechanical brake lamp switches have multiple circuits to open and close with the application of the brakes.

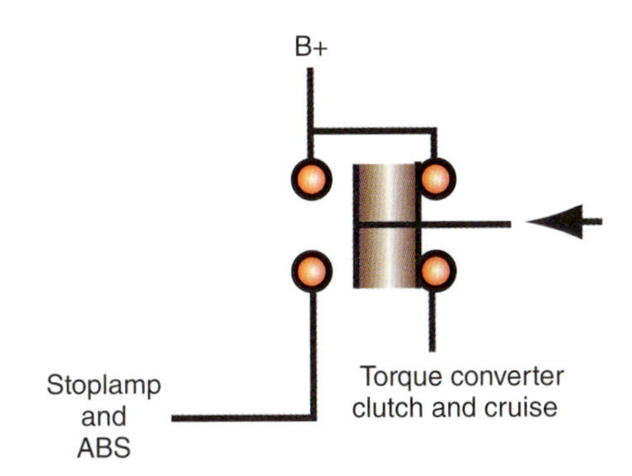

Figure 5-30 Stop lamp switches control the lamp circuit on the battery (B+) side of the circuit, and the switch contacts are always open.

The brake lamp contacts are usually connected to the brake lamps through the turn signal and hazard flasher switch. The switch contacts for the stop lamps are always normally open contacts (**Figure 5-30**). Another set of contacts is normally closed and will open when the brake pedal is depressed. This set of contacts supplies battery power to the cruise control and torque converter clutch (TCC). Both the TCC and cruise control must disengage when the brakes are applied. Some vehicles use a separate switch to control these systems.

The CHMSL and rear taillights are actually LED arrays on many models. The LED lamps come on almost instantaneously (which might give drivers that split second they need to stop) and consume small amounts of current, not to mention that the bulbs have a very long life.

Brake Pedal Position Sensor

Late-model vehicles use a brake pedal position sensor (**Figure 5-31**) in place of brake lamp switches. The advent of multiple uses for the brake input by the BCM, ABS, Transmission Control Module (TCM), and in particular, stability control systems required a finer input than the mechanical switch. The **brake pedal position (BPP) sensor** is a potentiometer, so it can inform the BCM that the brake lamps need to be illuminated, but it can also inform the computers how far and how fast the brakes have been applied. This information is very useful to the stability control system.

Stop Lamps and Bulbs

Stop lamps are included in the right and left taillamp assemblies. Vehicles also have a **center high-mounted stop lamp (CHMSL)** located on the vehicle centerline no lower than 3 inches below the rear window (6 inches on convertibles).

Some tail lamp systems are two-bulb assemblies with dual-filament bulbs that perform two functions (**Figure 5-32**). The stop-lamp circuit and the turn signal and hazard lamp

Figure 5-31 A brake pedal position sensor.

Figure 5-32 Taillamp bulbs. (A) bayonet style dual filament (B) bayonet style single filament (C) later style plug-in bulb.

circuit usually share a single dual-filament bulb, with the stop-lamp circuit connected to the high-intensity filament of the bulb. The taillamps are the separate, single-filament bulbs in a two-bulb assembly. Many late-model vehicles use a dual filament plug-in bulb even when only one filament is used (**Figure 5-33**).

In a two-bulb circuit, the stop lamps are wired through the turn signal and hazard switches. If neither turn signal is on, the current flows to both stop lamps (**Figure 5-34**). If the left turn signal is on, current for the right stop lamp is sent to the lamp through the turn signal switch and wire labeled 18 BR RD. The left stop lamp does not receive any current from the brake switch because the turn signal switch opens that circuit (**Figure 5-35**).

In a two-bulb circuit, the CHMSL can be wired in one of two ways. The first way is to connect the stop lamp circuit between the stop lamp switch and the turn signal switch

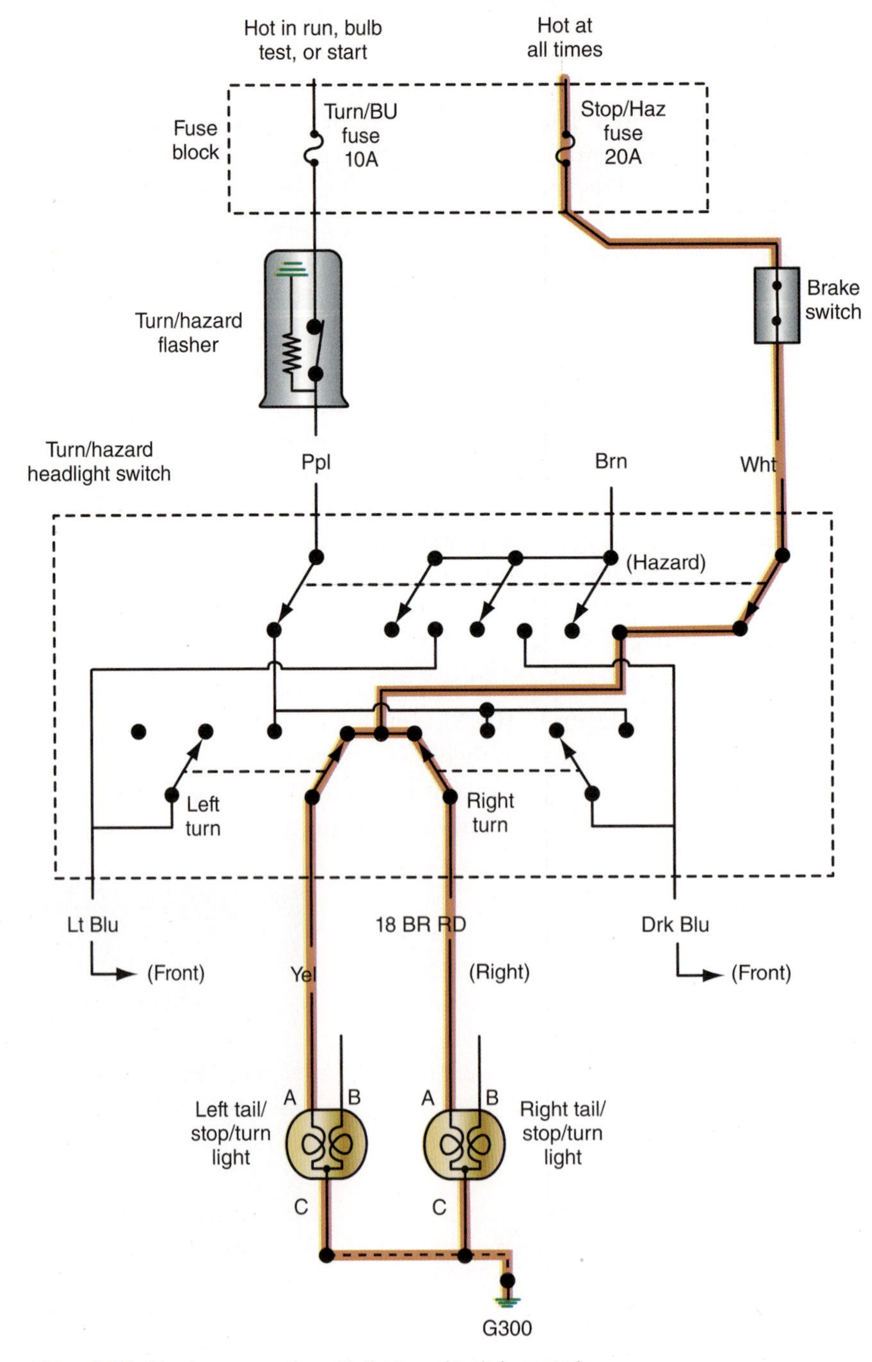

Figure 5-33 Stop lamp operation with the turn signals in neutral.

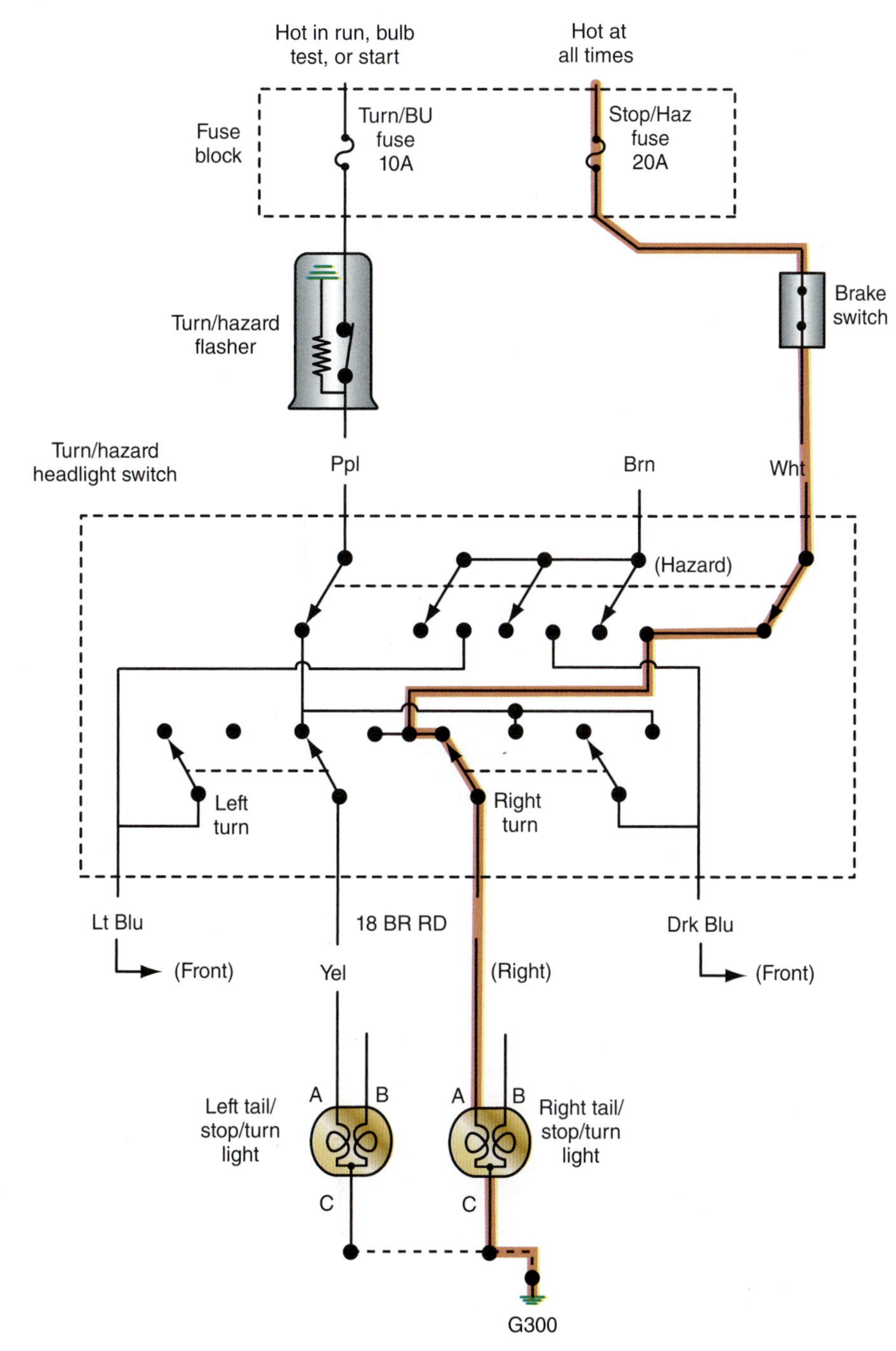

Figure 5-34 Stop lamp current path through the hazard flasher with turn signal switches in left-turn mode.

(**Figure 5-36**). However, this method increases the number of conductors needed in the harness. Therefore, most manufacturers prefer to install diodes in the wires that are connected between the left and right side bulbs (**Figure 5-37**). If the brakes are applied when the turn signal switch is in its neutral position, the diodes allow voltage to flow to the CHMSL. When the turn signal switch is placed in the left turn position, the left lamp must receive a pulsating voltage from the flasher. However, the steady voltage being applied to the right stop lamp would cause the left lamp to light continuously if the diode were not used. Diode 1 blocks the voltage from the right lamp, preventing it from reaching the left lamp. Diode 2 allows the voltage from the right stop lamp circuit to reach the CHMSL.

In a one-bulb system, one dual-filament bulb, one on each side of the vehicle, performs all of the rear warning light's functions. The high-intensity filament is for stop, signal, and

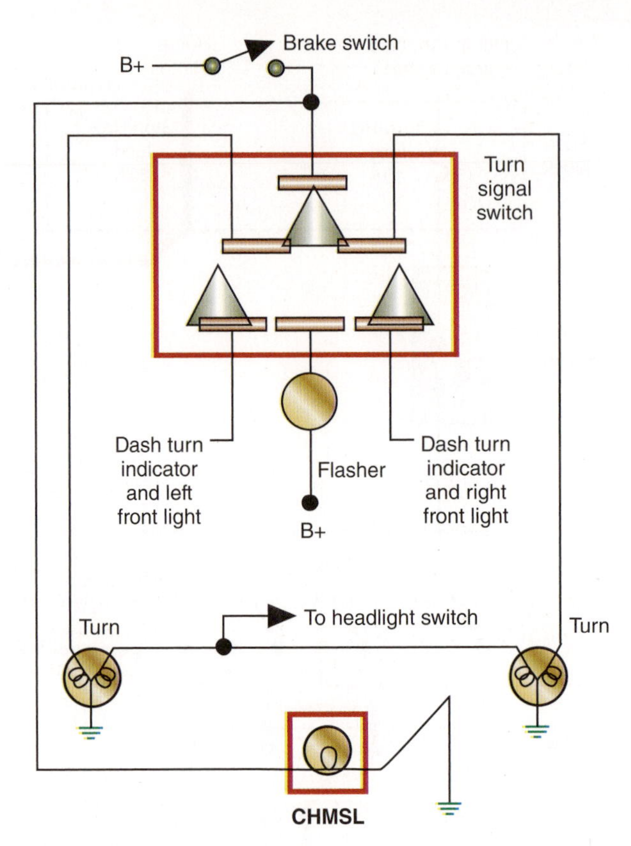

Figure 5-35 This CHMSL is connected directly to the brake switch and controlled by diodes.

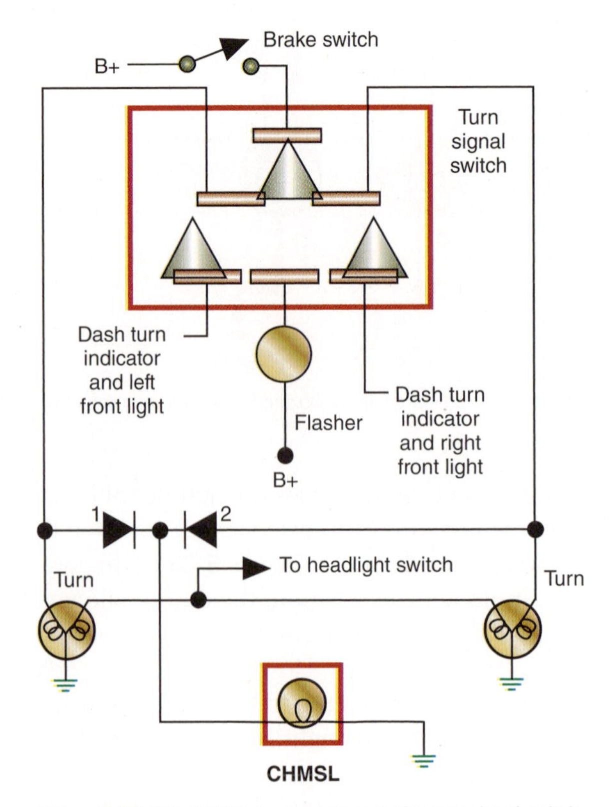

Figure 5-36 This CHMSL is wired through the turn signal switch and controlled by diodes.

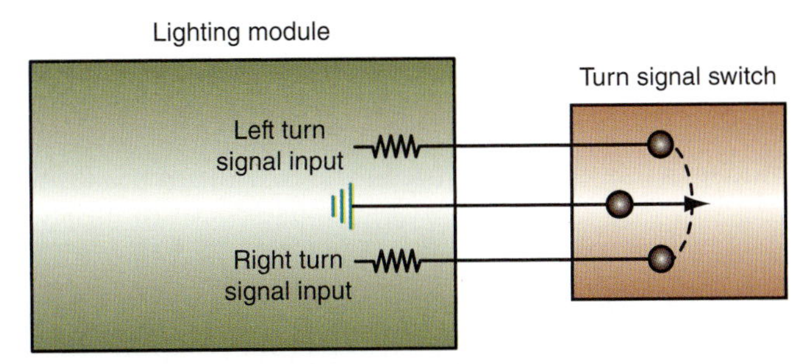

Figure 5-37 The turn signal switch on this vehicle is an electronic input to the lighting control module.

hazard. The low-intensity filament acts as the tail or marker light. The stop lamp is wired through the turn signal switch. With the brakes applied and assuming the left turn signals activated, power is routed through the turn signal to the right stop lamp. The left stop lamp is interrupted by the position of the turn signal switch. Power from a different fused circuit is routed through the turn signal switch and to the left-side dual-filament bulb. When the hazardous warning circuit is activated, front marker lights and both rear bulbs pulse on/off unless the brakes are applied. In that case, the front marker lights still blink but the stop lamp circuit is "powered" around the hazardous warning switch. This causes the rear stop lamps to be illuminated continuously as in normal braking operations.

The CHMSL in this system may be wired directly to the stop lamp switch bypassing the turn signal switch. Understanding this fact can assist the technician in determining which component is at fault: stop lamp fuse, stop lamp switch, turn signal switch, or one or more bulbs.

Turn Signal/Brake Lamp Operation with Lamp Module Control

The brake lamps, turn signals, and hazard flashers on many late-model vehicles are operated by a lighting module connected to the computer network. The BPP sensor and turn signal switch are actually inputs to the lighting module (Figure 5-37). When the turn signal or brake lamps are needed by the driver, the respective sensing circuit inside the lighting module is grounded. The module outputs power directly to the respective bulbs as shown in **Figure 5-38**. With this design, the bulbs are flashed electronically for a turn signal, or both rear taillamps are illuminated for brake lamps, depending on the need. This eliminates the need for complicated multi-position mechanical switches and the associated wiring.

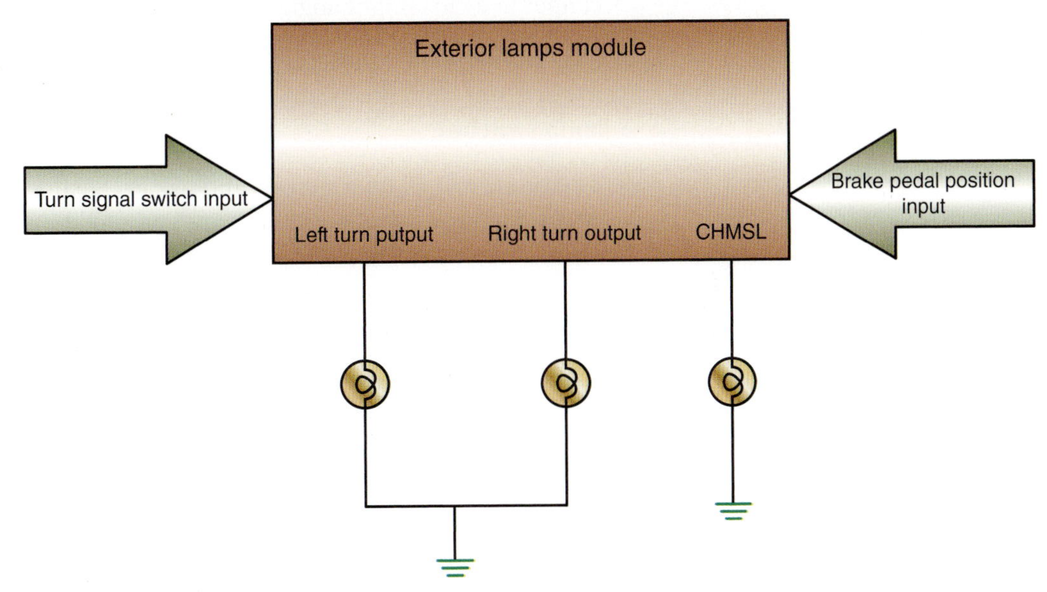

Figure 5-38 The turn signal switch on this vehicle is an electronic input to the lighting control module.

SUMMARY

- Hydraulic lines made of tubes and hoses transmit fluid under pressure from the master cylinder to each of the wheel service brakes.
- Brake hoses are the flexible links between the wheels or axles and the frame or body.
- The threaded parts that connect brake hoses and tubing together and to other brake components are called fittings.
- The metering valve keeps the front disc brakes from operating until the rear drum brakes have started to work.
- The proportioning valve reduces the hydraulic pressure at the rear drum brakes when high pressure is required at the front disc brakes.
- Brake systems also include switches for parking brakes, pad wear, stop lamps, and to warn of low brake fluid level in the master cylinder.
- The BPP sensor sends a signal to the lighting module concerning the brake lamp position. The lighting module then operates the external lamps.

REVIEW QUESTIONS

Essay

1. Explain why 100% copper line should never be used as brake lines.
2. Describe an ISO flare.
3. Why are brake hoses marked with raised ribs or colored stripes?
4. Describe how an internal brake hose problem can cause brake pull and/or brake drag.
5. Describe how a leak-free fitting is achieved with SAE flare fittings.
6. Describe why a double flare is a better choice than a single flare.
7. Explain how the ABS system has taken the place of many of the hydraulic control valves.
8. Describe the electronic brake pad wear indicator system used on some vehicles
9. Explain the operation of the BPP sensor.
10. Describe an ISO fitting.

Fill in the Blanks

1. Brake lines or tubing consist of _______________ tubes or pipes and flexible _______________ connected with fittings.
2. There are two common types of flares formed on the end of brake tubing, the _______________ flare and the _______________ flare.
3. SAE Standard J1047 requires that an 18-inch length of tubing must withstand an internal pressure of _______________ psi.
4. An ISO flare is not _______________ on itself as is a double flare.
5. _______________ fittings have become popular with both domestic and foreign vehicle manufacturers.
6. Straight _______________ fittings are usually found on the ends of brake hoses that attach to calipers or wheel cylinders.
7. A _______________ is used to attach a hose or tube to a port on a cylinder or caliper at a close right angle
8. Late-model vehicles use a _______________ _______________ position sensor in place of brake lamp switches.
9. The brake lamps, turn signals, and hazard flashers on many late-model vehicles are operated by a _______________ _______________ connected to the computer network.
10. The _______________ _______________ switch has replaced the pressure differential valve warning switch on most vehicles.

Multiple Choice

1. Technician A says that a 100 percent copper tube is often used as a brake line material. Technician B says there is a copper-nickel alloy brake tubing that meets SAE Standard J1047 and ISO 4038. Who is correct?
   A. A only
   B. B only
   C. Both A and B
   D. Neither A nor B

2. The flare on the end of a brake line has a bubble, or ball, shape. Technician A says that this is a bubble flare. Technician B says that this is an ISO fitting. Who is correct?
   A. A only
   B. B only
   C. Both A and B
   D. Neither A nor B

3. Technician A says fittings with an external taper is simply called a standard flare. Technician B says a fitting with an internal taper is called an inverted or LAP flare. Who is correct?
   A. A only
   B. B only
   C. Both A and B
   D. Neither A nor B

4. Technician A says that a metering valve delays the application of the rear drum brakes. Technician B says that the metering valve is only needed on four-wheel disc brake vehicles. Who is correct?
   A. A only
   B. B only
   C. Both A and B
   D. Neither A nor B

5. Technician A says on many vehicles the lighting module controls the brake lamps and turn signals. Technician B says that these vehicles use the turn signal switch as an input to the lighting module. Who is correct?
   A. A only
   B. B only
   C. Both A and B
   D. Neither A nor B

6. Technician A says that a proportioning valve provides greater pressure to the rear drum brakes during normal vehicle load. Technician B says that a proportioning valve provides lower pressure to the front disc brakes during initial brake application. Who is correct?
   A. A only
   B. B only
   C. Both A and B
   D. Neither A nor B

7. While discussing electrically actuated parking brakes, Technician A says these systems operate via electric actuators located on the rear brake calipers, or by electric cable actuators. Technician B says these actuators are controlled by their own control module. Who is correct?
   A. A only
   B. B only
   C. Both A and B
   D. Neither A nor B

8. Technician A says some vehicles use tubing sized in metric diameters. Technician B says some common metric diameters are 4.75 mm, 6 mm, 8 mm, and 10 mm. Who is correct?
   A. A only
   B. B only
   C. Both A and B
   D. Neither A nor B

9. While discussing mechanical brake lamp switches, Technician A says the stop-lamp switch is a normally open momentary contact switch. Technician B says the same circuit is used to alert the ABS to monitor wheel speed sensors during braking. Who is correct?
   A. A only
   B. B only
   C. Both A and B
   D. Neither A nor B

10. While discussing brake pad wear indicators, Technician A says that many brake pads utilize a simple scraper built into the brake pad itself that touches the rotor when the pads wear down. Technician B says the scraping sound goes away when the brakes are released. Who is correct?
    A. A only
    B. B only
    C. Both A and B
    D. Neither A nor B

CHAPTER 6
POWER BRAKE SYSTEMS

Upon completion and review of this chapter, you should be able to:

- Explain the relationship between atmospheric pressure and vacuum.
- Describe the relationship between atmospheric pressure and vacuum in a power brake vacuum booster.
- Describe the parts and operation of a vacuum power booster.
- Identify and describe the parts of a vacuum diaphragm assembly.
- List the three major kinds of vacuum boosters.
- Describe the parts and operation of the single diaphragm with a lever-reaction vacuum booster.
- Describe the parts and operation of the single diaphragm with a reaction-disc vacuum booster.
- Describe the parts and operation of the tandem diaphragm vacuum booster.
- Explain the parts and operation of the air and vacuum systems in a vacuum booster.
- Describe the parts and operation of a hydro-boost hydraulic power-assist system.

Terms To Know

Accumulator
Atmospheric pressure
Brake assist
Check valve
Diaphragm
Electrohydraulic brake (EHB)
Electronic brake system (EBS)
Hydro-boost
Lands
Pressure differential
Reaction disc (or reaction plate and levers)
Spool valve
Stroke sensor
Tandem
Tandem booster
Vacuum
Vacuum suspended
Valleys

INTRODUCTION

A power brake system is used on most cars to reduce the braking effort required from the driver. The power brake system reduces driver fatigue, increasing safety.

INCREASING BRAKE FORCE INPUT

Four methods can be used to reduce pedal pressure requirements or to boost the force applied to the master cylinder:

1. *Pedal Force.* The simplest way to increase braking force is for the driver to step on the pedal harder. This simple approach has definite limits, however. A 120-pound driver does not have the weight and probably not the leg strength of a 220-pound driver, but the braking requirements of the car do not change to compensate for the size and weight of the driver. Driver strength then becomes the limiting factor in how the car stops.

2. *Mechanical Advantage (Leverage).* Chapter 4 of this *Classroom Manual* explains the brake pedal ratio and how it provides leverage to increase force applied to the master cylinder. Mechanical limitations allow the pedal ratio to be increased only to a given point, however. The pedal arm can be only so long to fit into the car; and the longer the pedal arm, the greater the amount of pedal travel.
3. *Hydraulic Advantage (Force Multiplication).* Just as the pedal ratio multiplies force applied to the master cylinder, hydraulic piston size can be used to multiply hydraulic pressure applied by the master cylinder to the wheel cylinders and caliper pistons. These piston size and pressure relationships also are explained in Chapter 2 of this manual. Hydraulic force multiplication has its limits too. If wheel cylinder and caliper pistons are made larger for greater force, master cylinder piston travel may increase along with brake pedal travel. All brake systems use hydraulic multiplication to increase braking force, but it has limitations just as pedal leverage does.
4. *Power Boosters.* The fourth way to increase brake application force is to install a power booster in the system, and such boosters are the subject of this chapter. There are three general kinds of power boosters. One uses intake manifold vacuum acting on a diaphragm to help the driver apply pedal force to the master cylinder. The second type uses hydraulic pressure from a hydraulic pump to operate a hydraulic booster attached to the master cylinder.

The third power brake booster is not really a brake booster, but a system that controls brake hydraulics through the use of electronics. In this system, sensors on the brake pedal mechanism combined with signals shared with the ABS and vehicle stability system apply the brakes at each wheel. This is done through various configurations that share certain properties. One property is the sharing of sensors and controlling computers. The second is the installation of electronic valves (solenoids) within the brake's hydraulic lines or at each wheel. In either case, fluid pressure can be controlled to each wheel, resulting in better braking performance and vehicle control.

Power boosters are add-on devices that do not alter the basic brake system. They still allow braking, even if the booster fails or loses its power supply. All boosters have a power reserve to provide at least one power-assisted stop if power is lost. FMVSS 105 contains brake performance requirements for brake systems with the power boosters disabled. Because modern brake systems are designed to include the advantage of a power booster, pedal effort increases significantly if power is lost.

Atmospheric pressure is basically the weight of the atmosphere that is applying force. It amounts to approximately 14.7 pounds per square inch at sea level.

VACUUM PRINCIPLES

A **vacuum** is pressure less than atmospheric pressure.

To understand vacuum booster systems, the relationship between **atmospheric pressure** and **vacuum** must also be understood. The air around us has weight. For example, every 1-square-inch column of air extending from the Earth's surface to the edge of the atmosphere weighs about 14.7 pounds (**Figure 6-1**). The weight of this air is called atmospheric pressure. Atmospheric pressure varies with altitude and temperature; but at sea level and at 68°F, it is 14.7 psi. If you were to drive up in the mountains, you would find that atmospheric pressure gets lower. As you go up in altitude, the column of air is not as high, so there is less pressure.

Shop Manual
page 248

Atmospheric pressure is reduced by about 1 pound per each 1,000 feet of elevation above sea level. Vehicles operated in Denver, Colorado, are tuned slightly different from vehicles on the Pacific Coast.

Vacuum is a pressure lower than atmospheric pressure. When an engine is running, the intake strokes in the cylinders create low pressure. This low pressure draws in the mixture of air and fuel. In automotive work, we commonly call this low pressure a vacuum. A true vacuum, however, is a complete absence of air and is found only in the laboratory or out in space.

Atmospheric pressure and vacuum can be used as a strong force to make things move. **Figure 6-2** shows a piston that is free to move up and down in a cylinder. One end of the

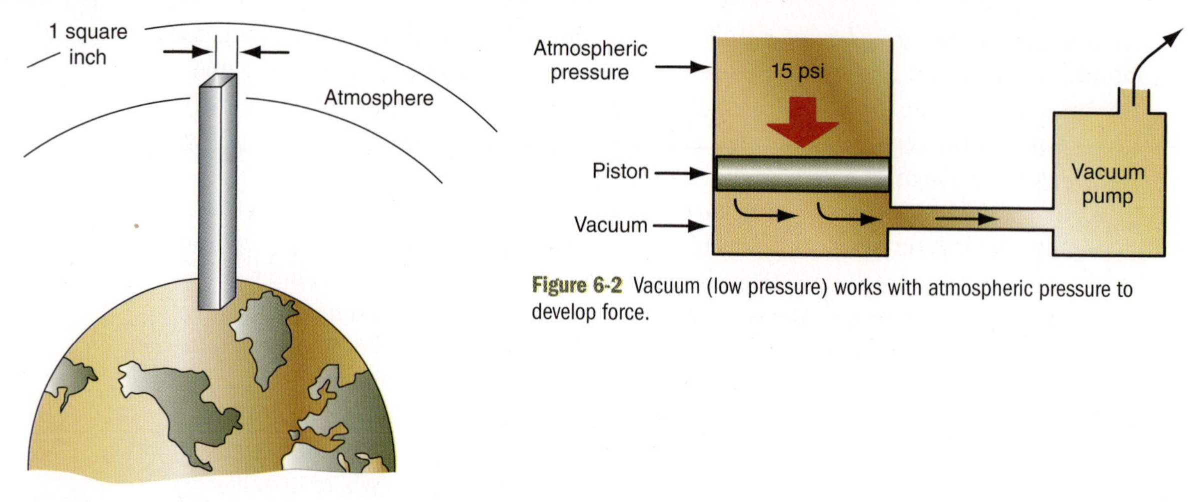

Figure 6-1 A 1-square-inch column of air, the height of the Earth's atmosphere, exerts 14.7 pounds of pressure on the Earth at sea level.

Figure 6-2 Vacuum (low pressure) works with atmospheric pressure to develop force.

cylinder is connected to a vacuum pump. The vacuum pump is used to lower the pressure under of the piston to below atmospheric pressure. When atmospheric pressure is applied on the other side of the cylinder, the piston will move down: toward the lower pressure.

The amount of force that is created depends on the **pressure differential** or the difference between the low pressure on one side of the piston and atmospheric pressure on the other. If a perfect vacuum exists on one side of the piston and atmospheric pressure is applied to the other side, the pressure differential equals:

$$14.7\,\text{psi} - 0\,\text{psi} = 14.7\,\text{psi}$$
$$(\text{atmosphere pressure}) - (\text{vacuum}) = (\text{pressure differential})$$

Remember, however, that a perfect vacuum does not exist in an intake manifold, and vacuum is not usually measured in psi. To get a realistic example of a typical pressure differential in a vacuum brake booster, two factors must be accounted for.

First, pressure may be measured as either absolute (psia) or gauge (psig) and in two different measurement units: pressure in pounds per square inch (psi) or vacuum in inches of mercury (in. Hg). Atmospheric pressure at sea level may be expressed as 14.7 psia or 29.9 in. Hg. Similarity, vacuum can be expressed as either 29.9 in. Hg below atmospheric pressure or –14.7 psig below atmospheric pressure.

AUTHOR'S NOTE The metric measurement for pressure is the kilopascal, abbreviated kPa. Atmospheric pressure (at sea level) is 100 kPa. At a perfect vacuum the measurement is 0 kPa. The reading on a scan tool with the engine running will show pressure above 0 kPa. The metric system of measurement does not go from psi to in. Hg so it can help in the understanding of vacuum and pressure.

However, most technicians do not normally rely on absolute pressure/vacuum measurements but use gauges in almost every instance. Gauges, vacuum, and pressure will register 0 in. Hg or 0 psi at sea level. In a complete vacuum a vacuum gauge would register 29.9 in. Hg, whereas a pressure gauge would register –14.7 psig. However, any math would have to use absolute measurements for accuracy.

A vacuum-suspended booster has two chambers. The front chamber is under vacuum, and the rear can be exposed to atmospheric pressure. A flexible diaphragm separates the two chambers. The amount of vacuum available in the intake manifold and the vacuum chamber will change as the engine load changes. To calculate the approximate booster force increase, the vacuum and pressure measurements must be in the same unit. Rounding 14.7 psia to 15 psia and 29.9 in. Hg to 30 in. Hg gives a rough conversion factor of –2 – because pressure and vacuum are on opposite sides of zero). To convert from psi to in. Hg multiply psi by – 2 and from in. Hg to psi divide in. Hg by – 2.

As an example, assume that the sea-level atmospheric pressure is 15 psia, the vacuum is 20 in. Hg, and the booster has 50 square inches of area.

$$15 \text{ psia (atmosphere pressure)}$$

$$20 \text{ in Hg}/-2 = -10 \text{ psi}$$

$$-10 \text{ psia} + 15 \text{ psia} = 5 \text{ psia}$$

$$15 \text{ psia (atmopshere pressure)} - 5 \text{ psia (vacuum booster pressure)} = 10 \text{ pounds of differential pressure}$$

$$10 \text{ pounds of pressure (psi)} \times 50 \text{ square inches} = 500 \text{ pounds of force}$$

This 500 pounds from a vacuum brake booster helps the driver apply the brakes.

VACUUM AND AIR SYSTEMS FOR POWER BOOSTERS

Enough vacuum and air must be delivered to the power booster for it to work correctly. Most power boosters have the same method of delivering air, with vacuum provided by either the intake manifold or an auxiliary vacuum pump.

Air Systems

Figure 6-3 shows the air system for a typical power booster. Air enters through passages in the pedal pushrod boot. The air passes through a fine mesh material called a silencer, which slows down the air and reduces any hissing sounds. The boot and air inlet are inside the car, so any noises could be heard by the driver. The air then passes through a filter to remove any dirt that could damage the valve. Air then flows into the power piston passages to the air valve.

Intake Manifold Systems

The vacuum for most power brake systems is supplied by the engine intake manifold. On older cars, a simple vacuum hose was attached from the intake manifold directly to the housing of the brake booster. A check valve was used to protect the booster against loss of vacuum. The problem with this system was that intake manifold vacuum decreases and increases as the engine is accelerated and decelerated. A check valve helps prevent these fluctuations in vacuum.

Many late-model cars that use intake manifold vacuum as the power source for the power brakes have a vacuum reservoir (**Figure 6-4**). The reservoir stores vacuum so it is always available, regardless of the changing vacuum in the intake manifold.

A **check valve** is installed between the manifold and the reservoir to prevent air from entering the reservoir during wide-open throttle. The check valve also is a safety device, protecting the system from losing vacuum in case of a leaking supply line or other failure in the vacuum supply. The check valve is typically mounted on the front of the booster where the vacuum hose is connected. The check valve can also prevent the complete loss of vacuum in case the engine stalls, providing at least one vacuum-assisted stop.

A **check valve** allows fluid or air to flow in one direction but not in the opposite direction.

Diesel engines do not create a vacuum in their intake manifold. Most diesel engines use hydraulically assisted power brake boosters. The pressure for the assist is developed in the power steering pump on some diesel engines.

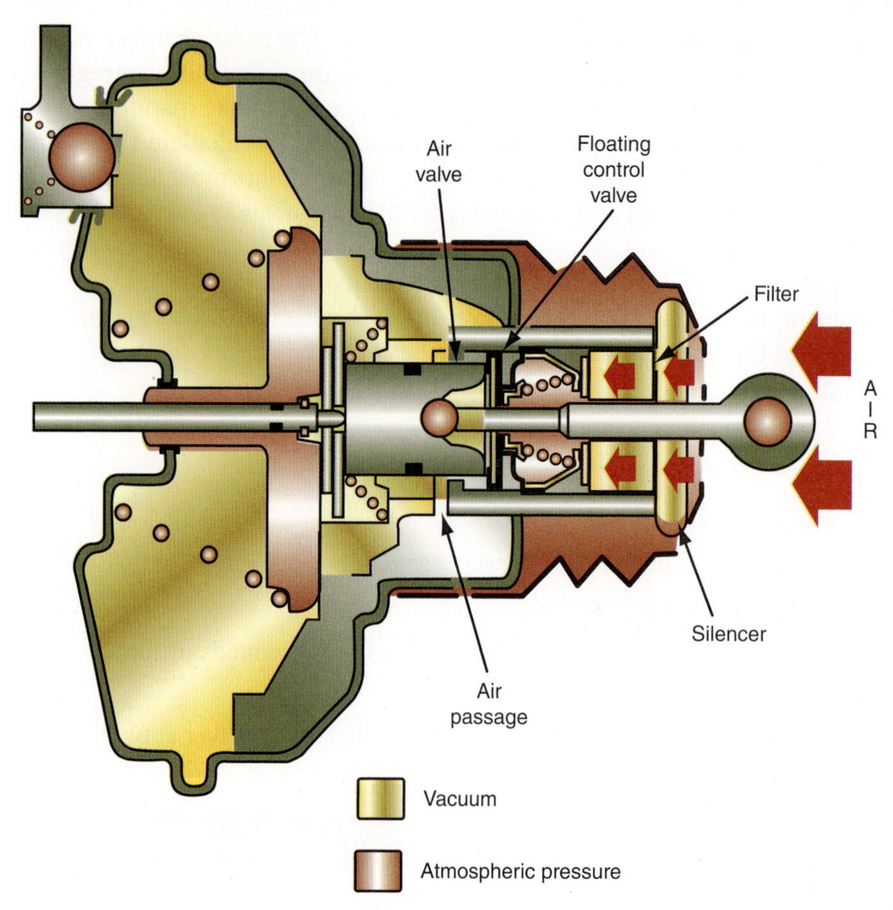

Figure 6-3 The air system for a vacuum brake power booster.

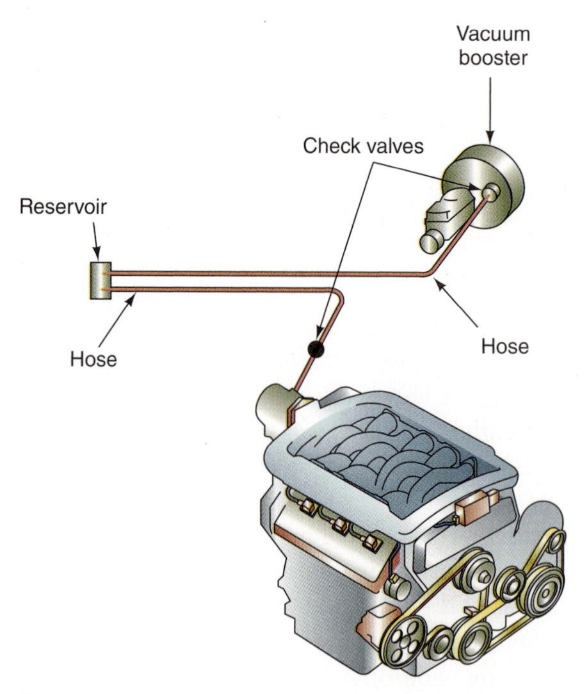

Figure 6-4 Many power brake systems have vacuum reservoirs to ensure a steady vacuum supply to the booster.

Vacuum Check Valves

Whether a power brake system has just a vacuum booster or a vacuum booster and a vacuum reservoir, it must include one or more vacuum check valves. The check valve uses a spring-loaded ball or disc to close the vacuum line if pressure in the line gets higher (closer to atmospheric pressure) than the vacuum in the booster or reservoir.

Figure 6-5 is a sectional view of a typical vacuum check valve. The valve has a small disc backed up by a spring. Manifold vacuum pulls and holds the disc off its seat. Vacuum is allowed into the booster. When vacuum drops below the calibration of the spring, the spring moves the disc against its seat. When the disc is seated, the valve closes and does not allow vacuum to leak out of the system. The check valve will keep vacuum in both sides of the booster when the engine is off. The vacuum check valve has a secondary function of keeping fuel vapors out of the vacuum booster. Without a check valve, vacuum in the booster could draw part of the air-fuel mixture into the booster when the engine is at wide-open throttle with little or no manifold vacuum. As an extra safety precaution, some systems have a charcoal filter between the manifold and the vacuum check valve to trap fuel vapors before they can get near the booster.

Most systems have a single check valve as part of the inlet fitting of the vacuum booster (see Figures 6-3 and **6-6**). Some vehicles, however, have an in-line check valve in the vacuum line. If the power brake system has a vacuum reservoir, as well as the booster, it will have two check valves. One valve is part of the booster inlet fitting; the other is between the intake manifold and the reservoir (see Figures 6-4 and **6-7**).

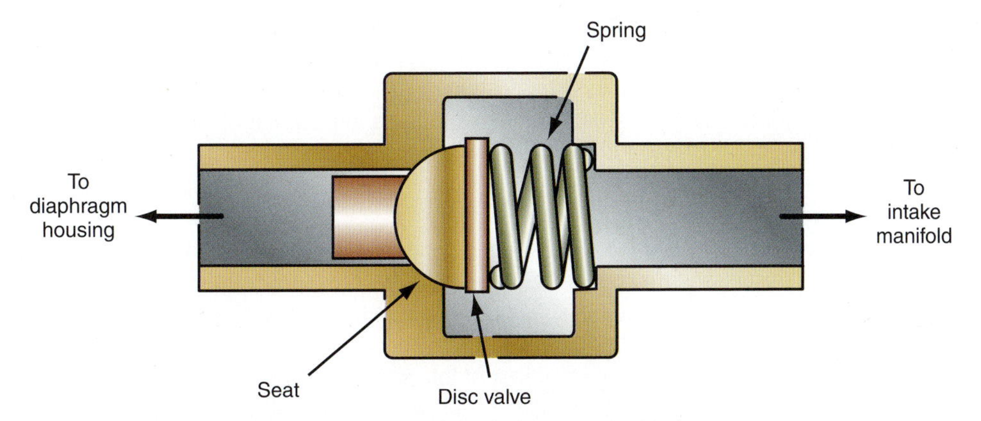

Figure 6-5 This cutaway view shows the parts of a typical vacuum check valve.

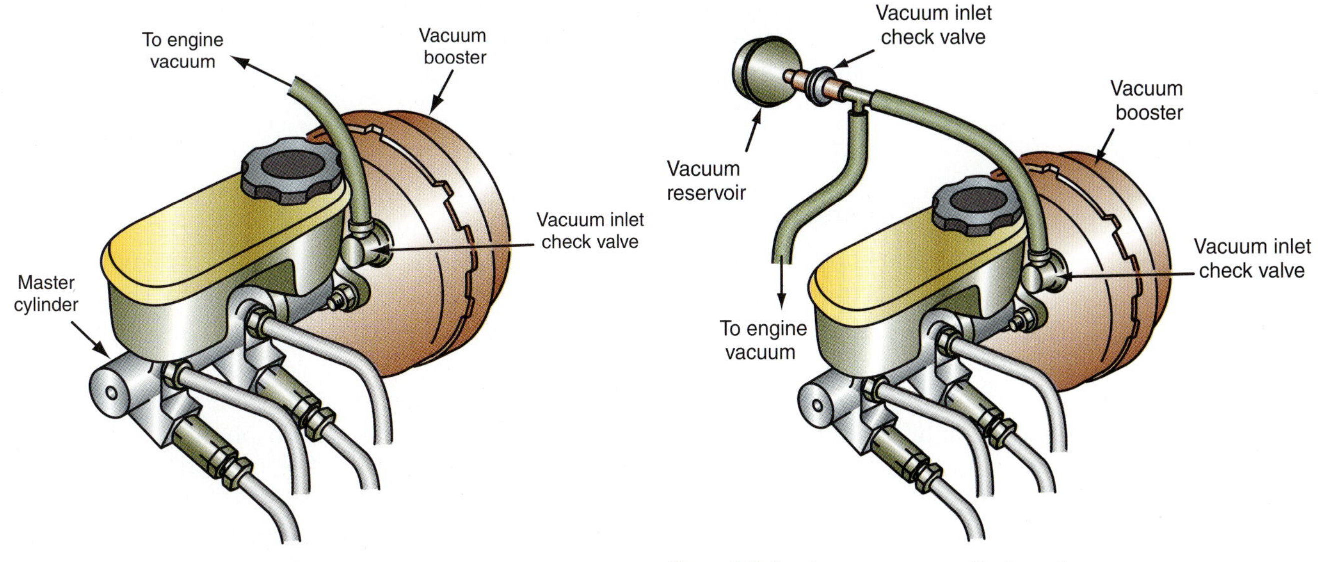

Figure 6-6 A typical vacuum brake booster.

Figure 6-7 Booster vacuum connection to engine.

AUTHOR'S NOTE A worn or mistuned engine can cause damage to the check valve by pushing carbon and combustion gases into the check valve. The customer may not even notice that the boost is not working right after engine start up. If the other vacuum hoses show carbon or heat damage, inspect the vacuum check valve.

Shop Manual
page 249

VACUUM POWER BOOSTERS

Vacuum-operated power boosters have different shapes, but they all work the same way. The booster is mounted between the master cylinder and the engine compartment bulkhead or fire wall. The booster is between the brake pedal pushrod and the master cylinder (**Figure 6-8**). A vacuum hose is connected from the engine intake manifold to the booster (**Figure 6-9**). The following sections describe the operation and construction of common vacuum boosters

A BIT OF HISTORY

Today, power brakes are universally standard equipment on all domestic and imported cars and light trucks. But this was not always so.

Modern power brakes appeared as extra-cost options on luxury cars in the early 1950s. The new, longer, lower, wider, heavier postwar cars brought drivers more power to go, and this required more power to stop. Power brakes, along with power steering and automatic transmissions, made driving these luxury liners a pleasure. Without these options, driving would have been a real chore.

Power brakes originally were a marketing feature and a driving convenience. At their introduction, one could also assert that power brakes were a safety feature that provided braking power for drivers who might not have the strength (or weight) to really stomp on the pedal for a hard stop. When disc brakes arrived on the scene a decade later, power boosters became a virtual safety necessity. The dual servo drum brakes of the 1950s and early 1960s used the self-energizing action of primary and secondary shoes to increase braking force when the driver pressed the pedal. Disc brakes, which began to appear in quantity by the mid-1960s, do not develop the self-energizing action of drum brakes. Although disc brakes develop greater braking energy at the wheels, they require greater application force from the driver. Disc brakes made some kind of power assist a necessity for almost all drivers.

Vacuum Booster Construction

Most vacuum boosters have essentially the same parts. **Figure 6-10** is a disassembled view of a typical vacuum booster. The parts are contained in a steel housing or shell that is divided into front and rear halves, held together with interlocking tabs. The rear housing has mounting studs to mount the unit on the fire wall.

The booster **diaphragm** is a large hemispherical shaped piece of rubber and is acted upon by atmospheric pressure and vacuum.

The **diaphragm** is a large, hemispherically shaped part made from rubber. The flexible diaphragm moves back and forth as it is acted on by atmospheric pressure and vacuum. The center of the diaphragm is supported by a metal or plastic diaphragm support.

The brake pedal is connected to the brake pedal pushrod, which is contained in the booster. One end of the pushrod sticks out of the center of the rear housing when the

Figure 6-8 A vacuum booster mounted on the vehicle. Note the area close around the booster.

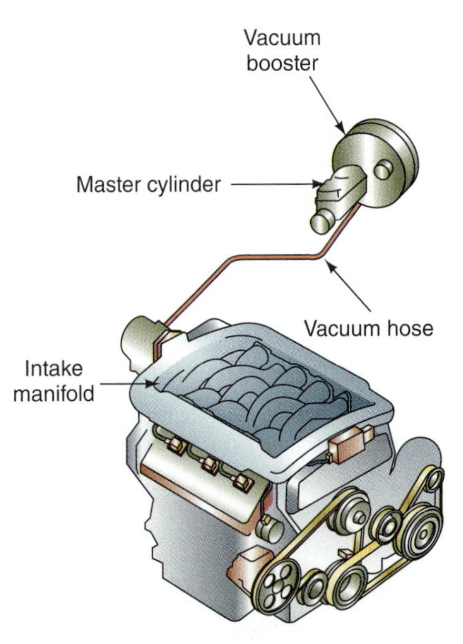

Figure 6-9 A vacuum hose from the intake manifold supplies vacuum to the booster.

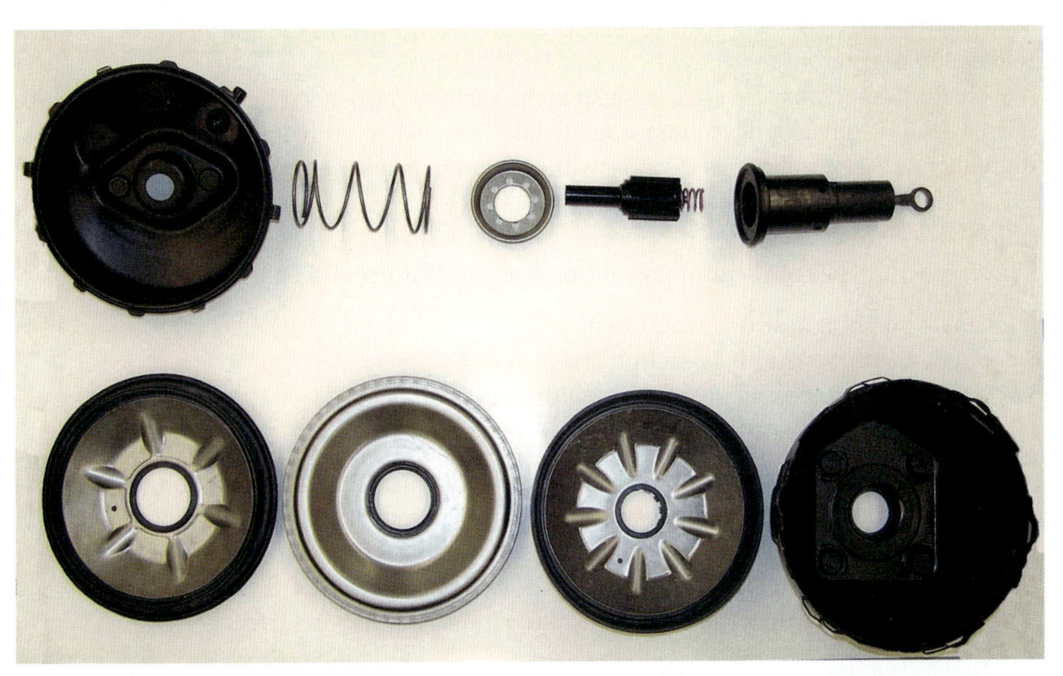

Figure 6-10 Disassembled vacuum booster. This is not a job for inexperienced technicians and is shown only to illustrate internal components.

booster is assembled. A rubber boot seals the area between the pushrod and the housing. The other end of the pushrod is attached to the power piston.

AUTHOR'S NOTE One thing that should be clarified here is the front and back of a brake booster. The rear or back of the booster is that part nearest the driver and that abuts the bulkhead. The front or forward part of the booster is where the master cylinder is installed.

The power piston is attached to the center of the diaphragm. The piston is often described as being suspended in the diaphragm. The power piston contains and operates the vacuum and air (atmospheric) valves that control the diaphragm. The power piston also transmits the force from the diaphragm through a piston rod to the master cylinder. Because the pedal pushrod is mechanically connected through the booster, any vacuum failure will not cause loss of braking action.

The piston rod also may be attached to another rod called a reaction retainer. A large coil spring is mounted between the front housing and the power piston and diaphragm assembly. The spring returns the diaphragm and piston to the unapplied position when the driver releases the brake pedal.

Two valves are located on the pedal side of the diaphragm. One, called the air valve, controls the flow of air at atmospheric pressure into one side of the booster. The other, called the vacuum valve, controls the buildup of vacuum in both sides of the booster. Both valves are connected to, and operated by, the pushrod.

Diaphragm Suspension

Diaphragms can be and are used in thousands of applications. The booster diaphragm separates the pressure chamber from the vacuum chamber.

The operation and construction of a vacuum booster and the booster diaphragm often are described as **vacuum suspended**. The brake boosters on almost all vehicles built since the mid-1970s have vacuum-suspended diaphragms. This means that when the brakes are released and the engine is running, vacuum is present on both sides of the diaphragm. When the pedal is pressed, atmospheric pressure is admitted to the rear of the diaphragm to develop booster force.

Vacuum reservoirs also are used with some vacuum-suspended boosters to ensure adequate vacuum with a lot of vacuum-operated devices such as emission controls and air conditioning. A reservoir is not as essential with a vacuum-suspended booster, however.

Types of Vacuum Boosters

Three general types of vacuum power boosters are (**Figure 6-11**):

1. Single diaphragm with a reaction disc
2. Single diaphragm with a reaction lever
3. Tandem booster (dual diaphragm) with a reaction disc

The single diaphragm with a reaction disc provides force (or reacts back) to the brake pedal through a rubber reaction disc. The single diaphragm with a reaction lever reacts back to the pedal through a lever arrangement. The tandem diaphragm has two diaphragms that work together to provide force. The following sections explain how a single diaphragm booster with a reaction disc works. The other two booster types operate similarly.

Vacuum booster operation can be divided into the following five stages:

1. Brakes not applied (released)
2. Moderate brake application
3. Brakes holding
4. Full brake application
5. Brakes being released

Brakes Not Applied. When the brakes are off, the return spring for the input pushrod holds the pushrod and the air control valve rearward in the power piston (**Figure 6-12**). The rear of the air control valve seats to close the atmospheric port. The plunger also compresses the air valve against its spring to open the vacuum port. This valve action closes the booster to the atmosphere and opens a passage between the front and rear of the booster chamber. Equal vacuum is present on both sides of the diaphragm. The power piston return spring holds the diaphragm rearward so no force is applied to the output pushrod and master cylinder.

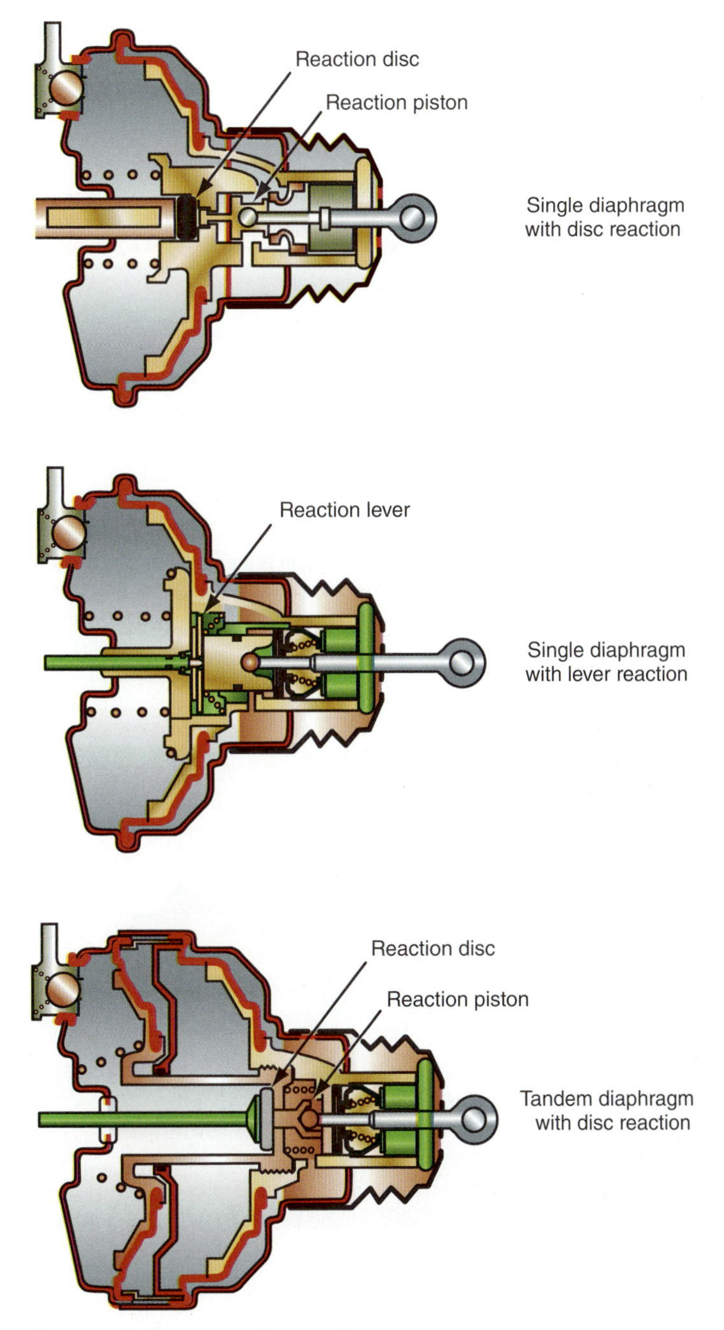

Figure 6-11 Three types of vacuum boosters.

Moderate Brake Application. When the driver applies the brakes, pedal pressure overcomes the input pushrod return spring to move the pushrod and air control valve forward (**Figure 6-13**). Spring pressure then moves the air valve to close the vacuum port to the rear chamber. As the air control valve continues to move forward, it opens the atmospheric port to the rear chamber. Vacuum (low pressure) still exists in the front chamber of the booster, and higher pressure (atmospheric) exists in the rear. The resulting pressure differential moves the diaphragm and power piston forward against the return spring to apply force to the master cylinder pushrod.

Brakes Holding. As long as the driver maintains unchanging foot pressure on the pedal, the input pushrod does not move. The diaphragm and power piston continue to move forward until the air control valve on the piston seats against the rear of the vacuum valve

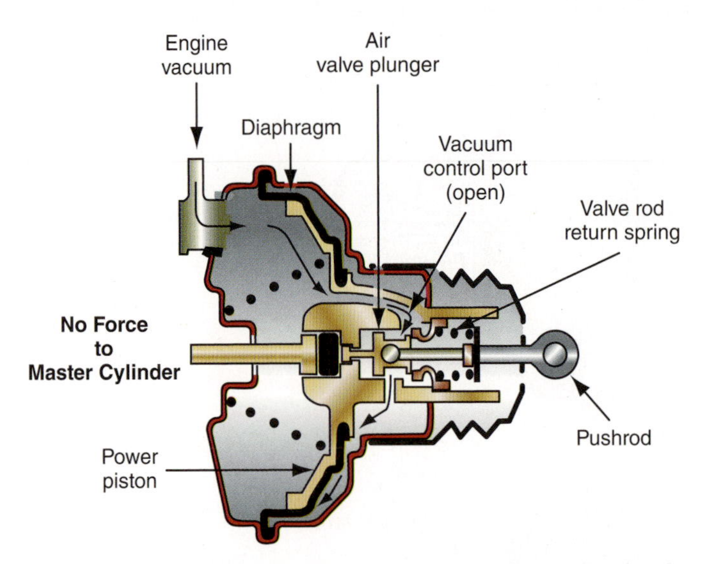

Figure 6-12 When brakes are released, vacuum is present on both sides of the diaphragm, and the air valve is closed to exclude atmospheric pressure.

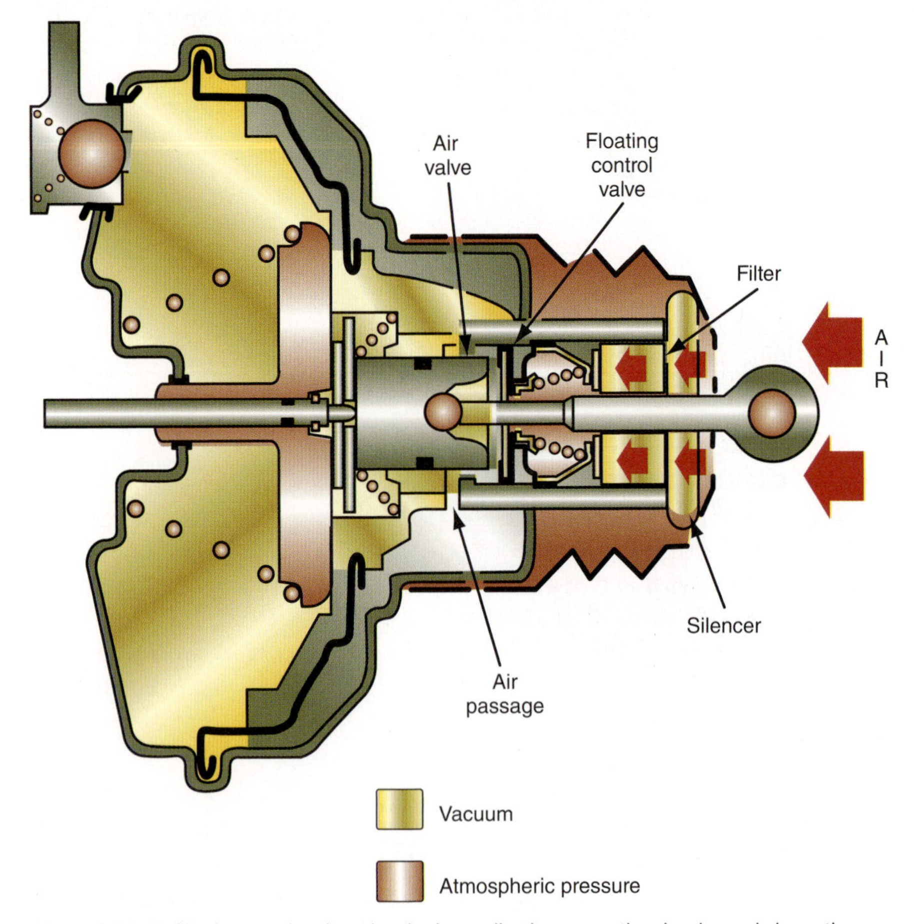

Figure 6-13 Air fills the rear chamber when brake application opens the air valve and closes the vacuum valve. The front chamber is kept in vacuum.

plunger to close the atmospheric port (**Figure 6-14**). This all happens very quickly, and as long as the atmospheric port is closed, the diaphragm does not move. It is suspended by a fixed pressure differential between the front and rear chambers of the booster. The booster always seeks the holding position when pedal force is constant or unchanging.

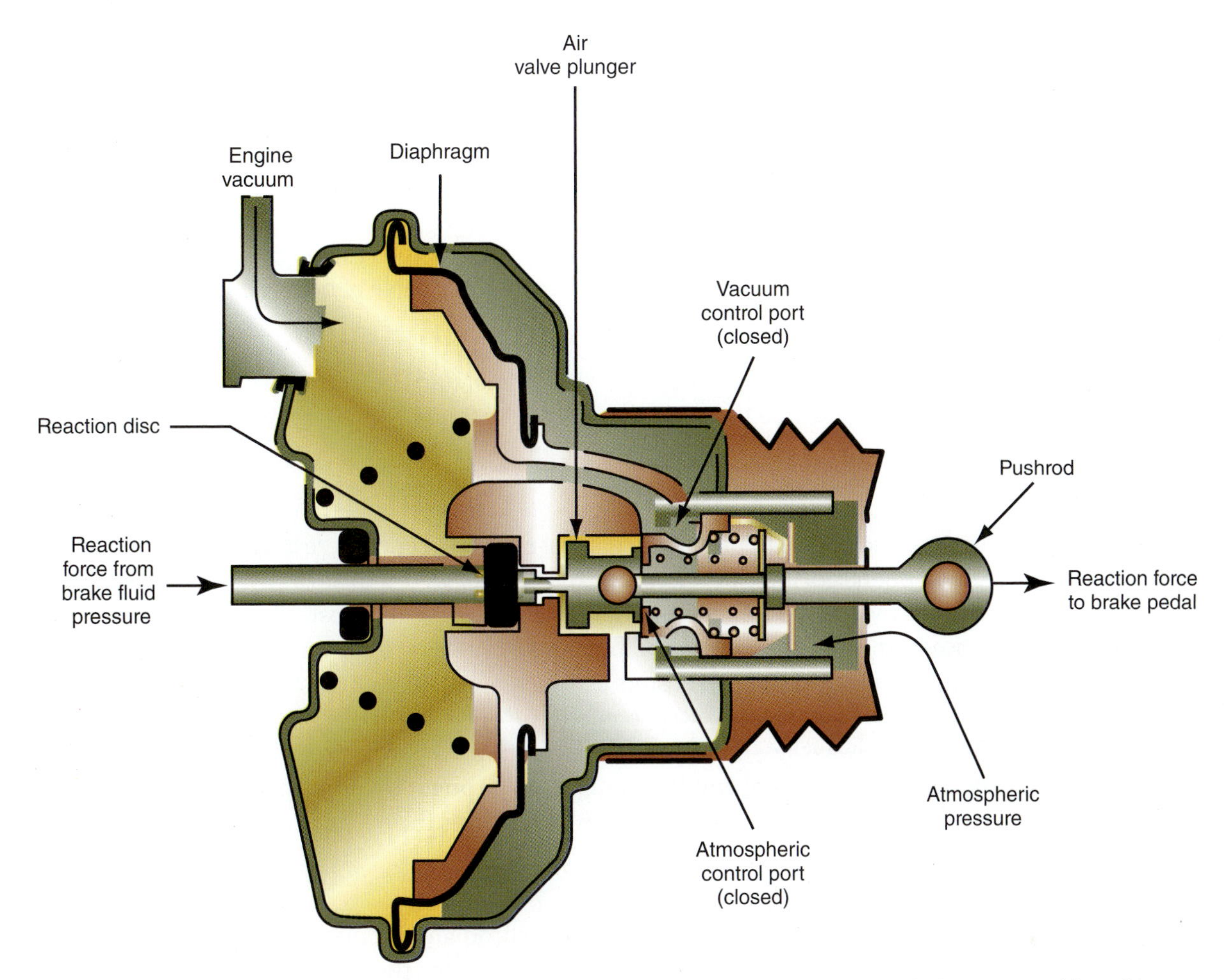

Figure 6-14 When the brakes are held at midpoint, both valves are closed to trap vacuum and atmospheric pressure and keep the diaphragm in a balanced position.

Full Brake Application. If the driver increases pressure on the pedal, she or he may push hard enough to force the air control valve plunger fully forward against the power piston. This action closes the vacuum port to the rear chamber and fully opens the atmospheric port. Maximum atmospheric pressure then exists in the rear chamber, and the booster supplies its maximum power assist. This condition sometimes is called the booster vacuum runout point. At this point, additional braking force can be applied to the master cylinder, but it must come entirely from foot pressure on the pedal. Because the booster is supplying its maximum force at this point, the driver will feel that the pedal becomes harder to press with additional foot pressure.

Brakes Being Released. When the driver releases the pedal, the input pushrod spring moves the pushrod and the air control valve rearward in the power piston. The rear of the air control valve closes the atmospheric port and opens the vacuum port. Vacuum then is applied to the rear chamber, and pressure equalizes on both sides of the diaphragm.

The large return spring moves the diaphragm, the power piston, and the output pushrod rearward. The booster returns to the released position, as was shown in Figure 6-12.

Brake Pedal Feel

A power brake booster must provide some kind of physical feedback to the driver. This feedback is called brake pedal feel. Without pedal feel, the booster would apply the power assist in sudden steps and cause very poor braking control. Also without pedal feel or feedback, when the booster reached the holding position, the driver would feel only the foot pressure applied to the pedal, not the actual force applied to the master cylinder. In fact, many early power brake systems had very poor pedal feel.

Pedal feel is provided by the **reaction disc and levers.**

Pedal feel is provided in vacuum boosters by a **reaction disc (or reaction plate and levers)**. Reaction simply means that as the driver's foot, the pedal, and the booster apply force to the master cylinder, an equal force develops in the opposite direction. The driver normally feels this equal and opposite reaction force as the resistance to application force. Simply put, when the brake pedal is applied as hard as possible, it stops moving. This means that the reaction force has equaled the maximum force that is capable of being applied to the pedal.

The earliest power brake felt as if the driver was pushing his or her foot into a sponge without any brake feel.

AUTHOR'S NOTE A simplistic means of understanding pedal feel is to press on a small coil spring with either your hand or your foot. The more you press, the more reaction force is generated to force your foot or hand back.

Using reaction force to provide pedal feel requires some interesting engineering. If the hundreds of pounds of force developed by the booster were applied as reaction, or feedback, to the driver's foot, the brakes could never be applied past the booster vacuum runout point. Theoretically, the booster could even force the pedal back to the unapplied position. To get around these problems, power boosters use the reaction disc or the reaction plate and levers to feed back only 20 percent to 40 percent of the booster force. The feedback force applied as pedal feel is always less than booster output force, but it also is always proportional. Whether the booster is applying 100 pounds or 400 pounds of force on the master cylinder, the feedback will be a constant percentage of whatever the output force is.

Reaction-Disc Booster

In a booster with a reaction disc, the input pushrod and vacuum valve plunger bear on a rubber disc (**Figure 6-15**). This reaction disc is located in the power piston and compresses under the force of the pedal. Its ability to compress lets it absorb reaction force back from the master cylinder when the brakes are applied. As the disc compresses and feeds back

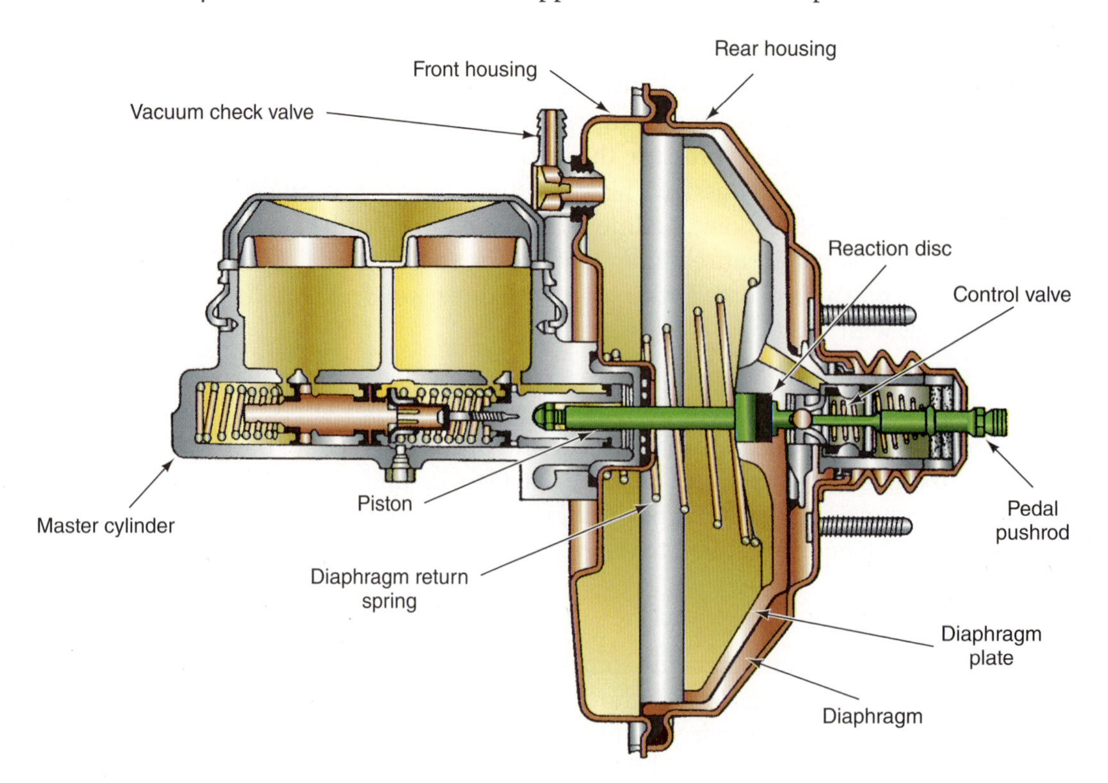

Figure 6-15 Reaction-disc vacuum booster.

reaction force to the pedal pushrod, it also modulates the action of the vacuum and air control valves to adjust pressure on the diaphragm. The harder the brake pedal is pressed, the more the disc compresses and the greater the feedback feel applied to the pedal.

Plate-and-Lever Booster

A plate-and-lever booster (**Figure 6-16**) uses a lever mechanism to react force back to the brake pedal. The connection between the pedal pushrod is through the reaction plate and levers in the power piston.

When the brakes are first applied, the fixed ends of each lever are in contact with the power piston. The other ends are spring loaded and free to move. The force back to the driver's foot on the brake pedal is kept low. As brake application continues, the springs deflect enough to allow contact between the movable ends of the levers and the vacuum and air valves. The operation of the air valve and vacuum valve begins to bleed out vacuum and adds atmospheric pressure to the brake pedal side of the diaphragm. The reaction force back to the brake pedal increases. The lever and reaction plate provide a resistance and feel to the pedal similar to a nonpower brake system.

Tandem Boosters

Automotive engineers are constantly working to save car weight and space. Lighter cars have been made possible by making components out of lighter materials and making components smaller. Master cylinders are one example. They are now much smaller than they were in the past and are made from lighter materials. Brake power boosters have not escaped this trend.

The amount of braking power from a vacuum booster is directly related to the area of the diaphragm. A booster with a smaller diaphragm would provide less power assist. Engineers

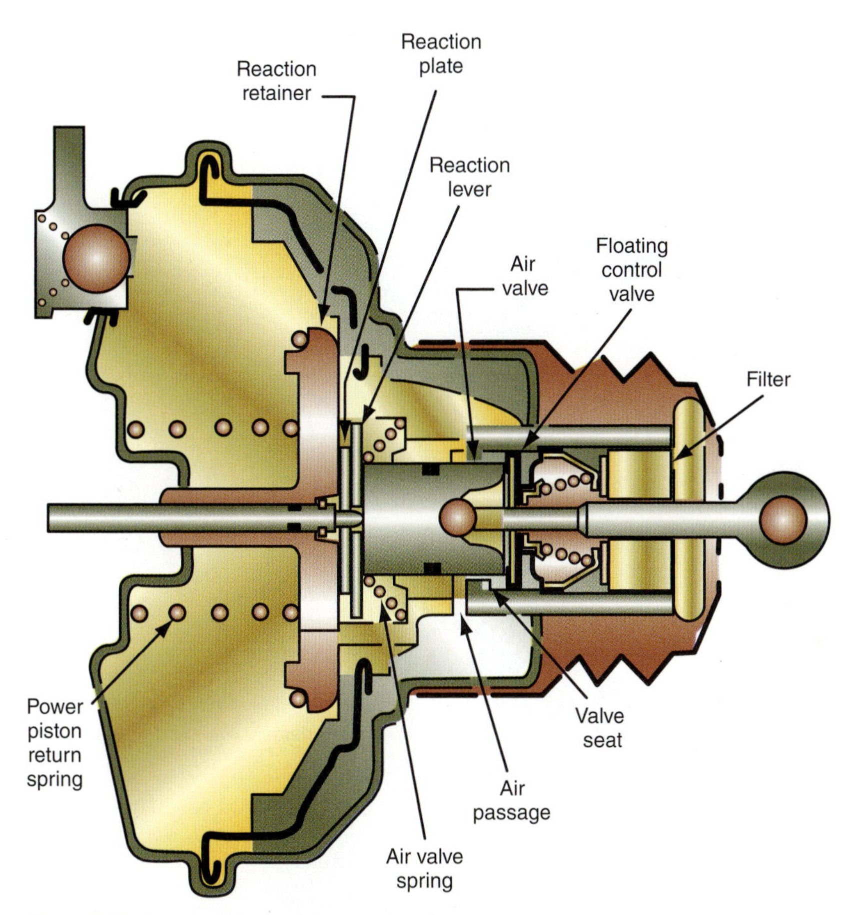

Figure 6-16 Reaction plate-and-lever vacuum booster.

Tandem refers to two or more devices placed one behind the other in line.

solved this problem by designing a smaller diaphragm housing and using two smaller diaphragms in **tandem**. The amount of force is proportional to the total area of both diaphragms. Two 10-square-inch diaphragms can provide the same power as one 20-square-inch diaphragm.

Figure 6-17 is a disassembled view of a **tandem booster**; **Figure 6-18** is a sectional view. The front and rear housings are the same as those of a single-diaphragm booster. The booster contains two separate diaphragms, however, both with support plates. Some units have a housing divider between the diaphragms and others do not. Both diaphragms

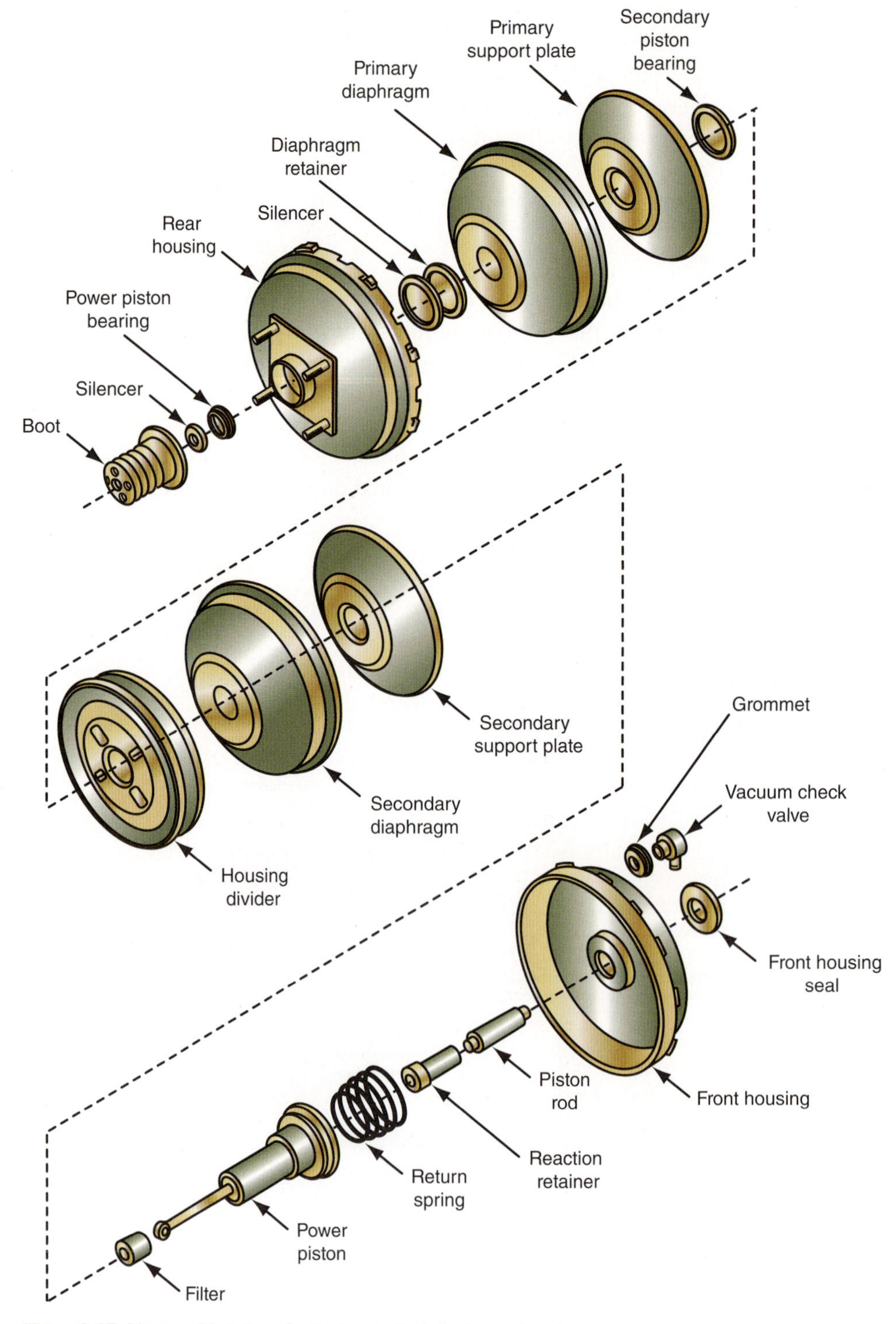

Figure 6-17 Disassembled view of a tandem or dual diaphragm booster.

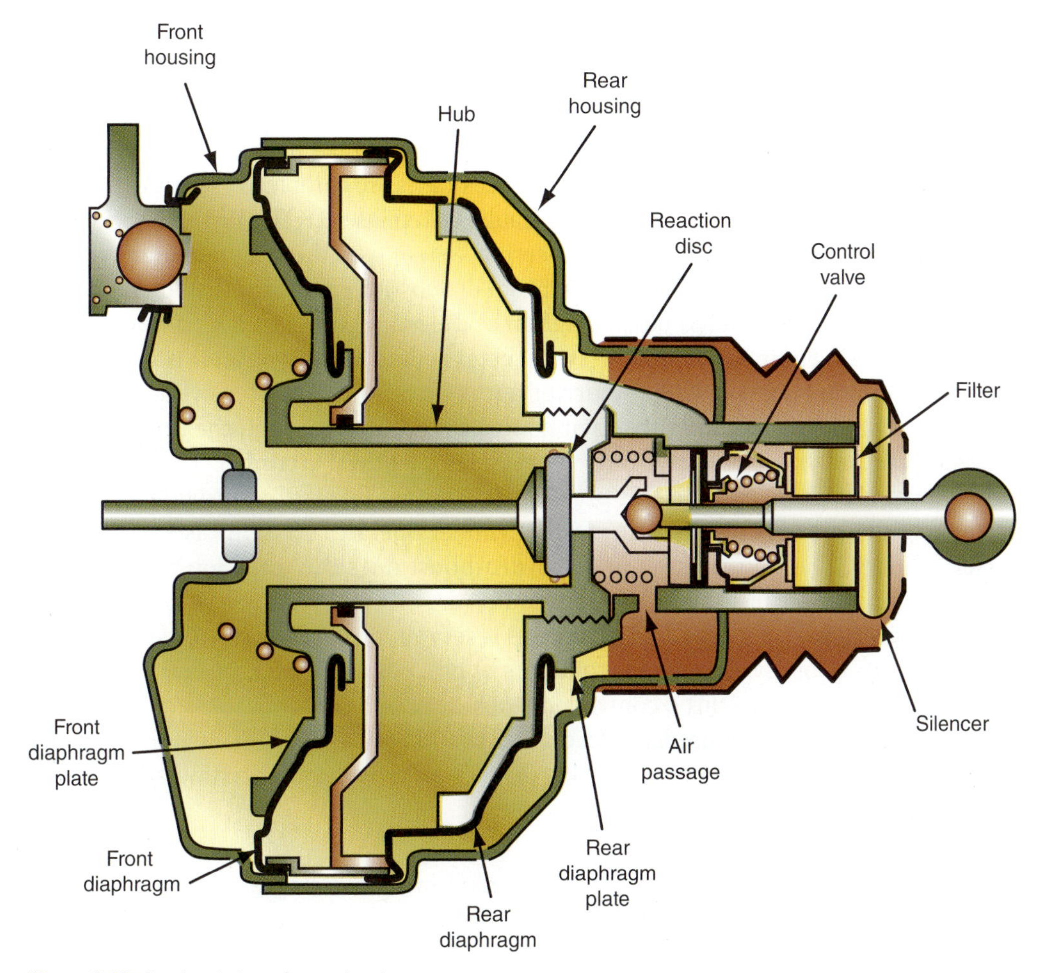

Figure 6-18 Sectional view of a tandem booster.

are connected to the power piston. A reaction disc or lever assembly is used, just as in a single-diaphragm booster.

The two diaphragms work the same way as a single diaphragm except that the air and vacuum valves must control vacuum and atmospheric pressure on both diaphragms at the same time. When the brakes are released, vacuum is on both sides of each diaphragm. During brake application, the air valve and vacuum valve operate to admit air on the brake pedal side of both diaphragms. With a vacuum on the master cylinder side of both diaphragms and atmospheric pressure on the opposite sides, power assist is developed.

HYDRAULICALLY ASSISTED POWER BRAKES

Shop Manual
page 249

Diesel engines do not produce intake manifold vacuum. Hybrid vehicles gas engine may not always be running. A way to provide for brake booster operation is to eliminate vacuum as a power source and use hydraulic power instead. The three kinds of hydraulic boosters are:

1. A mechanical hydraulic power-assist system operated with pressure from the power steering pump. This unit is a Bendix design called hydro-boost.
2. An electrohydraulic power-assist system with an independent hydraulic power source driven by an electric motor. These systems are used on vehicles with integral ABS systems.
3. An electrohydraulic system in which the brake fluid flow and pressure are controlled by solenoids.

Bendix developed the hydro-boost system to be used with the light-duty diesel engines.

Hydro-boost refers to the hydraulic power brake system that uses the power steering hydraulic system to provide boost for the brake system.

A **spool valve** is a cylindrical sliding valve that uses lands and valleys around its circumference to control the flow of hydraulic fluid through the valve body.

Hydro-Boost Principles

The hydro-boost power booster (**Figure 6-19**) fits in the same place as a vacuum booster, between the brake pedal and the master cylinder. Similar to a vacuum booster, the hydro-boost unit multiplies the force of the driver's foot on the brake pedal.

The hydraulic systems for steering and service brakes are completely separate from each other in a hydro-boost system. The brake hydraulic system uses DOT 3 or DOT 4 brake fluid as specified by the carmaker. The power steering system uses the manufacturer's specified power steering fluid. The fluids cannot be mixed or substituted one for the other.

Figure 6-20 is a simplified block diagram of the main parts of a hydro-boost system. The power steering pump develops pressure that is routed to the hydraulic booster assembly. The hydraulic booster helps the driver apply force to the master cylinder pushrod. The booster has a large **spool valve** that controls fluid flow through the unit. When the brakes are applied, the spool valve directs pressure to a power chamber. The boost piston in the power chamber reacts to this pressure and moves forward to provide force to the master cylinder primary piston. The master cylinder operates the same as a conventional master cylinder. The boost piston in the hydro-boost does the same job as the vacuum diaphragm and power piston in a vacuum unit.

An **accumulator** is a pressurized storage reservoir. Loss of the power steering drive belt on a hydro-boost system means total loss of brake boost. The accumulator provides sufficient boost for one to three controlled stops.

The accumulator can be used as a fluid shock absorber or as an alternate pressure source. A spring or compressed gas behind a sealed diaphragm provides the accumulator pressure.

Hydro-Boost Operation

Hydro-boost operation can be divided into the following five stages:

1. Brakes not applied (released)
2. Moderate brake application

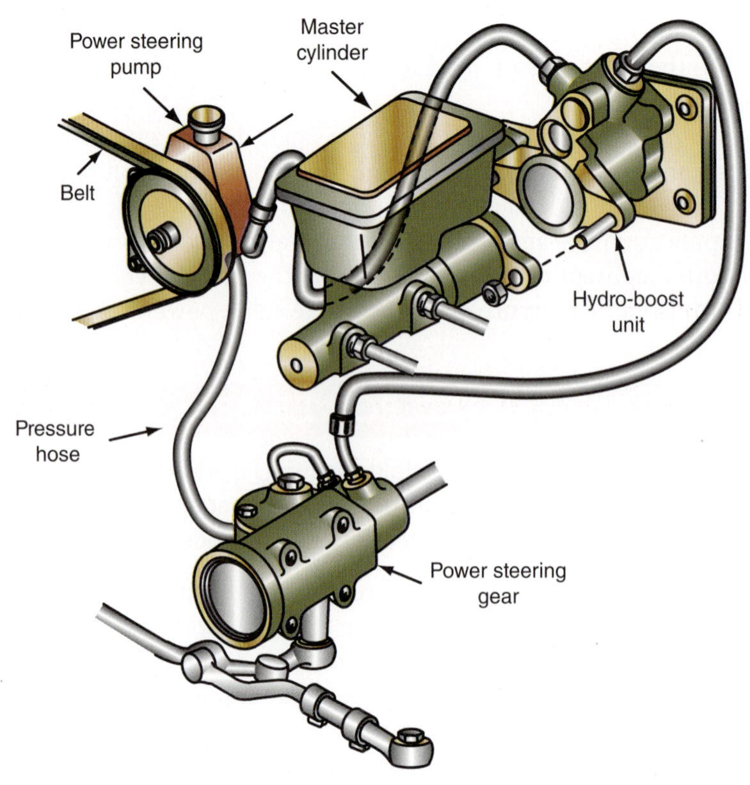

Figure 6-19 A hydro-boost unit is attached between the master cylinder and the firewall.

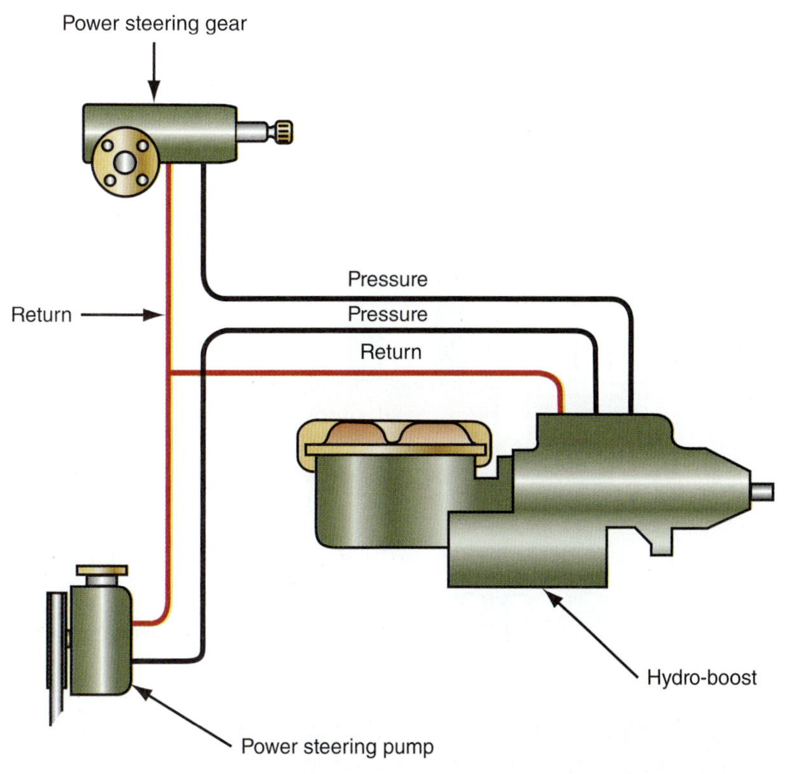

Figure 6-20 The hydro-boost system shared hydraulic power (pressure) with the power steering system.

3. Brakes holding
4. Brakes being released
5. Reserve brake application

Brakes Not Applied. **Figure 6-21** shows the booster in the unapplied position (brake pedal released). Fluid under pressure from the power steering pump enters the booster through the pump port in the housing. The fluid is directed through internal passages to the spool valve. The spool valve is held rearward by its spring so that the **lands** and **valleys** direct the fluid from the pump directly through the valve to the steering gear. The spool valve also opens a passage to vent the power cavity back to the reservoir. The power cavity has no pressure, so no force is applied to the output pushrod.

Lands are raised surfaces on a valve spool.

Valleys are annular grooves, or recessed areas, between the lands of a valve spool.

Moderate Brake Application. Moderate brake application causes the input pushrod to press on the reaction rod in the end of the power piston. The reaction rod moves forward and causes the lever to pivot on the power piston and to move the spool valve forward in its bore (**Figure 6-22**). Spool valve movement opens the pump inlet port to the power chamber and closes the vent port. Hydraulic pressure now increases in the power chamber and moves the power piston forward to operate the master cylinder.

As the spool valve moves forward, it also restricts fluid flow out to the steering gear. Closing the vent port and restricting flow to the steering gear causes pressure to rise in the brake booster. The farther the valve moves, the more the flow to the steering gear is restricted and the more the booster pressure increases. At maximum boost, hydraulic pressure in the brake booster can exceed 1,400 psi.

Brakes Holding. As long as the driver maintains unchanging foot pressure on the pedal, the input pushrod does not move. As pressure rises to this holding point, the power piston moves forward and makes the lever pivot on the lever pin, which moves the spool valve back toward the rear of its bore. This closes the fluid inlet port and reopens the bypass

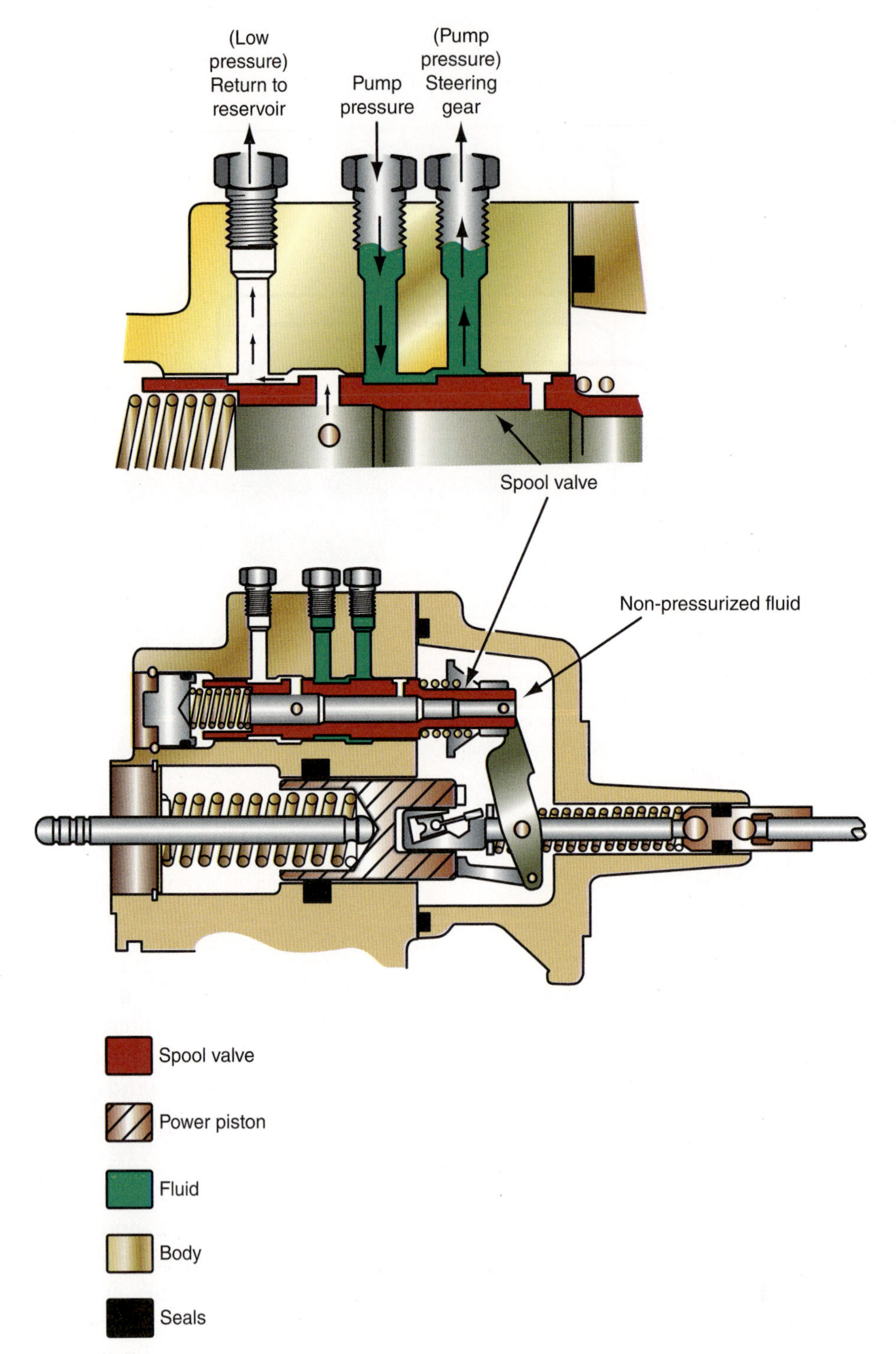

Figure 6-21 When the brakes are released power steering fluid flows past the spool and continues to the steering gear.

port to the steering gear. The vent port stays closed, however, so pressure in the power chamber reaches a steady state. The booster always seeks the holding position when pedal force is constant or is unchanging.

This all happens very quickly. As booster pressure increases, fluid flows through a small port to a space behind the reaction rod. This creates a counterforce that moves the rod and lever rearward, which also moves the spool valve rearward and moderates application force. The reaction rod provides the same kind of proportional feedback brake feel that the reaction disc or levers in a vacuum booster provide.

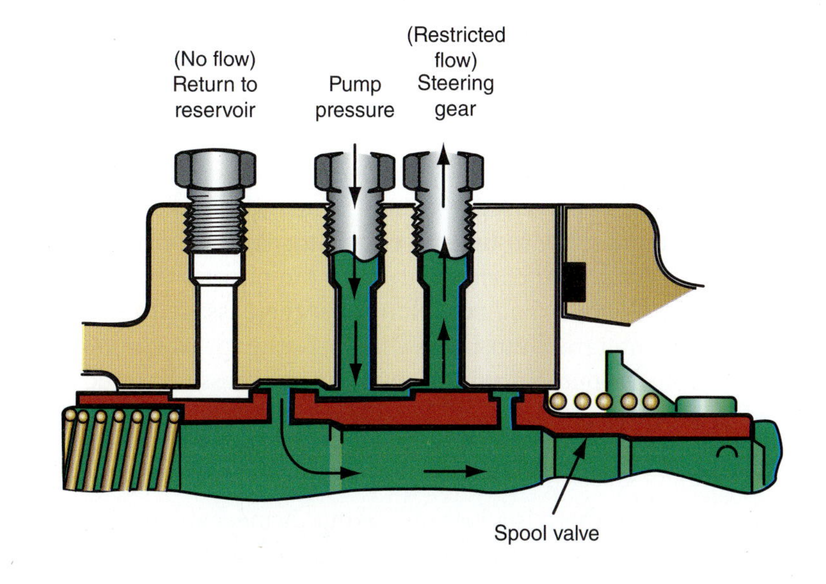

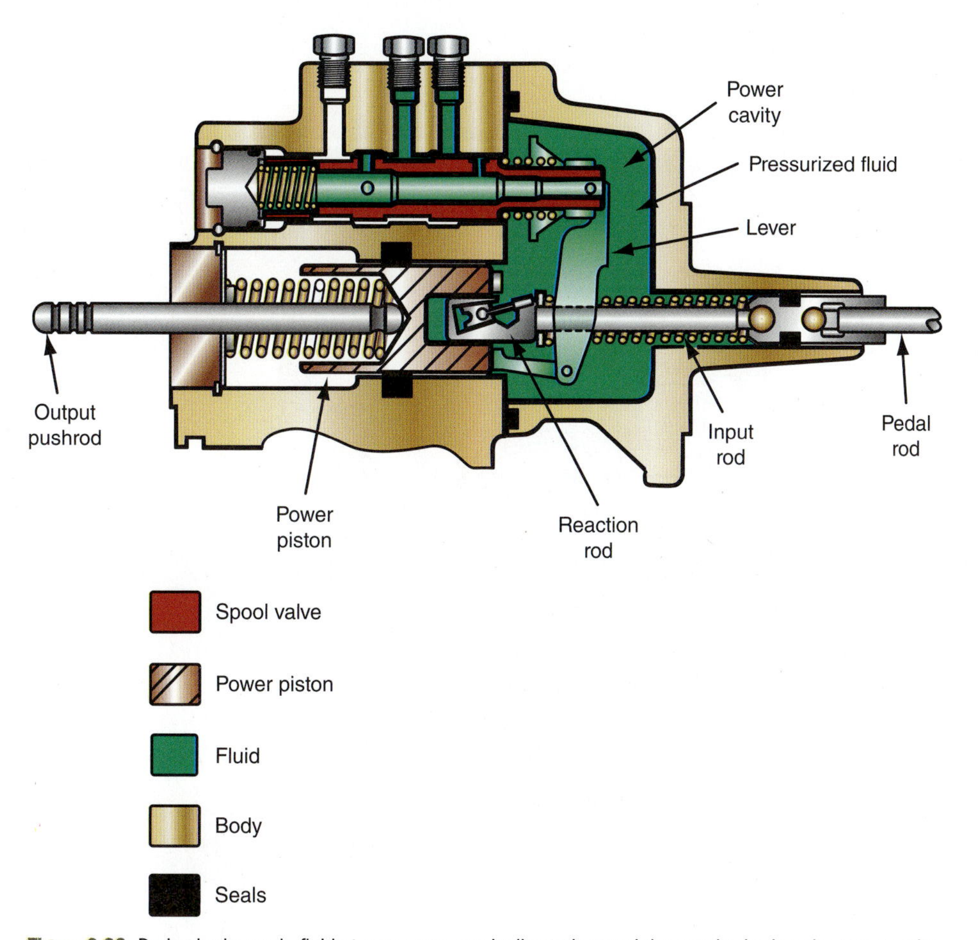

Figure 6-22 During brake apply, fluid at pump pressure is directed around the spool valve into the power cavity and charges the accumulator. Pressure is applied to the power valve. Fluid flow to the steering is restricted.

Brakes Being Released. When the brake pedal is released, the spool valve moves fully rearward to the unapplied position and vents pressure from the pressure chamber. Return springs on the power piston and lever quickly return the piston to the released position. Spool valve movement also blocks the fluid inlet port to the power chamber and lets fluid bypass the booster and flow to the steering gear. The hydro-boost returns to the released position as was shown in Figure 6-21.

Reserve Brake Application. A failure in the power steering system, such as a broken power steering hose, a broken power steering drive belt, pump failure, or a stalled engine could cause a loss of pressure to the hydro-boost system. The hydro-boost has a backup system, powered by an accumulator, that allows two or three power brake applications. When the pressure from the accumulator is needed, the gas or spring pressure inside the accumulator moves the pressurized hydraulic fluid to the brake booster.

Figure 6-23 shows the hydro-boost accumulator. During normal operation, the pump pressure fills the accumulator cavity in front of the piston. The piston compresses the accumulator spring. A check ball and plunger in the passage to the accumulator allow pressure in but prevent it from escaping. This keeps the spring compressed or charged. A loss of pressure from the pump opens a passage for the pressure stored in the accumulator to be routed to the power chamber to help the driver apply the brakes. The accumulator stores enough energy for two or three brake applications.

Loss of hydraulic power from the hydro-boost only means that the driver loses power assist. It does not mean there is a loss of braking. A mechanical connection exists from the brake pedal through the pedal rod, through the input rod, through the power piston to the output pushrod. The driver's pedal effort will increase, but the brakes will still work.

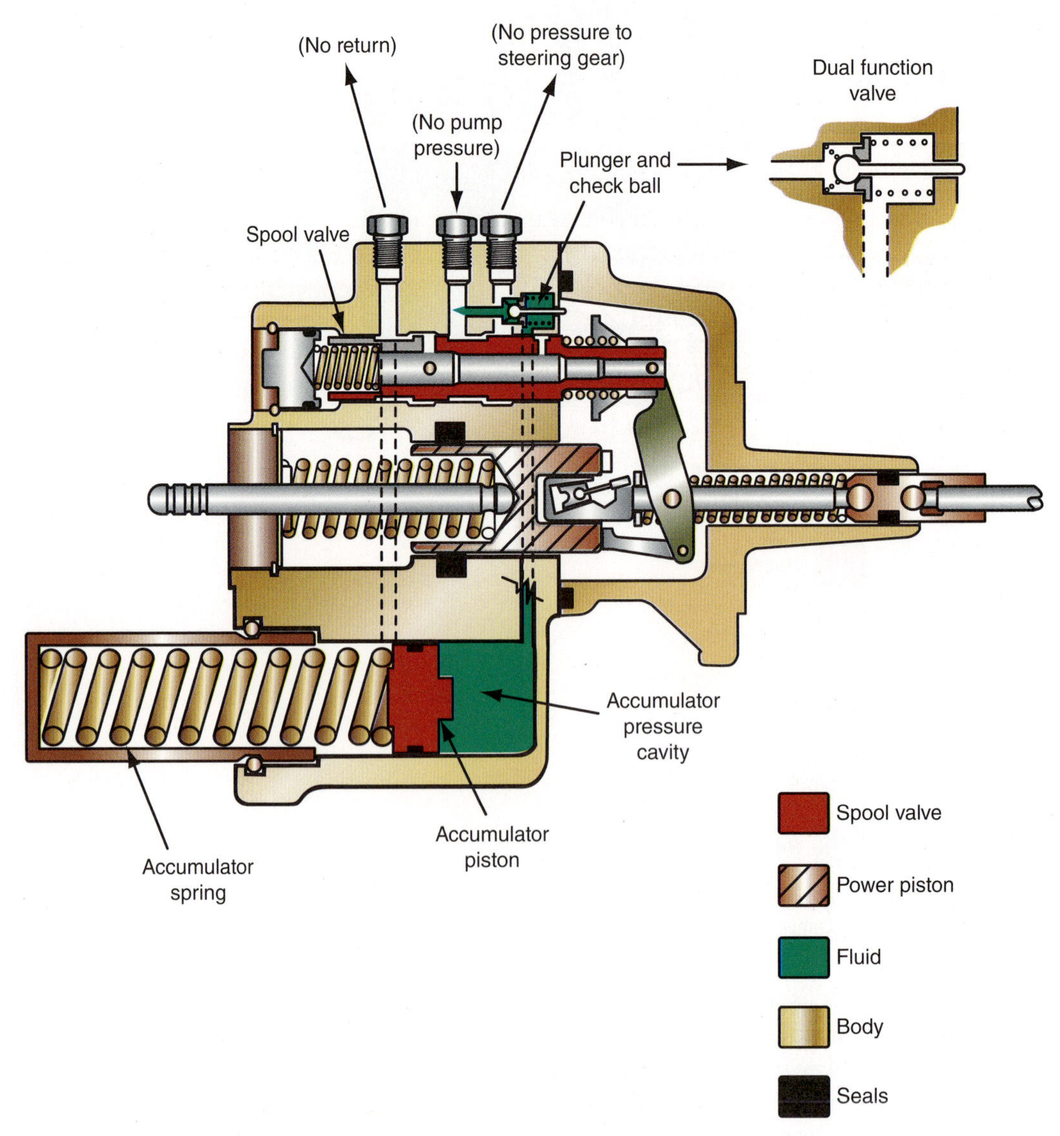

Figure 6-23 In case of pump failure, fluid under accumulator pressure is directed through the check valve to the power cavity. Fluid is preventing from entering the steering, pump, and return ports.

Integral ABS Electric Brake Booster

Integral ABS systems combine the master cylinder, electric pump motor, and hydraulic modulator into one unit. An electric pump motor pressurizes brake fluid and uses this pressurized brake fluid to supply hydraulic boost pressure for the master cylinder (**Figure 6-24**). These units have an accumulator to store hydraulic pressure. The accumulator allows for a few booster-assisted stops if the pump should fail. The unit has a pressure switch that activates the pump when the pressure falls too low. These units should never be serviced without first relieving the stored accumulator pressure.

Caution

Whenever an integral ABS system is serviced, relieve the pressure according to service procedures. This usually involves pumping the brake pedal with the key off until the pedal feels solid. This may take as many as 25 applications of the brake pedal. This is necessary, because the brake pressure stored in the accumulator can be as high as 1,500 psi.

Vacuum Booster with Brake Assist

The vacuum booster has recently been fitted with electronic sensors and actuators that will apply the brakes more quickly during panic stops (**Figure 6-25**). A **brake assist** (BA) unit is added to the vacuum booster. This unit detects the brake pedal application speed; in other words, how fast the driver is pushing on the pedal. During a panic stop, the driver naturally tries to move the pedal as fast as possible so the brakes will react quickly. When this condition is detected, the BA will more quickly activate the brake booster or signal the **electronic brake system (EBS)** controller. The brake fluid pressure is increased much more rapidly, and the brakes, in turn, react almost instantaneously. Although the time difference between non-BA units and assisted units is measured in fractions of a second, it may make the difference between vehicle control and a possible accident.

Electronic Brake Systems is a term for systems that combine electronic control with the conventional hydraulic braking.

Electrohydraulic Brake

As discussed before, an **electrohydraulic brake (EHB)** systems control hydraulic braking action through the use of electrical solenoids. EHB and BA units share information with ABS, TCS, and electronic suspension. The most common sensors are the yaw, wheel speed, and steering wheel angle. The most common electronic actuator is the hydraulic modulator, which is shared with ABS and TCS. The EHB have basically replaced the ABS and TCS and incorporated them into a single unit which also includes Vehicle Stability Control (VSC). In **Figure 6-26**, the EHB is compared to the BA unit discussed earlier. If the BA unit is removed (left side of figure), a conventional brake system with the ABS/TCS would be displayed. In this case, the BA unit is part of the EHB. The bottom of

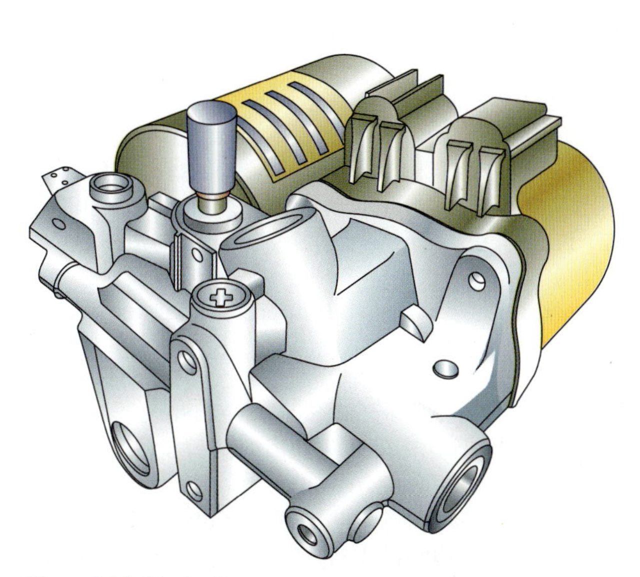

Figure 6-24 A hydraulic power booster assembly.

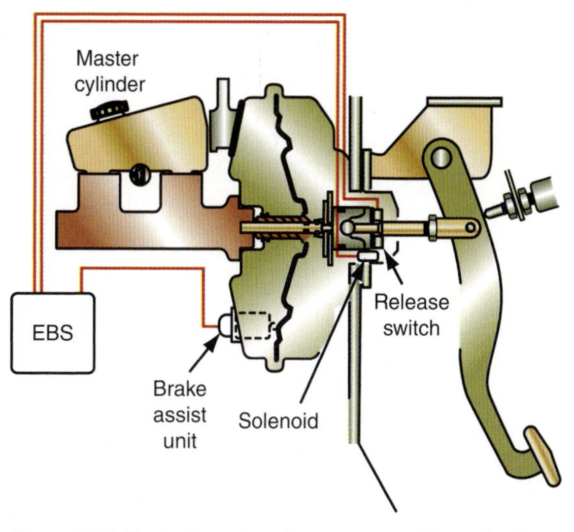

Figure 6-25 The brake assist unit measures pedal application speed and either actuates the vacuum booster or sends a signal to the EHB control module.

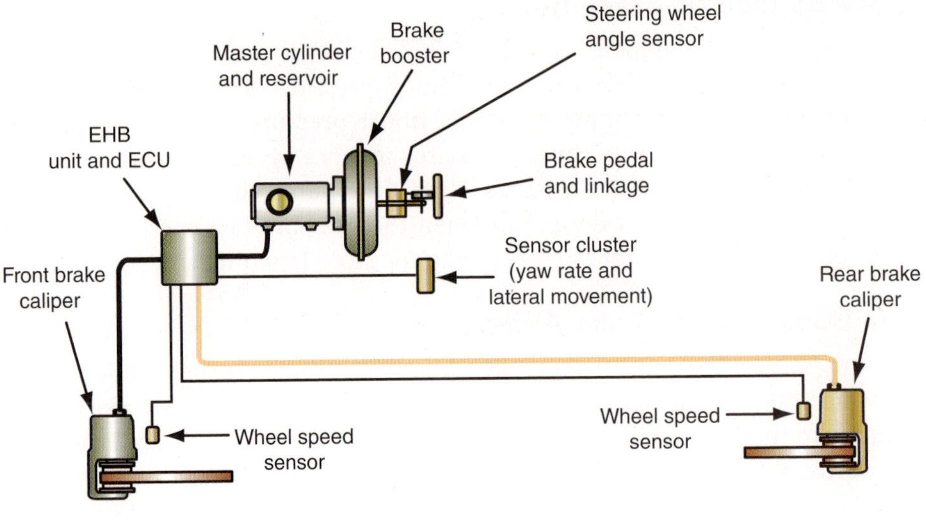

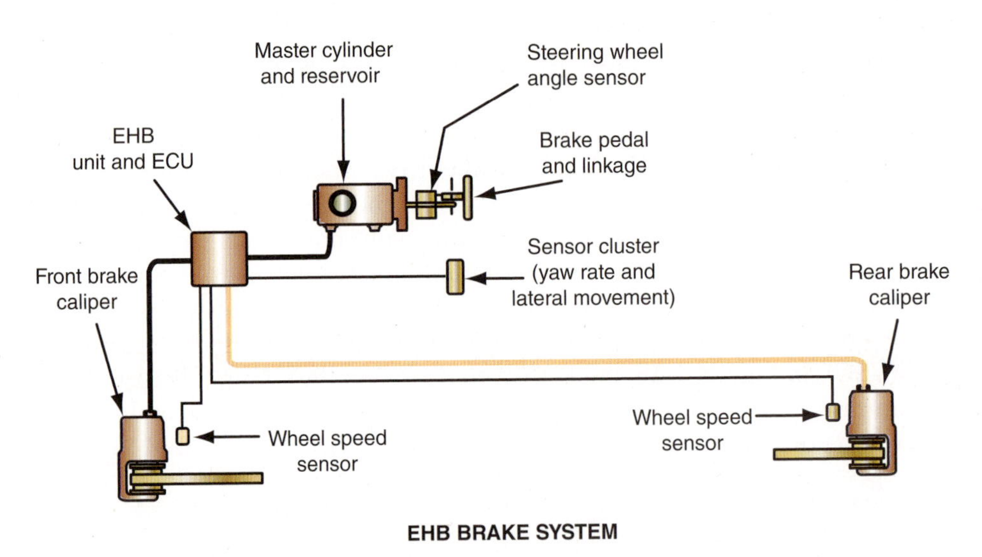

Figure 6-26 Note that the primary visual difference between the two layouts is the absence of the brake booster on this system. The EHB control module is more complex than the hydraulic model.

Figure 6-26 shows a complete setup for an EHB. Note that the brake components at the wheels are the same as the conventional brakes. The main components of an EHB are the electronic control unit and the electronic pedal module. A stroke sensor/module/master cylinder is shown in **Figure 6-27**. The main advantage of an EHB is shorter stopping distances and better vehicle control.

The **stroke sensor** tells the electrohydraulic controller how fast and how far the brake pedal has been applied to judge the driver's needs.

When the brake pedal is depressed on an EHB system, pedal movement and speed are detected by a **stroke sensor**, and the signal is sent to the EHB control module mounted on or near the hydraulic module. The pedal stroke sensor/master cylinder module also has components that will create a reaction or feedback to the driver's input.

The EHB control module collects data from the yaw and wheel speed sensor and commands the individual hydraulic valves within the hydraulic modulator to apply fluid to or release fluid from the brakes individually, as needed for the situation. In this manner, the EHB can control pressure to each wheel individually, preventing wheel lockup while maintaining maximum stopping power. In the opposite direction, if a wheel loses traction

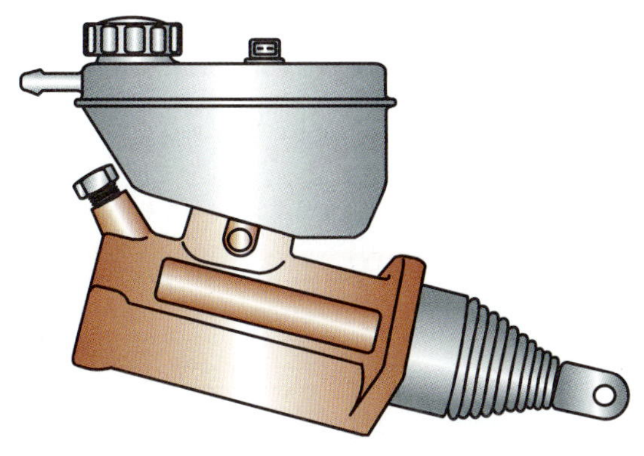

Figure 6-27 An EHB electronic pedal module with a master cylinder.

on acceleration, the EHB with the proper programming can act as a TCS. The EHB becomes an active brake system once it is married completely into other braking, steering, and ride control systems.

During normal operation of the EHB, the master cylinder supplies pressurized fluid to the modulator. Its function as the primary brake pressure-generator stops at that point, but it still provides the backup or redundancy in the event of electrical component failure. If the EHB should fail in part or completely, the driver will still have the ability to brake the vehicle using the brake's conventional operational hydraulics. As with the loss of the booster, the driver will have to apply more force, but the vehicle will be controllable and can be stopped.

SUMMARY

- Vacuum is simply air pressure below atmospheric pressure.
- Vacuum power brakes are by far the most common kinds of power brakes. They operate on power provided by the pressure differential between atmospheric pressure and vacuum.
- The basic vacuum power booster consists of a diaphragm in a housing along with vacuum and air valves.
- The air valve is the vacuum booster valve that controls atmospheric pressure on the vacuum diaphragm.
- The vacuum valve is the valve in the power booster that controls vacuum on the vacuum diaphragm.
- The power piston is a part of the power booster that controls the air and vacuum valves and transmits force to the master cylinder.
- The diaphragm return spring returns the diaphragm and power piston to the unapplied position when brakes are released.
- The three types of vacuum power boosters are the single diaphragm with lever reaction, the single diaphragm with disc reaction, and the tandem diaphragm with disc reaction.
- The air system for a vacuum booster consists of a silencer and air filter.
- A check valve and a reservoir prevent vacuum loss.
- Light-duty diesel trucks use hydraulically assisted power brake systems. These braking systems use hydraulic power developed in the power steering pump for power assist. One popular system is called hydro-boost.
- The BA unit measures brake pedal application speed and activates the booster quicker or signals the electronic brake module.
- The hydro-boost system uses a spool valve to control the flow of pressurized fluid to a power chamber.
- An accumulator stores energy for several brake applications if the hydraulic system fails.
- An EHB combines signals from sensors shared with ABS, TCS, steering, and suspension systems to control braking forces at individual wheels.

REVIEW QUESTIONS

Short-Answer Essays

1. Explain the three general kinds of power boosters used to increase braking force.
2. Brake pedal feel is provided in vacuum boosters by a reaction disc or a reaction plate and levers. Explain how this works.
3. What is the purpose of the power piston in a vacuum power unit?
4. Explain how the braking system can still apply the brakes if the booster fails completely.
5. Why are hydraulic power assist units used on vehicles with diesel engines?
6. Briefly explain how the hydro-boost power brake booster operates.
7. List the four methods that can be used to reduce pedal pressure requirements or to boost the force applied to the master cylinder
8. Explain the general operation of the BA system used on a vacuum booster.
9. Explain the purpose of the stroke sensor on an EHB system.
10. Explain the purpose of an accumulator in a hydraulic brake power assist system.

Fill in the Blanks

1. During normal operation of the EHB, the master cylinder supplies pressurized fluid to the ______________.
2. To understand vacuum booster systems, the relationship of atmospheric ______________ and ______________ must also be understood.
3. Atmospheric pressure varies with ______________ and ______________, but at sea level and at 68°F, it is 14.7 psi.
4. Without pedal feel, the booster would apply the power assist in ______________ steps and cause very poor braking control.
5. ______________ ______________ is provided in vacuum boosters by a reaction disc (or reaction plate and levers).
6. The amount of braking power from a vacuum booster is directly related to the ______________ of the diaphragm.
7. Because the pedal pushrod is mechanically connected through the ______________, any vacuum failure will not cause loss of braking action.
8. The hydro-boost system booster has a large ______________ valve that controls fluid flow through the unit.
9. Vacuum is air pressure less than ______________ pressure.
10. An ______________ is a pressurized storage reservoir.

Multiple Choice

1. Technician A says the pedal ratio multiplies force applied to the master cylinder. Technician B says hydraulic piston size can be used to multiply hydraulic pressure applied by the master cylinder to the wheel cylinders and caliper pistons. Who is correct?

   A. A only
   B. B only
   C. Both A and B
   D. Neither A nor B

2. Technician A says the amount of braking power from a vacuum booster is directly related to the area of the diaphragm. Technician B says a booster with a small diaphragm would provide more power assist. Who is correct?

   A. A only
   B. B only
   C. Both A and B
   D. Neither A nor B

3. Technician A says that when the brakes are released, vacuum is on both sides of the booster diaphragm. Technician B says that when the brakes are released, atmospheric pressure may be on one side of the booster diaphragm. Who is correct?

   A. A only
   B. B only
   C. Both A and B
   D. Neither A nor B

4. Technician A says the power piston contains and operates the vacuum and air (atmospheric) valves that control the diaphragm. Technician B says the power piston also transmits the force from the diaphragm through a piston rod to the master cylinder. Who is correct?

   A. A only
   B. B only
   C. Both A and B
   D. Neither A nor B

5. Technician A says that the EHB system is a direct power booster. Technician B says the EHB and the BA units share information with the ABS, TCS, and electronic suspension. Who is correct?

   A. A only
   B. B only
   C. Both A and B
   D. Neither A nor B

6. Tandem brake boosters are being discussed: Technician A says tandem diaphragm boosters use one large diaphragm. Technician B says that a tandem diaphragm can provide more brake boosting power in a smaller housing. Who is correct?

   A. A only
   B. B only
   C. Both A and B
   D. Neither A nor B

7. Technician A says that a hydro-boost system uses brake fluid for the booster. Technician B says the hydro-boost system uses power steering fluid for the brakes. Who is correct?

   A. A only
   B. B only
   C. Both A and B
   D. Neither A nor B

8. Power brake boosters are being discussed. Technician A says power brake boosters do not alter the basic brake system. Technician B says all boosters have a power reserve to provide at least one power-assisted stop if power is lost. Who is correct?

   A. A only
   B. B only
   C. Both A and B
   D. Neither A nor B

9. The BA system is being discussed. Technician A says that the BA system can build pressure in the brake system rapidly in case of a panic stop. Technician B says that the BA system can make a difference between vehicle control and a possible accident. Who is correct?

   A. A only
   B. B only
   C. Both A and B
   D. Neither A nor B

10. Technician A says if the EHB should fail completely, the vehicle will have to be towed to the nearest shop. Technician B says the driver can stop the vehicle but may have to press on the pedal harder than normal. Who is correct?

    A. A only
    B. B only
    C. Both A and B
    D. Neither A nor B

CHAPTER 7
DISC BRAKES

Upon completion and review of this chapter, you should be able to:

- Explain the advantages and disadvantages of disc brakes.
- Describe the basic parts of a disc brake assembly.
- Describe the different kinds of rotors and the types of hub-and-rotor assemblies.
- Describe how a caliper works to stop a vehicle.
- Name the kinds of friction material used on disc brake pads and explain the advantages and disadvantages of each.
- Identify the parts of a typical caliper.
- Describe the two principal kinds of caliper designs and the variations of each.
- Describe how a low-drag caliper works compared to other kinds of calipers.
- Describe the different kinds of parking brakes used with rear disc brakes.
- Explain how brake pad wear indicators operate.

Terms To Know

Aramid fibers
Automotive Friction Material Edge Code
Binder
Bonded lining
Brake caliper
Brake fade
Brake pad
Caliper support
Ceramic
Ceramic brake pad
Composite rotor
Curing agent
Drum-in-hat
Filler
Fixed caliper brake
Fixed rotor
Fixed seal
Floating caliper
Floating rotor
Friction modifier
Gas fade
Lathe-cut seal
Metallic lining
Mold-bonded lining
Organic lining
Pad hardware
Pad wear indicators
Phenolic plastic
Riveted lining
Rotor
Semi-metallic lining
Sliding caliper
Solid rotor
Square-cut piston seal
Steering knuckle
Swept area
Synthetic lining
Two-piece rotor
Unidirectional rotor
Ventilated rotor
Water fade

INTRODUCTION

Disc brakes are used on the front wheels of almost all cars and light trucks built since the early 1970s and on all four wheels of many vehicles.

Although no law or regulation requires disc brakes on the front wheels of cars and light trucks, the brake performance requirements of FMVSS 105 make front disc brakes virtually mandatory. Disc brakes provide greater stopping power than do drum brakes, with less brake fade and fewer problems such as grabbing and pulling. Disc brake efficiency is even more important on front-wheel drive cars, which require 60 percent to 80 percent of the stopping power at the front wheels.

This chapter explains the construction and operation of disc brake systems and begins with a summary of their advantages and disadvantages.

DISC BRAKE ADVANTAGES AND DISADVANTAGES

Disc brakes have several very important strong points and a few relatively small disadvantages. The principal advantages of disc brakes are strong fade resistance, self-adjustment, and reduced pulling and grabbing. Disc brake disadvantages are generally minor, but they include the lack of self-energizing (servo) action, noise, and poorer parking brake operation with complicated linkage.

Shop Manual
page 286

Fade Resistance

Remember that brake operation is a process of changing kinetic energy (motion) into thermal energy (heat) through the application of friction (**Figure 7-1**). When any brake installation reaches its limit of heat dissipation, **brake fade** sets in. Brake fade is simply the loss of braking power because of excessive heat that reduces friction between brake linings and the rotors or drums.

Brake fade is the loss of braking power because of excessive heat that reduces the friction between the brake and the rotor.

One major factor that makes disc brakes more fade resistant than drum brakes is that the friction surfaces are exposed to more airflow. Many brake rotors also have cooling passages or fins to dissipate heat even more (**Figure 7-2**).

Another factor that contributes directly to heat dissipation and fade resistance is the **swept area** of a disc brake rotor. With drum brakes, almost the entire swept area is in direct contact with the friction materials during the entire braking operation. With disc brakes, most of the swept area is cooling while only a small area directly contacts the friction materials at any given time (**Figure 7-3**). The greater the swept area, the greater

The **swept area** is the total area of the brake drum or rotor that contacts the friction surface of the brake lining.

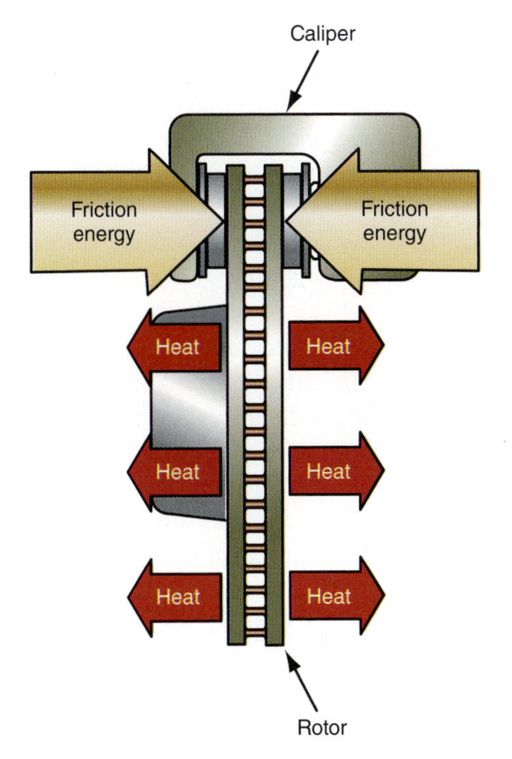

Figure 7-1 The clamping action of friction material in the rotating disc or rotor creates heat.

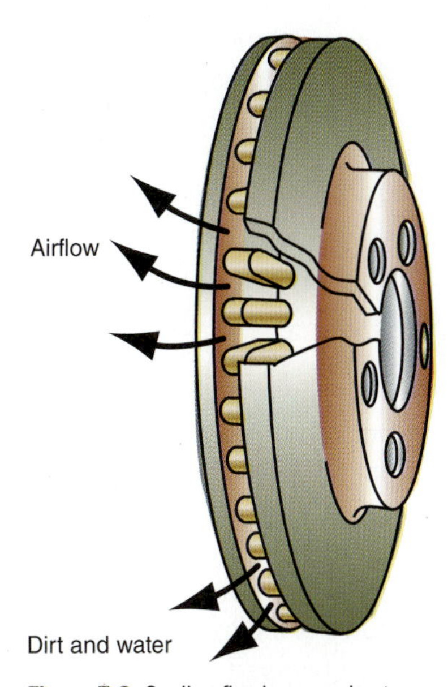

Figure 7-2 Cooling fins increase heat dissipation in many rotors, and centrifugal force helps to keep the rotor clean by throwing off dirt and water.

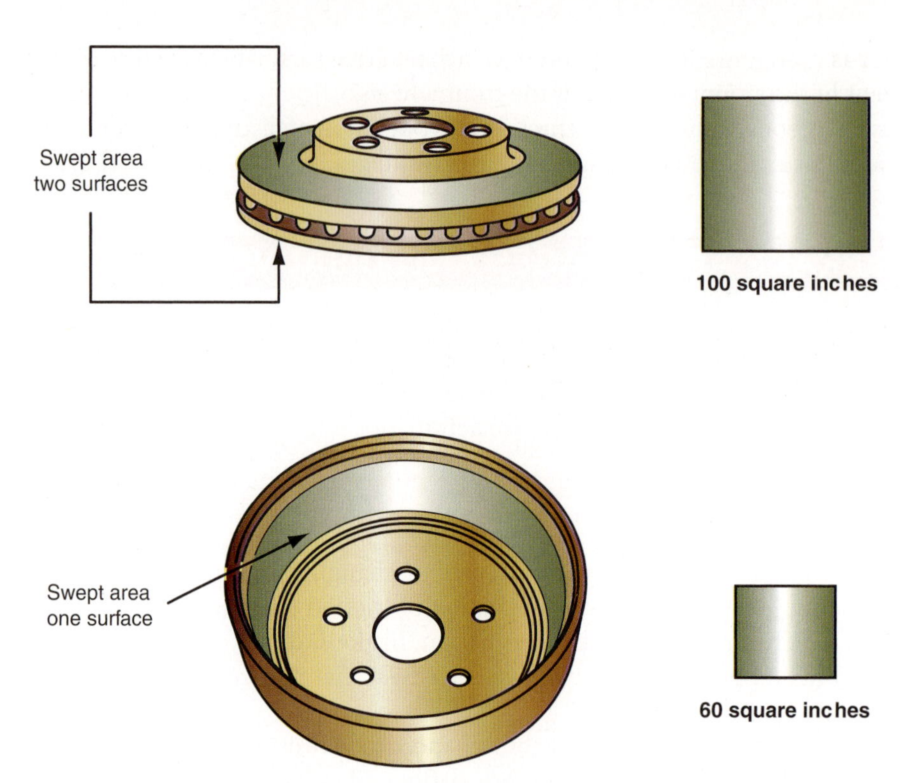

Figure 7-3 For any given size wheel, a disc brake always has 35 percent to 50 percent more swept area to dissipate heat.

the heat dissipation ability. Although a brake drum can have a relatively large swept area, the entire area is on the inner drum surface. The swept area of a disc brake comprises both sides of the rotor. For any given wheel size, the swept area of a disc brake will always be larger than the swept area of a drum brake. For example, a 10-inch-diameter rotor will have almost 50 percent more swept area than a 10-inch drum.

Freedom from mechanical fade is another disc brake advantage. Mechanical fade is a problem that can occur with drum brakes when the drum becomes very hot and expands outward, away from the brake shoes. The shoes then must travel farther to contact the drum surface with normal braking force. As a result, the pedal drops lower as the brakes are applied. The increased heat at the braking friction surfaces also reduces the coefficient of friction. The combined result is brake fade.

Disc brakes do not suffer from mechanical fade because the rotor does not expand away from the pads. If anything, the rotor expands very slightly toward the pads if it becomes very hot.

Water fade occurs when water is trapped between the brake pad and the rotor. Disc brakes are less prone to water fade than drum brakes because of their open design.

Gas fade occurs when a layer of hot gas and dust particles become trapped between the brake pad and rotor. Disc brakes are less prone to gas fade because their open design promotes cooling.

Disc brake design also reduces the effects of lining fade, water fade, and gas fade. Lining fade occurs when the linings are overheated and the coefficient of friction drops off severely. Some heat is needed to bring brake linings to their most efficient working temperature. The coefficient of friction rises, in fact, as brakes warm up. If temperature rises too high, however, the coefficient of friction decreases rapidly. Because disc brakes are exposed to more cooling airflow than drum brakes, lining fade is reduced due to this cooling air flow.

Water fade occurs when water is trapped between the brake linings and the drum or rotor and reduces the coefficient of friction. Severe water exposure can eliminate friction almost entirely and prevent braking until the friction surfaces dry themselves.

Gas fade is a condition that occurs under hard braking when hot gases and dust particles are trapped between the brake linings and the drum or rotor surfaces. These gases build up pressure that acts against brake force on the drum or rotor surface. More importantly, these hot gases actually lubricate the friction surfaces and reduce the coefficient of friction.

The basic design differences between disc and drum brakes make disc brakes much more self-cleaning to greatly reduce or eliminate water and gas fade. The rotor friction surfaces are perpendicular to the axis of rotation, and centrifugal force works to remove water and gas from the braking surfaces. In addition, the leading edges of the brake pads help to wipe the rotor surfaces clean, and many pad linings have grooves to drain water and help prevent gas buildup.

Drivers must beware when driving around construction or road repair sites on clear days. Sufficient water may be on the road to cause water fade and cause a rear-end collision in the stop-and-go traffic at these sites.

Disc Brake Self-Adjustment

Brake adjustment is the process of compensating for lining wear and maintaining correct clearance between brake linings and the drum or rotor surfaces. Disc and drum brakes both require adjustment, but unlike drum brakes, disc brakes do not need cables, levers, screws, struts, and other mechanical linkages to maintain proper pad-to-rotor clearance for the service brakes. On some vehicles, the emergency brake is incorporated into the rear calipers. These brakes have a mechanical function to turn the caliper piston out to apply the pads for the emergency brakes. Most late-model vehicles have the emergency brake shoes separate from the caliper, mounted into the rear brake rotor.

When disc brakes are applied, the caliper pistons move out far enough to apply braking force from the pads to the rotors. When the brakes are released, the caliper pistons retract only far enough to release pressure (**Figure 7-4**). The pads always have only a few thousandths of an inch clearance from the rotor, regardless of lining wear. In later sections of this chapter, brake caliper operation and disc brake self-adjustment are explained in detail.

Brake Servo Action

Drum brakes have an operating feature by which they develop mechanical leverage as the lining contacts the drum. One end of the lining contacts the drum before the other does and becomes a pivot point as friction increases quickly. The brake shoe becomes a self-energizing lever and adds its own mechanical leverage to hydraulic force to help apply the brakes. This self-energizing action, called duo-servo, is shown in **Figure 7-5**.

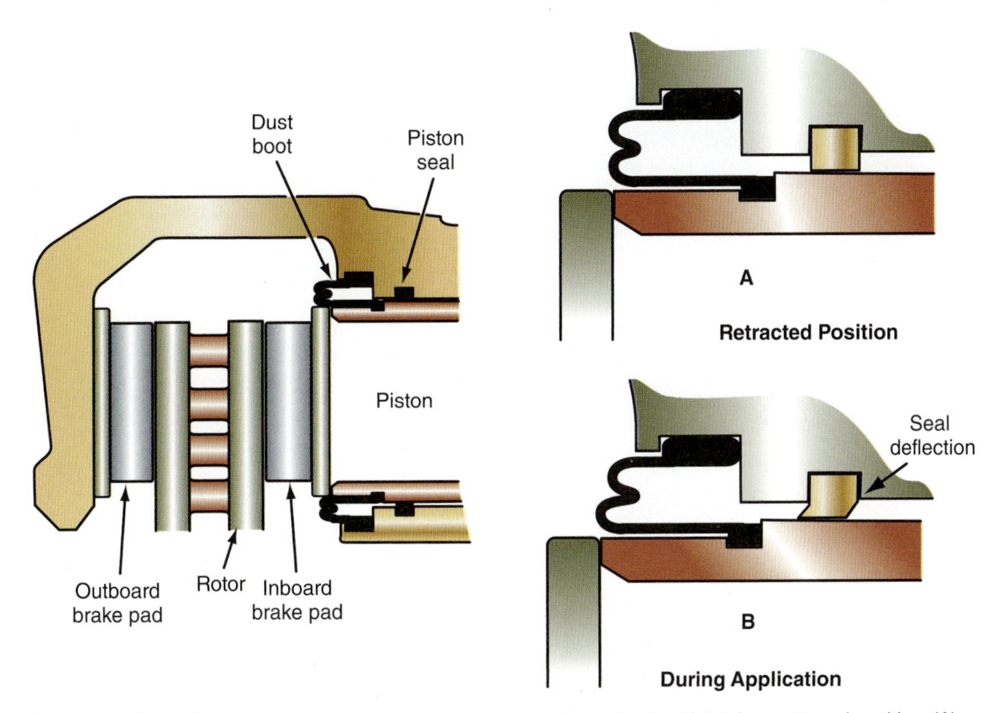

Figure 7-4 When the brakes are released, the caliper piston seal holds the piston in a retracted position. (A) When the brakes are applied, the seal warps (B) and then returns to the relaxed position retracting the piston and providing clearance between the pads and the rotor.

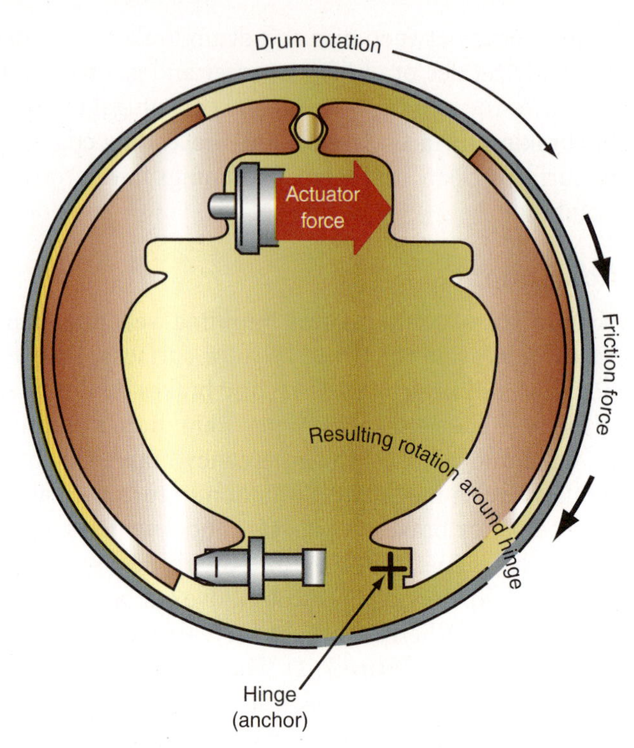

Figure 7-5 Drum rotation adds leverage to the brake shoe as it contacts the drum. This is called self-energizing action.

AUTHOR'S NOTE The self-energizing or servo action of drum brakes created enough braking force that many older vehicles with four-wheel drum brakes did not have power brakes. Disc brakes require more stopping pressure because they are not self-energizing and they have a smaller contact area.

As more and more vehicles became smaller and predominantly front wheel drive, the need for heavy braking action diminished. Most brakes became the leading-trailing type. Leading-trailing brakes servo action is limited to the leading shoe, since the anchor pin is located at the bottom of the brake assembly. The trailing shoe has little effect because the rotation of the brake drum is actually pushed back into the wheel cylinder (**Figure 7-6**).

A servo action is using one component to increase the input force so the second component of the assembly is applied with greater force.

Because disc brakes do not develop mechanical servo action, they less prone to severe grabbing or pulling. Slight pull may occur in a front disc brake installation because of a sticking caliper piston or problems in hoses or valves. Because equal hydraulic pressures are applied to both surfaces of the rotor during braking, no distortion of the rotor occurs regardless of the severity or duration of application. This is another reason that grabbing and pulling are reduced with disc brakes.

AUTHOR'S NOTE Many times the customer will define a slight pulling to one side as a drift; for example, "the vehicle tends to drift toward one side as I apply the brakes slowly." Such a slight pull or drift may be caused by a brake or alignment problem on either side of the vehicle.

Lack of servo action in disc brakes may eliminate some operational problems, but it also means that disc brakes require higher application force from the hydraulic system than drum brakes require. The greater size ratio between master cylinder pistons and

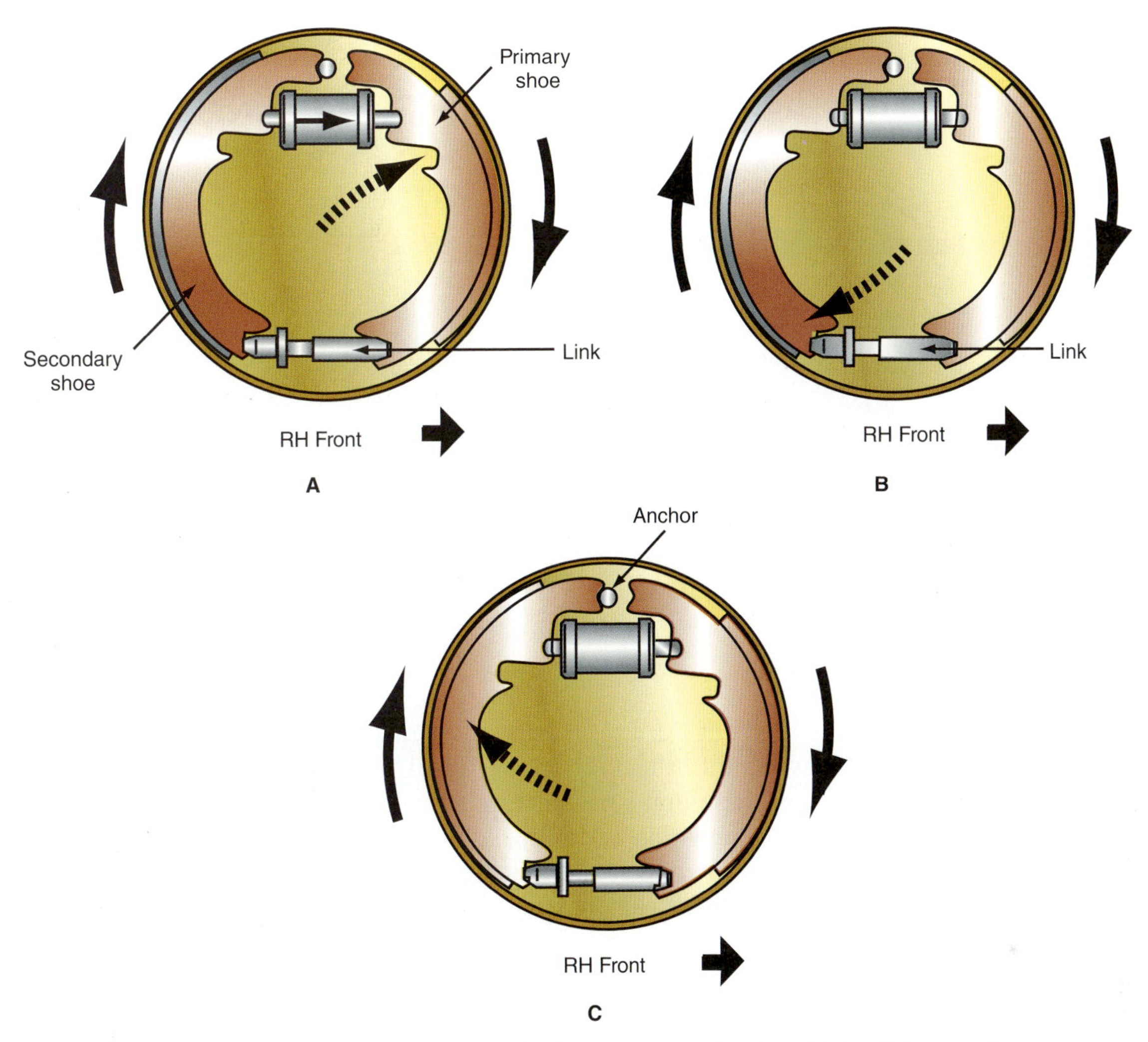

Figure 7-6 Self-energizing action of the primary shoe applies force to the secondary shoe. This is called servo action.

caliper pistons, along with universal use of power boosters with disc brakes, provides the necessary braking force without requiring undue pedal effort from the driver.

Disc Brake Noise

A common complaint about disc brakes is that they are noisy. The metallic brake pad used by many manufacturers is the loudest of all. Noise usually occurs when the brakes are applied and released and most commonly results from slight high-speed rattling of the pads. Modern disc brakes built for sale in North America use various devices and techniques to minimize noise. Anti-rattle springs and clips hold pads securely in calipers and help to dampen noise. Adhesive to hold pads to pistons and caliper mounting points further dampens squeaks and rattles. On some installations, metal or plastic shims between pads and calipers and pistons help to reduce vibration and noise.

Most disc brake silencers do not eliminate the squeaking noise. They cause the frequency to change so the human ear will not hear it.

Simple squeaks and rattles may be annoying to the driver, but they are generally not a problem for brake operation. When the noise becomes a continuous grinding or loud scraping sound as the brakes are applied, it often indicates that the linings have worn to the metal surfaces of the pads. These sounds mean that it is time for brake service before vehicle safety is jeopardized further.

Many disc brake pads have audible wear indicators, which are small steel pins or clips that rub on the rotor to create a constant squeal when the pads have worn to their minimum thickness. However, with the radio on and traffic noise the sound created by the

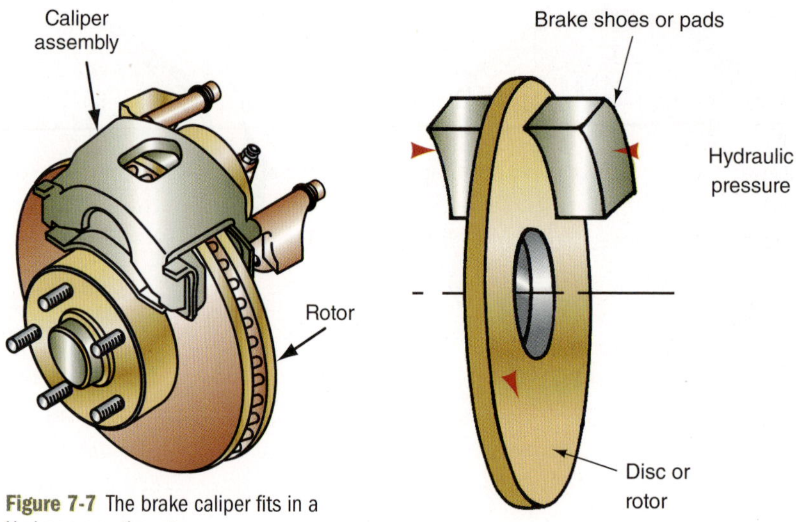

Figure 7-7 The brake caliper fits in a U-shape over the rotor.

Figure 7-8 Pads, lined with friction material, are forced against the rotating disc to stop the car.

wear indicators is masked out. As a result, the driver may not be aware of the problem until the rotor and possibly the caliper piston are ruined. In an effort to combat this problem, many vehicles have disc brake pads equipped with an electronic wear indicator (**Figure 7-7**). A ground wire within a warning light circuit is installed onto the inboard pad. When the pad wears down, the wire touches the rotor, grounding the circuit and lighting a warning light in the instrument panel (**Figure 7-8**).

Disc Brake Parking Brake Disadvantage

Drum brakes provide a better static coefficient of friction than do disc brakes. The brake linings grab and hold the drums more tightly than disc pads can hold a rotor. The servo action of drum brakes contributes to this feature, as does the larger area of brake shoe linings compared to disc brake pads.

Because most brake installations have discs at the front and drums at the rear, the parking brake weaknesses of disc brakes are not a problem. Four-wheel disc brake installations, however, must have some way to mechanically apply the rear brakes. With movable (floating or sliding) calipers, this is usually done with a cam-and-lever arrangement in the caliper that mechanically moves the piston to develop clamping force. Many late-model rear disc brakes with sliding calipers also have small drum-type parking brakes built into the rotors.

DISC BRAKE CONSTRUCTION

Shop Manual
page 307

The principal parts of a disc brake are a rotor, a hub, and a caliper assembly (Figure 7-7). The rotor provides the friction surface for stopping the wheel. The wheel is mounted to the rotor hub by wheel nuts and studs. The hub houses wheel bearings that allow the wheel to rotate. The rotor has a machined braking surface on each side.

The hydraulic and friction parts are housed in a caliper that straddles the outside diameter of the rotor. When the brakes are applied, pistons inside the caliper are forced outward by hydraulic pressure. The pressure of the pistons is exerted through the pads or shoes in a clamping action on the rotor. A splash shield on most installations helps to keep water and dirt away from the rotor and caliper and directs airflow to the rotor for improved cooling. Figure 7-8 illustrates basic disc brake operation.

Rotors, Hubs, and Bearings

The disc brake **rotor** has two main parts: the hub and the braking surface (**Figure 7-9**). The hub is where the wheel is mounted and contains the wheel bearings. The braking surface is the machined surface on both sides of the rotor. It is machined carefully to provide two parallel friction surfaces for the brake pads. The entire rotor is usually made of cast iron, which provides an excellent friction surface. The rotor side where the wheel is mounted is the outboard side. The other side, toward the center of the car, is the inboard side.

The rotor is the rotating part of a disc brake that is mounted on the wheel hub and contacted by the pads to develop friction to stop the car. It is also called a disc.

The size of the rotor braking surface is determined by the diameter of the rotor. Large cars, which require more braking energy, have rotors measuring 12 inches in diameter and larger. Smaller, lighter cars can use smaller rotors. Generally, manufacturers want to keep parts as small and light as possible while maintaining efficient braking ability.

The rotor is protected from road splash on the inboard side by a sheet metal splash shield that is bolted to the steering knuckle (**Figure 7-10**). The outboard side is shielded by the vehicle wheel. The splash shield and wheel also are important in directing air over the rotor to aid cooling.

A BIT OF HISTORY

The kind of disc brakes that we know today were originally called spot brakes because caliper action clamped a friction pad against a rotor. Braking action takes place at the "spot" of the brake pad. The earliest automotive disc brakes, however, actually used a pair of circular discs lined with friction material. These systems consisted of large finned housings mounted on the hubs or axles and in turn held the wheels and tires. Two lined pressure plates inside each housing were forced outward hydraulically during brake application. The plates—or discs—did not rotate but clamped against the inside of the rotating housing to provide braking force. A ramp-and-ball mechanism on each disc provided servo action to aid brake application. These early disc brake systems appeared on the subcompact Crosley and heavyweight Chrysler Imperial models of 1949. Although they provided improved braking action, they were complicated, costly, and prone to trouble. They also added too much unsprung weight to each wheel. The Crosley version lasted one model year. Chrysler discontinued its system in 1955. Ten years later, modern disc (spot) brakes were taking over the industry.

Figure 7-9 Major parts of a typical rotor and hub assembly.

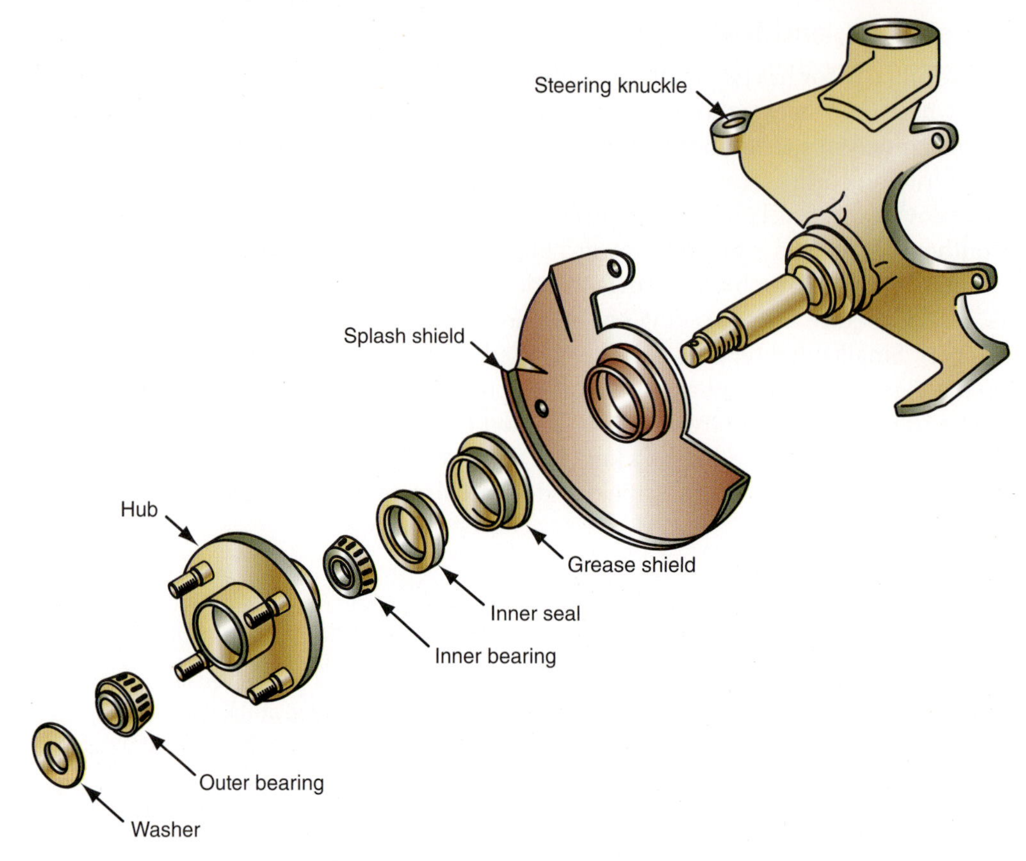

Figure 7-10 The splash shield is mounted on the steering knuckle inboard of the rotor.

A **fixed rotor** has the hub and rotor cast as a single part.

A **floating rotor** and its hub are two separate parts. They are also known as two-piece rotors.

Fixed and Floating Rotors

Rotors can be classified by the hub design as fixed (with an integral hub) or floating (with a separate hub). A **fixed rotor** has the hub and the rotor cast as a single part (**Figure 7-11**) and is commonly known as a one-piece rotor.

Floating rotors and their hubs are made as two separate parts (**Figure 7-12**) and may be known as a **two-piece rotor**. The hub is a conventional casting and is mounted on

Figure 7-11 The wheel hub and the rotor are cast as a single part in a fixed or integral rotor.

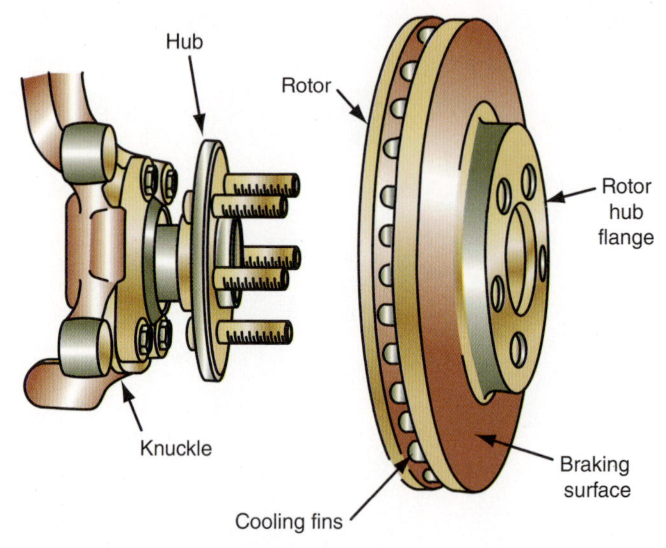

Figure 7-12 A rotor that is cast as a separate part and fastened to the hub is called a floating rotor.

wheel bearings or on the axle. The wheel studs are mounted in the hub and pass through the rotor center section. This kind of rotor is called a hubless or floating rotor. One advantage of this design is that the rotor is less expensive and can be replaced easily and economically when the braking surface is worn beyond machining limits.

Composite Rotors

Traditionally, brake rotors, as were drums, were manufactured as a single iron casting. The development of floating, two-piece rotors and the need to reduce vehicle weight led to the development of **composite rotors**. Composite rotors are made of different materials, usually cast iron and steel, to reduce weight. The friction surfaces and the hubs are cast iron, but supporting parts of the rotor are made of lighter steel stampings. The steel and iron sections are bonded to each other under heat and high pressure to form a one-piece finished assembly (**Figure 7-13**). Composite rotors may be fixed components with integral hubs, or they may be floating rotors mounted on a separate hub. Because the friction surfaces of composite rotors are cast iron, the wear standards and refinishing methods are generally the same as they are for other rotors.

Shop Manual
page 322

Composite rotors do require a special adaptor when machining on a bench lathe.

AUTHOR'S NOTE The latest material being utilized for high-end vehicles is for carbon ceramic rotors. A few manufacturers have utilized Formula One braking for their supercars such as the Mercedes-Benz SLR, Aston Martin DBS, and the Chevrolet Corvette ZR1. These brakes provide the advantages of reducing weight and increasing life of the rotor, but costs are still prohibitive.

A **solid rotor** is usually a floating rotor.

Solid and Ventilated Rotors

A rotor may be solid or it may be ventilated (**Figure 7-14**). A **solid rotor** is simply a solid piece of metal with a friction surface on each side. A solid rotor is light, simple, cheap, and easy to manufacture. Because solid rotors do not have the cooling capacity of ventilated rotors, they usually are used on small cars of moderate performance or on rear disc brake installations.

In general, solid rotors cannot be machined to the extent that ventilated rotors can be, if at all.

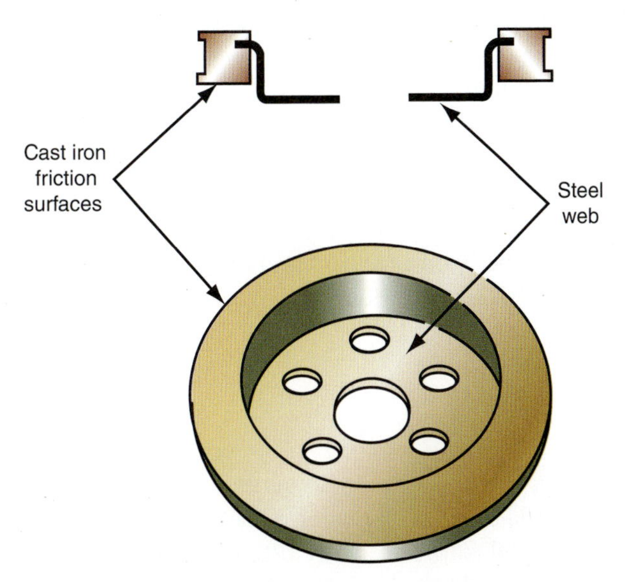

Figure 7-13 A composite rotor combines a steel web with a cast iron friction surface, which saves weight.

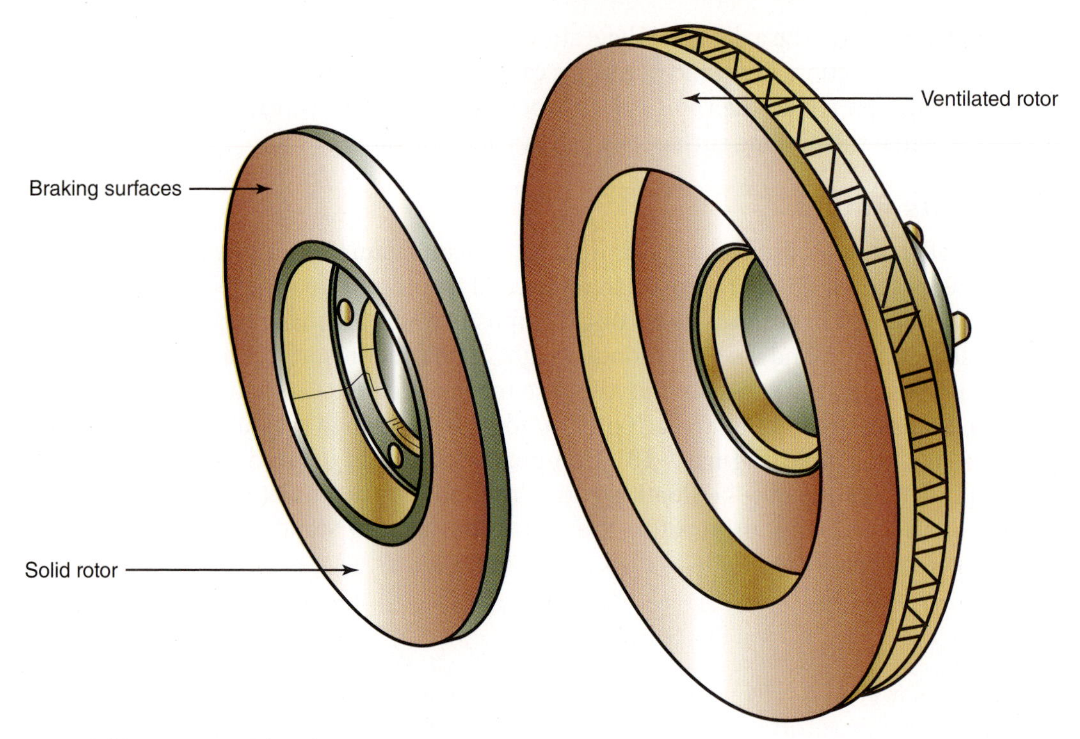

Figure 7-14 A solid rotor, left, and a large ventilated rotor, right.

A **ventilated rotor** has cooling fins cast between the rotor surfaces.

A **ventilated rotor** has cooling fins cast between the braking surfaces to increase the cooling area of the rotor. When the wheel is in motion, the rotation of these fins in the rotor also increases air circulation and brake cooling.

Although ventilated rotors are larger and heavier than solid rotors, these disadvantages are more than offset by their better cooling ability and heat dissipation. Some heavy ventilated rotors may have weights between the fins, or in an area near the fins, ground away to balance the rotor.

A **unidirectional rotor** can also be called a **directional rotor**, because the rotor is designed to turn in a certain direction.

Some ventilated rotors have cooling fins that are curved or formed at an angle to the hub center. These fins increase centrifugal force on rotor airflow and increase the air volume that removes heat. Such rotors are called **unidirectional rotors** because the fins work properly only when the rotor rotates in one direction. Therefore, unidirectional rotors cannot be interchanged from right to left on the car and the fins of the installed rotor must point forward when viewed from the top.

A few high-performance sports cars have solid rotors with holes drilled through the friction surfaces. These drilled rotors are not made to increase cooling so much as they are to release water and hot gases from the rotor surface that can cause water or gas fading. Drilled rotors are typically very light and can have very short service lives. Therefore, they are used mostly on race cars and dual-purpose, high-performance cars. Rotors that are "hash-cut" are actually less prone to stress cracking than rotors that have been drilled.

AUTHOR'S NOTE Drilled or slotted rotors have become increasingly common on passenger cars. While the gas and water release properties are still engineering correct, slotted and drilled rotors are more often selected more for the cost-effective "cool" look they add to a dressed-out car.

Rotor Hubs and Wheel Bearings

Tapered roller bearings, installed in the wheel hubs, are the most common bearings used on the front wheels of rear-wheel drive (RWD) vehicles and the rear wheels of front-wheel drive (FWD) cars. The tapered roller bearing has two main parts: the inner bearing cone and the outer bearing cup. A tapered roller bearing has the rollers set at an angle to the centerline of the bearing assembly. The rollers are held in place by the bearing cage mounted to the inner race. The race is the precision machined area where the rollers run. The bearing cup acts as the outer race and can usually be removed from the bearing assembly. The bearing fits into the outer cup or race, which is pressed into the hub. This provides two surfaces, an inner cone and outer cup, for the rollers to ride on. See Chapter 3 for details on tapered roller bearings and their service.

Almost every FWD vehicle has a sealed double-row ball bearing pressed into the front hubs (**Figure 7-15**). Some vehicles also use a sealed wheel bearing and hub that is serviced as an assembly. The bearing usually requires no periodic service and is replaced when necessary. The FWD axle stub slides through the inner bearing race. A typical sealed bearing may last several hundred thousand miles, and some will last the life of the vehicle. The rear wheels on front-wheel drive vehicles also have pressed-in, single-roll roller bearings, or bearings serviced with the hub as an assembly like the front wheels.

Shop Manual
page 327

Brake Pads

Each brake caliper contains two or four **brake pads**. Pads are the disc system braking friction surfaces. They perform the same function as the shoes in a drum brake system. The brake pads are positioned in the caliper on the inboard and outboard sides of the rotor. The caliper piston, or pistons, forces the brake pad linings against the rotor surfaces to stop the car.

Brake pads are the disc brake system friction surfaces.

Fundamentally, a brake pad is a steel plate with a friction material lining bonded or riveted to its surface (**Figure 7-16**). The simple appearance of a brake pad, however, hides some sophisticated engineering that goes into its design and manufacturing. Chapter 2 of this *Classroom Manual* introduced the subject of friction materials and summarized the common requirements of both drum brake linings and disc brake pads. Friction materials

Figure 7-15 Most FWD vehicles have a pressed-in, double-roll sealed ball bearing.

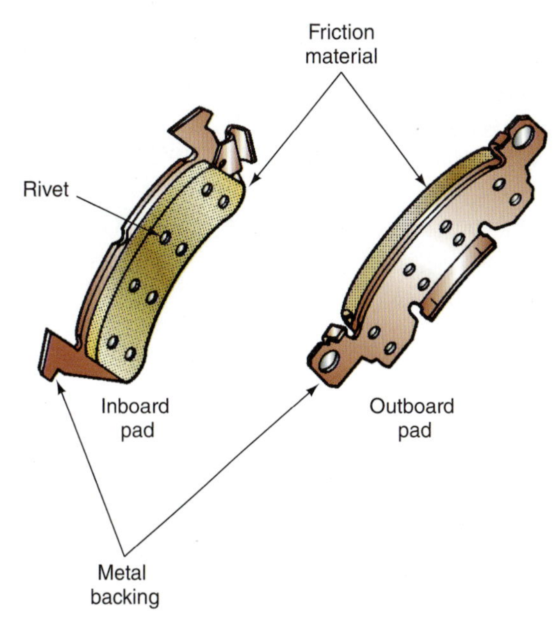

Figure 7-16 Parts of a disc brake pad.

are classified as organic (nonmetallic), semi-metallic, fully metallic, and—in some advanced systems—synthetic.

Brake Friction Materials (Drum and Disc)

The friction material used for disc brake pads is generally harder than that used on drum brake linings because the friction surface is smaller and higher pressures are used to push the pads into contact with the rotor. The friction material used for drum brake lining is generally softer than that used on disc brake pads because the friction surface is larger.

Asbestos has excellent friction qualities and long service life. Therefore, it was the most common brake lining material for years. The health hazards of asbestos have led to its drastic reduction or removal from most brake friction materials. Today's basic types of disc pad lining materials are organic, semi-metallic, metallic, synthetic, and ceramic.

Organic linings are made from non-metallic fibers bonded together to form a composite material.

Organic Linings. Organic linings usually wear faster than do semi-metallic linings, but they have the benefit of breaking in faster. **Organic linings** are made from nonmetallic fibers bonded together to form a composite material. For decades, asbestos was the main ingredient in organic linings. Asbestos is actually an inorganic mineral but not truly a metal. Regardless of its chemistry, the organic name stuck to friction material based on asbestos.

Health regulations have caused asbestos content in organic linings to drop from as high as 75 percent to below 25 percent. The goal is to eliminate asbestos from all brake linings, and it has been achieved for friction materials made in North America. The problem is imported brake linings can contain asbestos material. Another factor is that some of the compounds used to replace asbestos may be as dangerous to health as asbestos itself.

Today's organic brake linings contain the following kinds of materials:

Friction materials help provide the frictional stopping power, such as graphite, powdered metal, and nut shells.

Curing agent refers to a class of materials used in brake linings to accelerate the chemical reaction of the binders and other materials.

- Friction materials and **friction modifiers** are the materials that help provide the frictional stopping power. Some examples are graphite, powdered metals, and even nut shells.
- **Fillers** are secondary materials added for noise reduction, heat transfer, and other purposes.
- **Binders** are glues that hold the other materials together.
- **Curing agents** accelerate the chemical reaction of the binders and other materials.

Organic linings have a high coefficient of friction for normal braking; they are economical; they are quiet; they wear slowly; and they are only mildly abrasive to rotors. Organic linings fade more quickly than other materials, however, and they do not operate well at high temperatures. High-temperature organic linings are available for high-performance cars, but they do not work as well at low temperatures and they wear faster than common organic linings.

Semi-metallic friction material can seriously damage or destroy some types of rotor.

Semi-metallic linings contain about 50% iron and steel, are more fade resistant than organic materials but require a higher braking effort.

Semi-metallic linings. Semi-metallic materials are made from a mixture of organic or synthetic fibers and certain metals molded together; they do not contain asbestos. **Semi-metallic linings** are harder and more fade resistant than organic materials but require higher brake pedal effort.

Most semi-metallic linings contain about 50 percent iron and steel fibers. Copper also has been used in some semi-metallic linings and, in smaller amounts, in organic linings. Concerns about copper contamination of the nation's water systems has led to its reduced use in brake linings, however.

AUTHOR'S NOTE A few vehicles, and not necessarily high-end ones, will work quietly only with the manufacturer's brand of brake pads. Aftermarket pads, particularly semi-metallic pads, make a lot of noise even when first installed. This is usually caused by the composition of the rotor's friction area, the friction area finish, and the makeup of the pads themselves. At times the only possible solution to the noise problem is to buy pads from the local dealership. Luckily this problem applies to only a few models of a few brands.

Semi-metallic linings operate best above 200°F to 250°F (90°C to 120°C) and actually must be warmed up to bring them to full operating efficiency. Consequently, semi-metallic linings often have poorer operating characteristics than organic linings at low temperatures.

Semi-metallic linings often are blamed for increased rotor wear, but this is not entirely true. Early semi-metallic linings were more abrasive than current materials, which may cause no more wear with the properly matched rotors than organic materials. Also, the better heat transfer characteristics of semi-metallic linings can reduce rotor temperatures and help to counteract abrasiveness. Many small, FWD cars built since the early 1980s have smaller front brakes that require the better high-temperature friction characteristics and heat transfer abilities of semi-metallic linings.

Currently, semi-metallic linings are used only on front disc brakes of passenger cars and light trucks. The lighter braking loads on rear brakes, particularly on FWD cars, may never heat semi-metallic linings to their required operating efficiency. Semi-metallic linings also have a lower static coefficient of friction than organic linings, which makes them inferior for parking brake use.

Metallic Linings. Fully metallic materials were used for many years in racing, particularly in the heyday of drum brakes. **Metallic lining** is made from powdered metal that is formed into blocks by heat and pressure. These materials provide excellent resistance to brake fade but require high brake pedal pressure and create the most wear on rotors and drums. Metallic linings work very poorly until they are fully warmed. On the other hand, metallic linings tend to heat quickly and develop heat fading more quickly. Improved high-temperature organic linings and semi-metallic materials have made metallic linings almost obsolete for late-model automotive use. Metallic linings are extremely noisy, which must be considered when talking about choices to a customer.

Metallic linings are made from powered metal, and are very hard on brake rotors.

Synthetic Linings. The goals of improved braking performance and elimination of the disadvantages of current lining materials has led to a new generation of **synthetic lining** friction materials. They are called synthetic because that term generally describes nonorganic, nonmetallic, and nonasbestos materials. Two principal kinds of synthetic materials show promise as brake linings: fiberglass and **aramid fibers**.

Synthetic lining may be used on disc or drum brakes.

Aramid fibers are a family of synthetic materials that are five times stronger than steel but are very light weight.

Fiberglass was introduced as a brake lining material to help eliminate asbestos. As does asbestos, it has good heat resistance, good coefficient of friction, and excellent structural strength. The disadvantages of fiberglass are its higher cost and its reduced friction at very high temperatures. Overall, fiberglass linings perform similarly to organic linings, but the higher costs have confined their use primarily to rear drum brake linings.

Aramid fibers are a family of synthetic materials that are five times stronger than steel, pound for pound, but weigh little more than half what an equal volume of fiberglass weighs. Friction materials made with aramid fibers are manufactured similarly to organic and fiberglass linings. Aramid fibers have a coefficient of friction similar to semi-metallic linings when cold and close to that of organic linings when hot. Overall, the performance of aramid linings is somewhere between organic and semi-metallic materials but with much better wear resistance and longevity than organic materials.

Ceramic Brake Pads. Many automotive manufacturers are now supplying brake pads of **ceramic** materials as the pad of choice. Akebono is supplying many vehicle manufacturers with their OEM **ceramic brake pads**. In addition, some aftermarket vendors such as Raybestos Brakes' "Quiet Stop" and NAPA's "Ceramix" are two brands of ceramic pads used to replace semi-metallic types. Combining ceramic material and copper fibers is the typical formula for ceramic pad manufacture. Many of the drawbacks of semi-metallic pads have been reduced or eliminated entirely.

Ceramic brake pads are made from ceramic and copper and make less noise and do not damage rotors as badly as semi-metallic pads do.

Ceramic brake pads are a combination of ceramic material and copper or some other metal fibers.

The steel in semi-metallic pads tends to make noise, wear rotors, and create lots of brake dust, which affects the appearance of the wheel and rims. Ceramic and copper composite pads make much less noise, are not as damaging to the rotors, and create virtually no dust. These benefits are much more satisfactory to the owner/driver. Ceramic pads are usually slotted vertically, horizontally, or diagonally to change the noise frequency to a range beyond human hearing. The slots also provide an escape route for gases and dust, thereby reducing brake fades at higher temperatures. Some ceramic pads have chamfered leading and trailing edges, which also help to reduce noise. Porsche offers ceramic pads and rotors on some of its top-of-the-line vehicles.

Regardless of wear pattern, all of the pads on an axle should be replaced at the same time to prevent uneven braking and wear.

Ceramic brake pads do have a few disadvantages over semi-metallic. The main disadvantage is primarily the result of improper pad/rotor configuration. Some operators may wish to add "high-performance" ceramic pads in lieu of the specified semi-metallic pads. Many times, this configuration would warp or wear the rotor(s) very quickly, even during normal driving and braking. The high-performance ceramic pad does not dissipate heat as quickly as semi-metallic pads and the heat is trapped in the rotor, caliper, and brake fluid. As mentioned earlier, there are aftermarket ceramic pads for many cars, but they are made to directly replace those specified by the vehicle manufacturer. The technician or service writer needs to caution the customer on the use of ceramic brake pads. Many times the customer only wishes to keep the wheels and rims looking neat.

Another item the technician and customer must consider is the cost of ceramic pads. Typically, a set of ceramic brake pads cost three to four times the cost of standard semi-metallic or organic pads. The labor cost should be the same. In most cases, ceramic brake pads are overkill braking for the typical driver and vehicle. However, the appearance of the vehicle may be of more concern to the customer than the price of the parts.

Friction Material Selection

The friction material of a new brake pad can be identified from a code printed on the edge of the pad or shoe (**Figure 7-17**). This code is called the **Automotive Friction Material Edge Code**. Letters and numbers in the code identify the manufacturer, the material, and, of most importance, the cold and hot coefficients of friction. In the example shown in Figure 7-17, the first letters "NRSS" identify the manufacturer. The numbers "12041" identify the lining material, and the last two letters "FF" identify the cold and hot coefficients of friction, respectively.

From a service standpoint, the friction codes are probably the most important. The technician may be able to compare them to a decoding chart in brake parts catalogs to determine the friction material and whether it is recommended for a specific vehicle. The following chart gives definitions of different DOT Automotive Friction Material Edge Codes (**Table 7-1**). Note that EE coded pads have roughly the same coefficient of friction of steel on steel or bare caliper piston against steel rotor. A pad with the edge HH would probably not work well on a conventional vehicle when noise, brake wear, and cost are considered. The two different temperatures given are considered as cold and hot.

The codes on some inexpensive brake pads often cannot be read because of the printing and paint used.

It is important to use the recommended friction material when replacing brake pads or drum brake shoes. The incorrect type of friction material can affect the stopping characteristics of the car. These codes, however, indicate only the coefficient of friction. They do not address lining quality or its hardness.

Hard and *soft* are terms applied to linings within a general category of material. Thus, any particular organic lining may be considered as a hard or a soft organic material. Overall, organic linings have been considered softer than semi-metallic linings, and

Figure 7-17 Automotive Friction Material Edge Code.

Table 7-1 AUTOMOTIVE FRICTION MATERIAL EDGE CODE

DOT Edge Code	Coefficient of Friction (C.F.) @ 250°F and @ 600°F	Fade Probability
EE	0.25 to 0.35 both temps	0% to 25% @ 600°F
FE	0.25 to 0.35 @ 250°F temp 0.35 to 0.45 @ 600°F temp	2% to 44% fade at 600°F
FF	0.35 to 0.45 @ both temps	0% to 22% fade at 600°F
GG	2.45 to 0.55	Very rare
HH	0.55 to 0.65	Carbon/carbon only. Glow at about 3000°F

semi-metallic linings are considered softer than fully metallic linings. Typically, a hard lining has a low coefficient of friction but resists fade better and lasts longer than a soft lining. A soft lining has a higher coefficient of friction but fades sooner and wears faster than a hard lining. Additionally, a soft lining is less abrasive on rotor and drum surfaces and operates more quietly than a hard lining.

It also is important to know that carmakers sometimes specify different friction materials for the inboard and outboard disc pads. Original equipment installations may have an organic disc pad on one side of the caliper and a semi-metallic disc pad on the other. The reason for using these different coefficients of friction can be in the difference of cooling from the inboard to outboard pad, noise reduction, or simply the results of testing for a particular rotor and pad operation. It also is common to use linings with a lower coefficient of friction on the rear brakes than on the front brakes to minimize rear brake lockup. Always follow the manufacturer's recommendation for brake lining friction to maintain the original design performance of the braking system.

A BIT OF HISTORY

In the 1950s, it was not uncommon for the technician or DIYer to install the lining onto the old brake shoe web. You could buy the asbestos linings and the rivets at most parts stores. The old lining would be knocked off and the rivets removed with a hammer and chisel. The new lining was then riveted to the web generally using only a hammer and a punch. The ends of the lining would be chamfered a little (using a hand file) to prevent the lining from grabbing the drum. The result was a satisfactory brake shoe for then, but now the result would be much different because of the average driving speeds and the almost universal use of power brakes. Now the lining would probably be ripped from the web at the first near-panic stop.

Although the coefficient of friction and hardness of linings can vary quite a bit for the same kind of material, these general rules for the coefficient of friction will assist in comparing different materials:

Organic—cold: 0.44; warm: 0.48
Semi-metallic—cold: 0.38; warm: 0.40
Metallic—cold: 0.25; warm: 0.35
Synthetic—cold: 0.38; warm: 0.45

When troubleshooting a brake performance problem, a noise problem, or a problem with premature rotor wear, it often is a good idea to install new brake pads that match the original equipment exactly. This establishes a baseline of what should be original brake performance for further diagnosis.

Friction Material Attachment

Both disc brake and drum brake linings are attached by rivets or adhesive bonding or a combination of riveting and bonding to the metal backing (**Figure 7-18**).

Riveted linings are attached to the metal backing plate with rivets.

Riveted Linings. In a riveted brake pad or shoe assembly, the **riveted lining** is attached to the steel pad or shoe by copper or aluminum rivets. Riveting allows a small amount of flexing between the pad and lining to absorb vibration and reduce noise. Riveting also is very reliable, and rivets maintain a secure attachment at high temperature and high mileage.

Rivets require that about one-third to one-quarter of the lining thickness remain below the rivet for secure attachment. This places the rivet head closer to the lining friction surface and reduces the service life or mileage of the pad. In some cases, the holes above the rivet heads can trap abrasive particles that can score the rotor.

Historically, riveting was the most common method for lining attachment. Since bonding was perfected, however, rivets are used more with semi-metallic linings than with organic linings.

Bonded Linings. Bonding is a method of attaching the friction material to the pad or shoe with high-strength, high-temperature adhesive.

Bonded brake linings are attached to the steel backing with adhesive.

Bonded linings can provide longer service life because more material is available for lining wear before the steel pad or shoe contacts the rotor or drum. This feature can be misleading and provides false comfort to many motorists, however. If a driver neglects lining wear until a disc pad hits the rotor, it often severely scores the rotor and destroys it. In contrast, minor scoring from rivet heads can often be resurfaced.

Bonded linings do not have the flexibility between the pad and shoe lining that riveted linings have. Therefore, they can be noisier than riveted linings. A noise complaint often can be fixed by replacing bonded linings with riveted linings that have the recommended friction characteristics.

Mold bonded linings are molded to the steel backing plate that has had adhesive applied,

Mold-Bonded Linings. In a disc brake pad assembly with **mold-bonded lining**, an adhesive is applied to the pad, and the uncured lining material is poured onto the pad in a mold. The assembly is then cured at high temperature to fuse the lining and adhesive to the pad. Holes drilled in the pad are countersunk from the rear so that the lining material flows through the holes and rivets itself to the pad as it cures. Many high-performance pads are made this way to avoid the stress-cracking problems associated with rivets while providing very secure pad attachment. Some pad manufacturers mold bond the friction material to the pad by using small hooks machined into the pad backing plate.

Figure 7-18 The linings can be bonded (left) or riveted (right) to the disc brake pads.

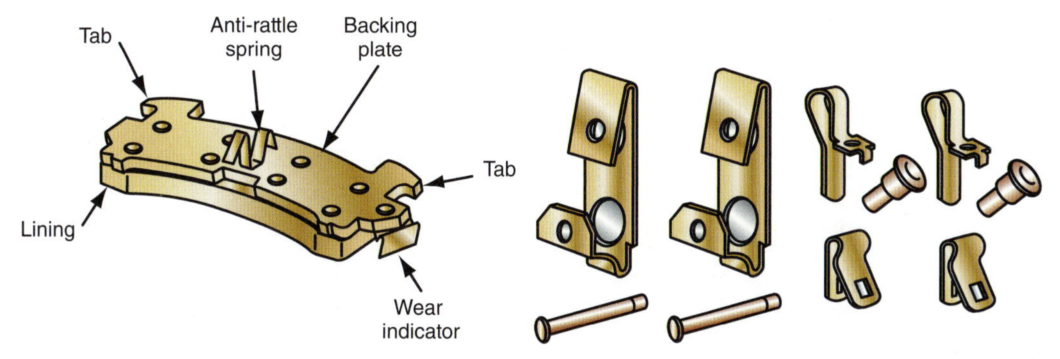

Figure 7-19 Tabs on the metal backing plate locate the pad in the caliper.

Figure 7-20 Typical anti-rattle and retaining hardware for brake pads.

Pad-to-Caliper Attachment

Most brake pads are held in the caliper by locating tabs formed on the end of the metal backing plate (**Figure 7-19**). The pads also may be retained by retaining pins that go through holes in the metal pad backing.

Many pads have anti-rattle or support clips, which are spring steel clips that hold the pads in position to keep them from rattling when the pads are out of contact with the rotor. These small parts are called **pad hardware** and should be replaced when the pads are replaced. **Figure 7-20** shows a selection of typical pad hardware.

Pad hardware refers to miscellaneous small parts, such as anti-rattle clips and support clips, which hold brake pads in place and keep them from rattling.

AUTHOR'S NOTE Test driving a brake repair and a clicking noise is heard each time the brakes are applied and released. Check the installation of the pad hardware. Chapter 7 of the *Shop Manual* gives the details.

Brake Pad Wear Indicators

Many current brake systems have **pad wear indicators** or sensors to warn the driver that the lining material has worn to its minimum thickness and that the pads require replacement. The most common types are audible contact sensors and electronic sensors.

Wear sensors warn the driver that the brake pads will need attention soon.

The audible sensor is the oldest type. The audible system uses small spring clips on the brake pads and lining that make a noise when the linings are worn enough to be replaced. The spring clips are attached to the edge of the brake shoe or into the shoe from the pad side. They are shaped or positioned to contact the rotor when the lining wears to where it should be replaced. When the linings wear far enough, the sensor contacts the rotor (**Figure 7-21**) and makes a high-pitched squeal to warn the driver that the system needs servicing. This squeal usually goes away when the brakes are applied. Unfortunately, many drivers are deaf to the sound of these wear indicators when the radio is on or there is a lot of traffic noise or they just ignore noise.

AUTHOR'S NOTE In defense of some drivers, we experienced an incident where the brake pads were completely gone, the pad metal backing was rubbing on the rotors, and on one side the backing was gone to the point where the caliper piston was almost out of the bore. I questioned the owner, a fellow faculty member, as to why the car was not brought in when it started to make a lot of noise during braking. The owner stated that there was never any noise, and that the car just got to the point where it did not want to stop. As soon as the owner left and before we put the car on the lift, I drove it around the parking lot several times and applied the brakes both in easy and in panic modes just to prove a point. But the owner was right. The car did not stop well, and there was absolutely no noise that would be associated with extremely worn pads. So sometimes when the owners say there was no noise, they may be correct.

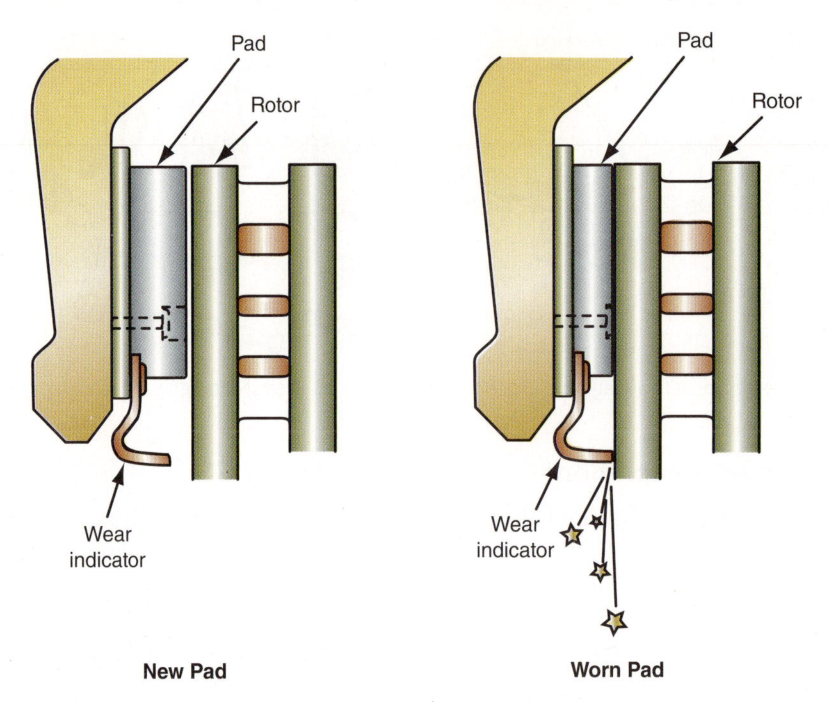

Figure 7-21 As the linings wear, the wear indicator eventually hits the rotor and creates noise to warn the driver that new pads are needed.

Electronic sensors provide a warning lamp or message on the instrument panel to inform the driver of brake pad wear or circuit problems. The electronic sensor uses pellets embedded into the friction material for circuit completion. Early systems used a grounding logic to complete a parallel electrical circuit and turn on the warning lamp (**Figure 7-22A**). If a circuit wire became grounded, this too would turn on the warning lamp. An open circuit, however, could not be detected and may disable the detection circuit. Later systems were designed to use an open logic to do the same thing. Open logic allows the system to detect both opens and grounds in the circuit (**Figure 7-22B**).

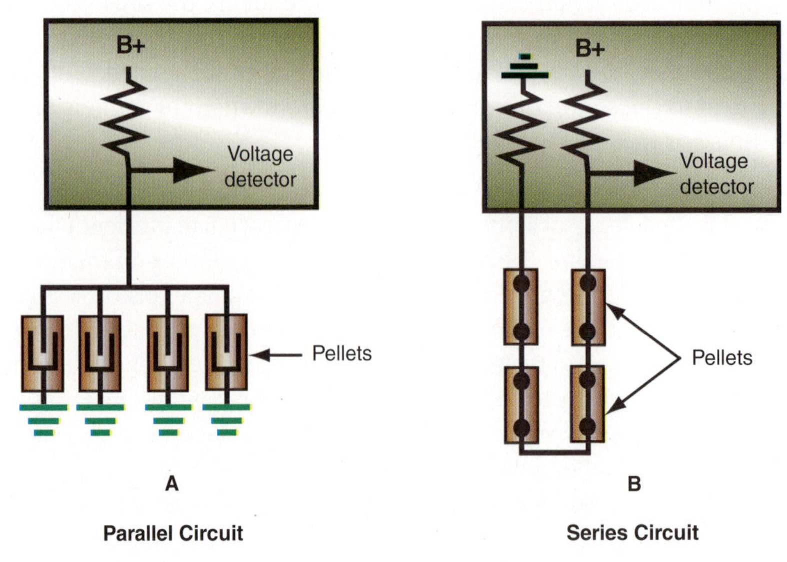

Figure 7-22 Electronic wear sensors are used in both parallel (A) and series (B) circuits.

A BIT OF HISTORY

Automotive disc brakes were developed for racing, and the first practical disc brake system was used in 1953 on a C-type Jaguar that won the 24-hour endurance race at Le Mans. The effectiveness of the brakes in this event led automotive engineers to develop disc brakes for passenger cars by the 1960s and 1970s.

Figure 7-23 Note the depth of the slot on the old pad (left) compared to the one on the new pad (right).

A wear indicator that is commonly overlooked is the narrow slot running across the pad (**Figure 7-23**). It may not have been the intention of the design engineers, but that little slot provides a quick visual indication of brake lining remaining. In most cases, it can be observed without removing the tire-and-wheel assembly. The more shallow the groove depth, the less lining remaining.

CALIPER CONSTRUCTION AND OPERATION

The disc **brake caliper** converts hydraulic pressure from the master cylinder to mechanical force that pushes the brake pads against the rotor. The caliper is mounted over the rotor (**Figure 7-24**). Although there are many design differences among calipers (described in a later section of this chapter), all calipers contain these major parts:

- Caliper body or housing
- Internal hydraulic passages
- One or more pistons
- Piston seals
- Dust boots
- Bleeder screw

The **brake caliper** converts hydraulic pressure to mechanical force that applies the brake pads.

Shop Manual page 288

Figure 7-24 The caliper is bolted to the caliper support, which is bolted to the steering knuckle.

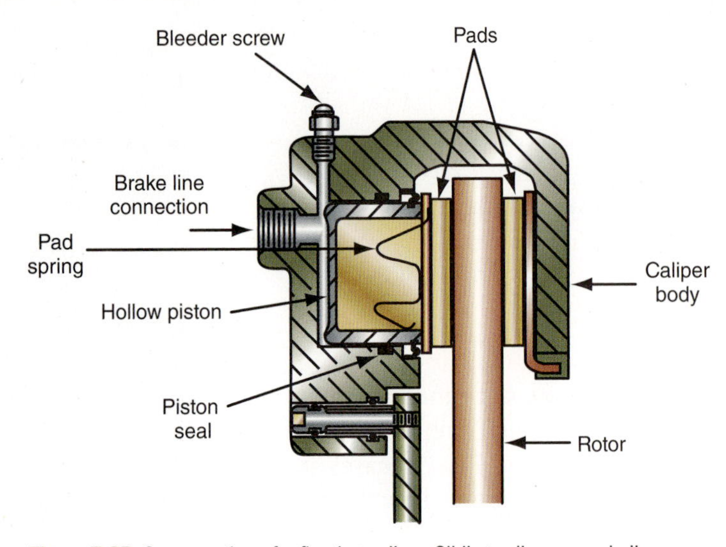

Figure 7-25 Cross section of a floating caliper. Sliding calipers are similar.

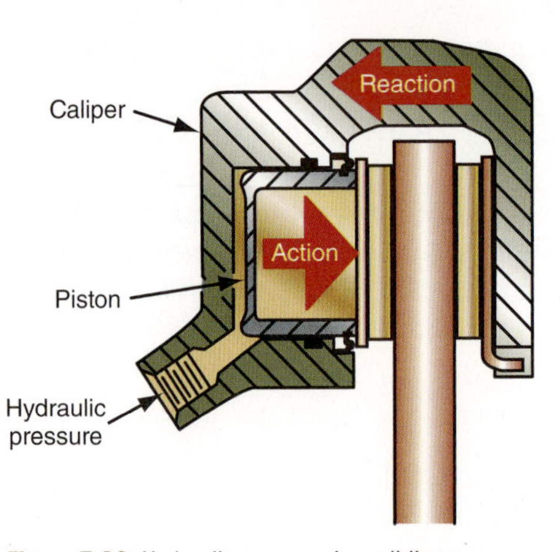

Figure 7-26 Hydraulic pressure in a sliding or floating caliper forces the piston and one pad in one direction and the caliper body and the other pad in the opposite direction. This is an application of one of the Laws of Motion: For every action there is an equal and opposite reaction.

Figure 7-25 is a sectional view of a caliper. A brake line from the master cylinder is attached to the caliper body. When the brakes are applied, fluid under pressure from the master cylinder enters the caliper. The caliper has at least one large hydraulic piston located in a piston bore. During braking, fluid pressure behind the piston increases. Pressure is exerted equally against the bottom of the piston and the bottom of the cylinder bore. The pressure applied to the piston is transmitted to the inboard brake pad to force the lining against the inboard rotor surface. Depending on caliper design, fluid pressure may be routed to matching pistons on the outboard side of the caliper, or the pressure applied to the bottom of the cylinder can force the caliper to move inboard on its mount (**Figure 7-26**). In either case, the caliper applies mechanical force equally to the pads on both sides of the rotor to stop the car.

Caliper Body

The caliper body is a U-shaped casting that wraps around both sides of the rotor (**Figure 7-27**). Almost all caliper bodies are made of cast iron. Single-piston caliper bodies are usually cast in one piece, but caliper bodies with pistons on the inboard and outboard sides of the rotor are usually cast in two pieces and bolted together.

A **steering knuckle** is the outboard part of the front suspension that pivots on the ball joints and lets the wheels turn for steering control.

Front calipers are mounted on **caliper supports** or adapters that may be an integral part of the **steering knuckles**. The steering knuckle provides a spindle for the wheel to rotate on and provides connections to the vehicle's suspension steering systems. Although

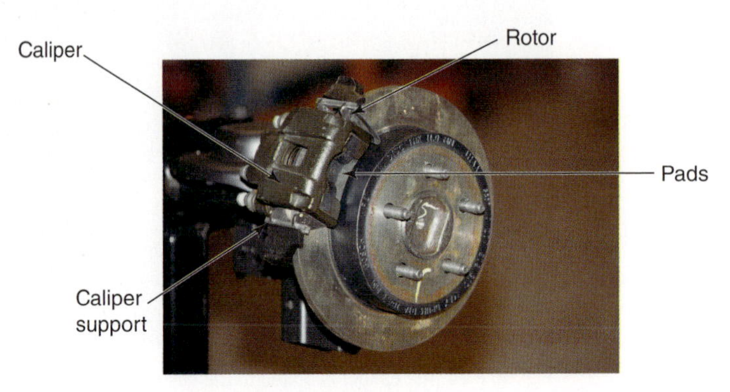

Figure 7-27 The brake caliper straddles the rotor.

one-piece steering knuckle and caliper support forgings are the simplest and most economical way to manufacture these components, most front caliper supports are separate pieces that bolt to the steering knuckle. Rear calipers are mounted on caliper supports that bolt to the rear suspension or the axle housing.

Most caliper bodies have one or two large openings in the top of the casting through which the lining thickness can be inspected. Although these openings are handy inspection points, that is not their primary purpose. Caliper bodies are cast with these openings to reduce weight and, of more importance, to minimize large masses of iron that can cause uneven thermal expansion of the assembly. A few caliper bodies are without these openings because they are not needed structurally.

Caliper Hydraulic Passages and Lines

The fitting for the brake line is located on the inboard side of the caliper body, as is the bleeder screw. The bleeder screw is almost always at the top of the caliper housing. A movable caliper with one or two pistons on the inboard side has hydraulic passages cast into the inboard side of the body. A fixed caliper with pistons on both the inboard and the outboard sides requires crossover hydraulic lines to the outboard side. These usually are cast into the caliper halves and sealed with O-rings when the caliper body is bolted together (**Figure 7-28**). Some fixed calipers, however, have external steel passages to carry fluid to the outboard pistons.

Caliper Pistons

Depending on its design, a caliper may have one, two, or four pistons, with single-piston calipers being the most common by far. The piston operates in a bore that is cast and machined into the caliper body (**Figure 7-29**). The piston is the part that actually converts hydraulic pressure to mechanical force on the brake pads. To do its job, the piston must be strong enough to operate with several thousand pounds of pressure, and it must resist corrosion and high temperatures.

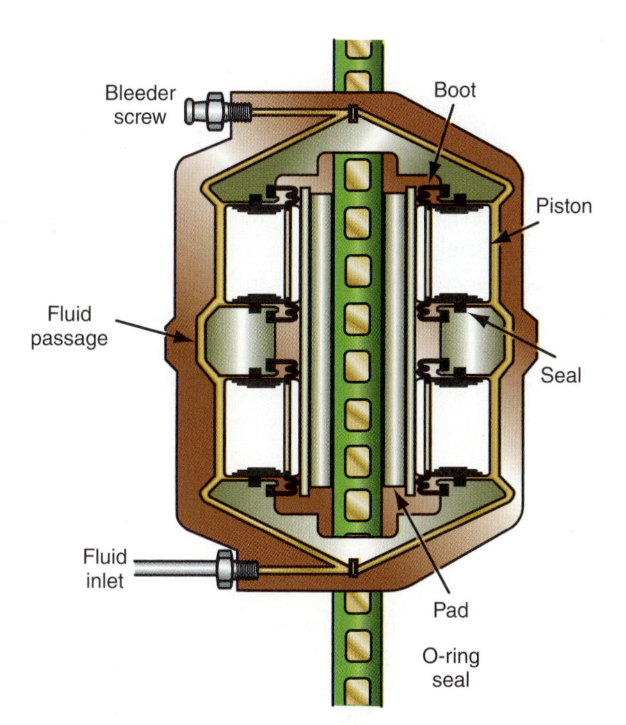

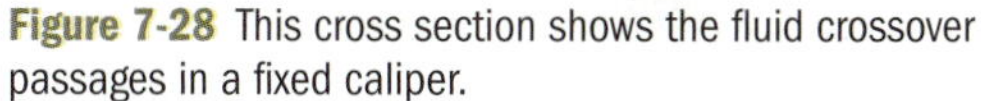
Figure 7-28 This cross section shows the fluid crossover passages in a fixed caliper.

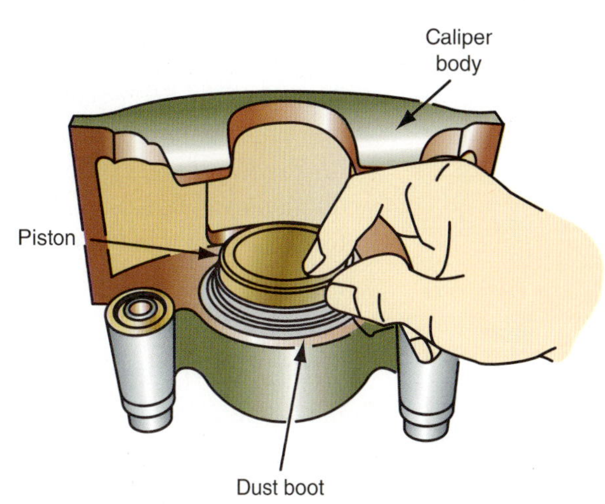

Figure 7-29 The piston operates in a machined bore in the caliper body.

The inner side of the piston contacts the brake fluid in the caliper bore, and the outer side bears against the steel brake pad. The brake pads operate at temperatures above the boiling point of brake fluid, and the piston must help to insulate the fluid from this extreme heat. The piston surface that contacts the pad is hollow or cup shaped, which reduces weight and reduces the area available to absorb heat from the pad. The ability to absorb heat and not transfer it to the fluid also is an important design consideration for the piston materials. Caliper pistons typically are made of steel, cast iron, **phenolic plastic**, and aluminum.

Phenolic plastic is plastic made primarily from phenol, a compound derived from benzene; phenol is also called carbolic acid.

Chrome-plated cast-iron and steel pistons are the most common. Iron pistons were used in many early disc brakes, but steel is more common in late-model designs. Steel pistons are strong and thermally stable, but they are heavier than desired in late-model brake assemblies, and they can conduct excessive heat to the brake fluid.

To reduce both weight and heat transfer, manufacturers turned to phenolic plastic pistons in the mid-1970s. Today, phenolic plastic pistons are original equipment on about half the light-duty vehicles sold in North America. Phenolic plastic pistons are strong, light, excellent insulators, immune from corrosion, and economical to make. In addition, the plastic surface is not as slippery as chrome plating and grips the piston seals better than a steel piston does.

Phenolic pistons do have some disadvantages; among them is a tendency to wear faster and score more easily than steel pistons. Early designs also tended to stick in their bores because of caliper bore varnish and corrosion. Improved dust boots and seals have reduced that problem, however. Phenolic plastic pistons also may be damaged if mishandled.

Aluminum would seem to be a good material for caliper pistons because of its light weight. In fact, aluminum has been used for pistons in some high-performance brakes where weight is critical, but it has serious drawbacks that limit its use in passenger car and light truck brakes.

Aluminum expands faster that iron or steel, and aluminum pistons must be made with more clearance in their caliper bores. This additional clearance, in turn, increases the possibility of leakage, as does aluminum's greater tendency toward scoring and corrosion. The major problem with aluminum caliper pistons, however, is their ability to transfer heat. Aluminum is a very poor thermal insulator and increases the danger of boiling the brake fluid.

AUTHOR'S NOTE Some racing calipers are equipped with as many as 12 pistons made of stainless steel or titanium. Some specialty calipers are nickel-coated forged steel. The large number of pistons can help even out and increase the force applied to the brake pad. Several smaller pistons can have more surface area than one large piston. As noted earlier, these brake systems can also use carbon ceramic brake rotors. The expense of some racing equipment can be prohibitive on a production car, but many of today's technology, just like disc brakes themselves, came from the racing industry.

Fixed-seals do not move.

A **lathe-cut seal** or **square-cut piston seal** is a fixed seal for a caliper piston that has a square or irregular cross section; it is not round like an O-ring.

Caliper Piston Seals

Brake calipers require seals to keep fluid from escaping between the pistons and their bores, but caliper piston seals also perform other functions. Disc brakes do not have return springs to move the pads out of contact with the rotor. This is accomplished by the caliper piston seal. Movable calipers use **fixed seals**, also called **square-cut piston seals** or **lathe-cut seals**.

Fixed Seals. A fixed seal, used with a movable caliper, is installed in a groove in the inner circumference of the caliper bore. The piston fits through the inside of the seal and is free to move in the seal. The outer circumference of the seal remains in a fixed position in the caliper bore. Many seals are square or rectangular in cross section, but others have different cross-sectional shapes. Because all seals are not identical, it is important to be sure that replacement seals match the shape of the originals.

During braking, the piston seal is deflected or bent by the hydraulic pressure. When the pressure is released, the seals relax or retract, pulling the pistons back from the rotors. **Figure 7-30** is a cross section of a piston and seal that shows the seal action from relaxed (before application) to an applied position and back to a relaxed position. The seal flexing releases pressure from the rotor but maintains only very slight clearance of 0.006 inch or less between the pad and the rotor. The seal fits closely around the piston and holds it in position. Because the seal is installed at the outer end of the caliper bore, it keeps dirt and moisture out of most of the bore and away from the piston. This minimizes corrosion and damage to the caliper bores and pistons.

As the brake linings wear, the piston can move out toward the rotor to compensate for wear. The seal continues to retract the piston by the same amount, however. Thus the piston can travel outward, but its inward movement is restricted by the flexibility of the seal. This action provides the inherent self-adjusting ability of disc brakes, and pedal height and travel remain constant throughout the life of the brake linings. However, the increased cavity will require a larger volume of brake fluid.

Steel pistons used with fixed seals have very close clearances in their bores. Manufacturers typically specify 0.002 inch to 0.005 inch of clearance. Phenolic plastic

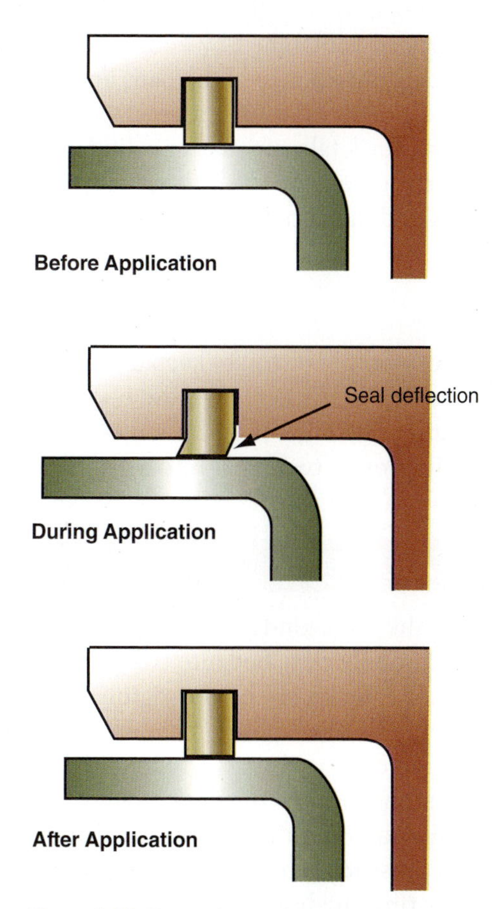

Figure 7-30 These cross sections show how a seal deflects during brake application and then returns to the relaxed position to retract the piston.

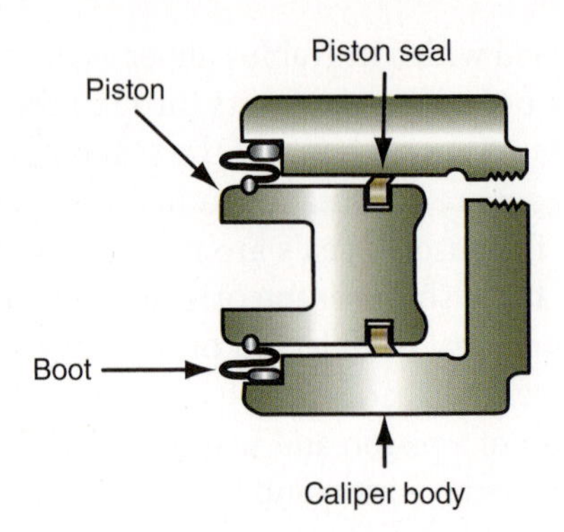

Figure 7-31 A rubber boot keeps dirt and moisture out of the caliper bore.

pistons require slightly more clearance to allow for more expansion. They are typically installed with 0.005 inch to 0.010 inch of clearance. The close fit of the pistons in their bores and the close running clearances between pads and rotors keep pistons from cocking in the calipers and minimize piston knock back because of rotor runout or warping. Close piston-to-bore clearances also keep the seals from flexing too much and rolling in their grooves.

Low-Drag Caliper Seals. Low-drag calipers increase the clearance between the brake pads and rotors when the brakes are released. This increased clearance reduces friction and improves fuel mileage. In a low-drag caliper, the groove for the fixed seal has a tapered outer edge. This lets the seal flex farther as the brakes are applied. This increased flexing outward as the brakes are applied is matched by equal inward flexing as the brakes are released. The result is more clearance between the pads and rotors.

Low-drag calipers require a quick take-up master cylinder as described in Chapter 4 of this *Classroom Manual*. The quick take-up master cylinder provides more fluid volume with the initial pedal movement to take up the greater pad-to-rotor clearance. The combination of low-drag calipers and quick take-up master cylinder maintain normal brake pedal height and travel.

Caliper Dust Boots

A rubber boot fits around every caliper piston to keep dirt and moisture out of the caliper bore (**Figure 7-31**). The opening in the center of the boot fits tightly around the outer end of the piston. The outer circumference of the boot may be attached to the caliper body by a retaining ring in the caliper or by tucking it into a groove inside the bore.

TYPES OF DISC BRAKES

Although all disc brakes have the same kinds of common parts—calipers, rotors, pistons, pads, and so on—they are commonly classified into two groups in terms of caliper operation: fixed and movable. Fixed calipers are bolted rigidly to the caliper support on the steering knuckle or on the rear axle or suspension. Fixed calipers have pistons and cylinders (bores) on the inboard and outboard sides. Hydraulic pressure is applied equally to the inboard and outboard pistons to force the pads against the rotor. Moveable calipers slide or float on the caliper support. Sliding or floating calipers have a piston only on the inboard side, and hydraulic pressure is applied to the piston to force the inboard pad against the rotor. At the same time, hydraulic pressure on the bottom of the caliper bore forces the caliper to move inboard and clamp the outboard pad against the rotor with equal force. Remember: For every action, there is an opposite and equal reaction.

Fixed Calipers

Fixed caliper disc brakes are the oldest type of disc brake. Their use on production cars in high volume dates to the mid-1960s. Probably the best-known **fixed caliper brake** is the four-piston Delco Moraine brake used on Chevrolet Corvettes from 1965 through the early 1980s. Fixed caliper brakes from other suppliers such as Bendix, Budd, and Kelsey-Hayes were used by other domestic carmakers through the mid-1970s, and some late-model imported vehicles continue to have fixed caliper brakes.

A **fixed caliper brake** has a brake caliper that has piston(s) on both the inboard and the outboard sides, is bolted to its support, and does not move when the brakes are applied.

AUTHOR'S NOTE Ford and other light- and medium-duty truck manufacturers (F-350 and above) were using fixed, four-piston calipers up through the mid-1990s. This was primarily because of the length of the pads used on these trucks. Compared to lighter trucks and passenger cars, the pads were about twice as long and could not be evenly applied along their full length by one piston and a sliding caliper without making the single piston very large.

A fixed caliper is bolted to its support and does not move when the brakes are applied. A fixed caliper must have pistons on both the inboard and the outboard sides. The most common are four-piston brakes, with two pistons on each side of the caliper (**Figure 7-32**). Two-piston fixed caliper brakes were common on some lighter imported vehicles, and a few three-piston designs also were manufactured (with one large inboard and two small outboard pistons).

The pads for a fixed caliper brake are held in the caliper body by locating pins. They are mounted more loosely than the pads of a movable caliper brake because the outboard pad cannot be attached solidly to the caliper body. Anti-rattle clips and springs reduce vibration and noise, however.

Fixed calipers must have equal piston areas on each side so that equal hydraulic pressure will apply equal force to the pads (**Figure 7-33**). When the brakes are released, the pistons are retracted by their seals as explained earlier in this chapter.

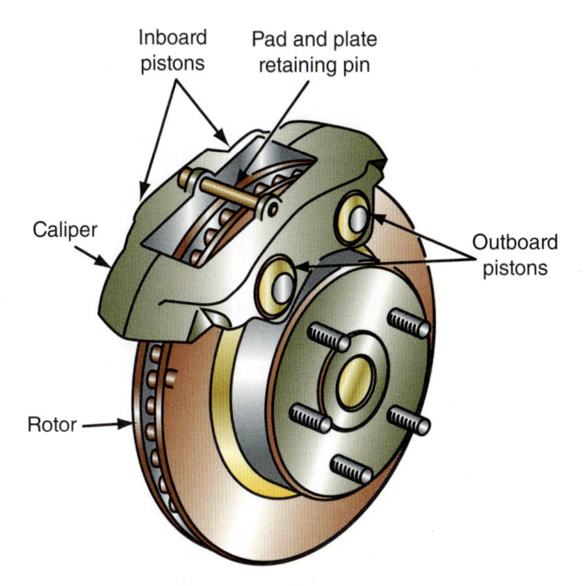

Figure 7-32 A fixed caliper is mounted rigidly over the rotor and has one or two pistons on the inboard and outboard sides to apply the brake pad.

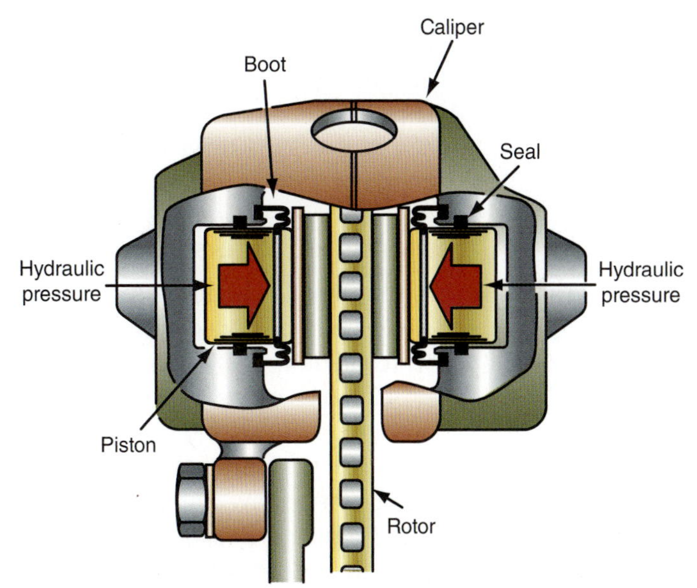

Figure 7-33 Equal hydraulic pressure on both sides of a fixed caliper applies equal force to the inboard and outboard pistons.

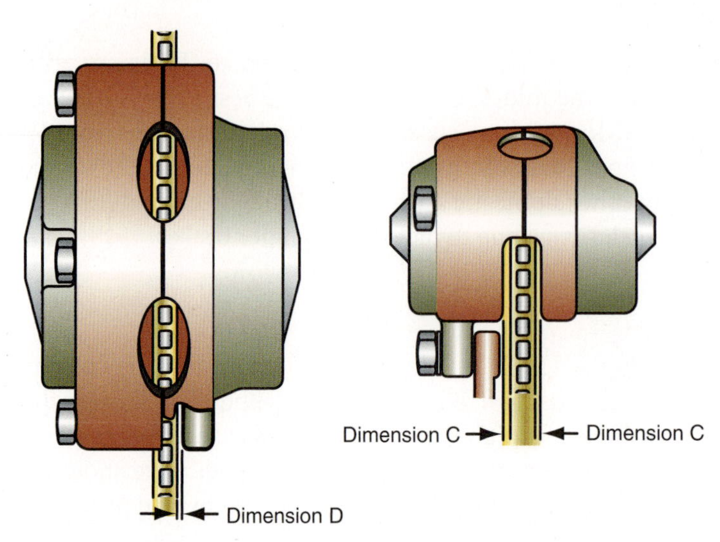

Figure 7-34 A fixed caliper must be parallel to the rotor surface (Dimension D) and spaced equally on each side of the rotor (Dimension C).

Because the body of a fixed caliper brake is mounted rigidly to the vehicle, it must be aligned precisely over the brake rotor. Piston travel must be perpendicular to the rotor surface, and the pads must be exactly parallel to the rotor (**Figure 7-34**). If the caliper is misaligned or cocked on its support, the pads will wear unevenly and the pistons may stick in their bores. Just as important, the caliper must be centered over the rotor so that the inboard and outboard pistons move equal distances to apply the pads. If the caliper is offset to one side or the other, unequal piston travel can create a spongy brake pedal feel and uneven pad wear. Moreover, it may be more difficult to bleed all air from the caliper during service. Shims often are used to align fixed calipers on their supports.

Fixed calipers are large and heavy, which provides good heat dissipation and strength to resist high hydraulic pressures. Because a fixed caliper is mounted rigidly to the vehicle, it does not tend to flex under high temperature and pressure from repeated use. Fixed calipers can provide consistent braking feel under repeated, hard use.

The use of fixed calipers has declined over the past 40 years, and they are not found on many late-model, lightweight cars. The greater weight of a fixed caliper is its biggest disadvantage, and that caliper weight increases the percentage of unsprung weight on the steering knuckle. Fixed calipers also are more expensive to make than floating or sliding calipers, as well as harder to service. Because fixed calipers are built with a two-piece body, they require more time to service and leaks have more opportunities to develop around O-rings and crossover line fittings. In summary, cost, weight, and complexity have made fixed caliper disc brakes less common on late-model vehicles.

Floating Calipers

A **floating caliper** has a one-piece caliper body and one large piston on the inboard side.

By far, most late-model cars and light trucks have disc brakes with movable calipers. Movable caliper brakes are further subdivided into **floating calipers** and sliding calipers, which identify the ways in which the calipers are mounted on their supports.

Floating calipers began to appear on domestic and some imported vehicles in the late 1960s. A floating caliper brake has a one-piece caliper body and usually one large piston on the inboard side. Some medium-duty trucks, particularly Fords, and some GM front-wheel drive cars have floating caliper brakes with two pistons on the inboard side (**Figure 7-35**).

The caliper is mounted to its support on two locating bolts or guide pins that are threaded into the caliper support (**Figure 7-36**). The caliper slides on the pin in a sleeve or bushing. The bushing may be lined with Teflon or have a highly polished surface for

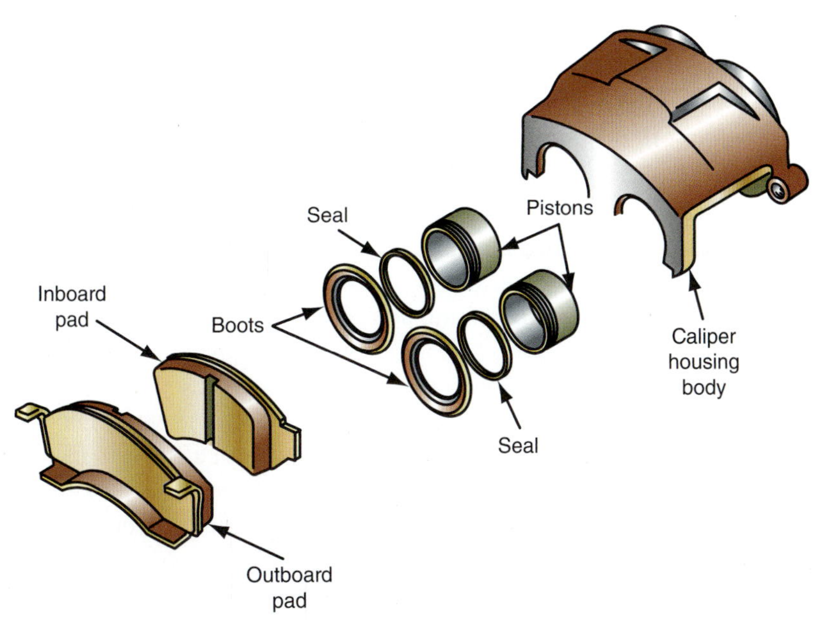

Figure 7-35 Some late-model floating calipers have two pistons.

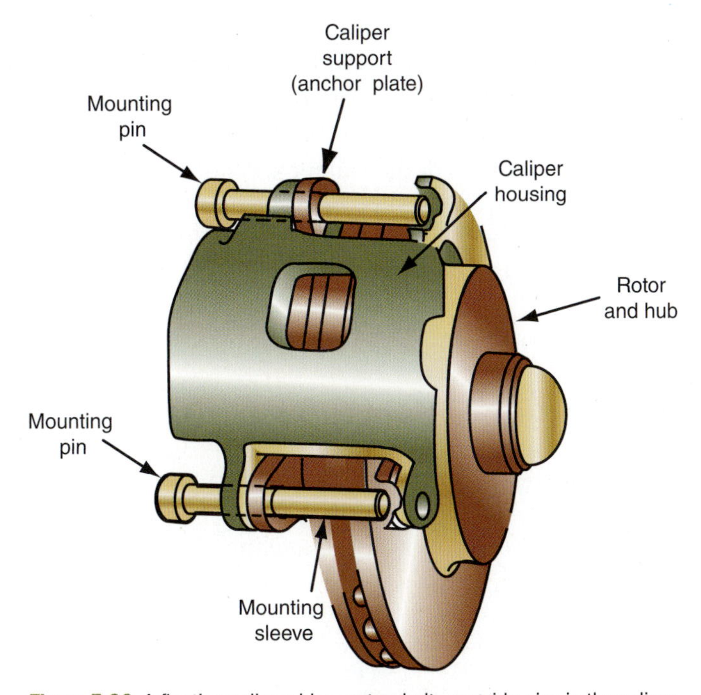

Figure 7-36 A floating caliper rides on two bolts or guide pins in the caliper support.

low friction. The pins let the caliper move in and out and provide some flexibility for lateral movement to help the caliper stay aligned with the rotor. The pads are attached to the piston on the inboard side and to the caliper housing on the outboard side.

When the brakes are applied, the fluid pressure behind the piston pushes the piston outward in its bore. The pressure is transmitted directly to the inboard pad, which is forced against the inboard rotor surface. The pressure applied to the bottom of the cylinder bore forces the caliper to slide or move on the guide pins toward the inboard side. This movement causes the outboard section of the caliper to apply equal pressure against the back of the outboard pad, forcing it against the outboard rotor surface.

In a floating caliper, the piston and the bottom of the caliper bore are both hydraulic pressure surfaces of equal size, and equal pressure is applied to both the piston and the caliper. Hydraulic pressure against the bottom of the caliper bore creates a reaction force that moves the caliper body inward as the piston moves outward. As a result, the rotor is clamped between the piston on one side and the caliper body on the other.

When the driver releases the brakes, the pressure behind the piston drops. The seal relaxes, or springs back, and moves the piston back. As the piston moves back, the caliper relaxes and moves in the opposite direction on the guide pins to the unapplied position. The piston seals provide the self-adjusting action and required pad-to-rotor clearance described earlier in this chapter.

Floating calipers (and sliding calipers, described in the next section) are lighter, easier to service, and cheaper to build than fixed calipers. Because of their simpler design, they also are less likely to develop leaks. The flexible mounting of a floating caliper provides some beneficial self-alignment of the caliper with the rotor. If the flexibility becomes excessive, however, the pads can wear at an angle or become tapered, which decreases pad life.

Sliding Calipers

sliding calipers have a one piece caliper and a piston on the inboard side and move on slides instead of pins.

Sliding calipers operate on exactly the same principles as floating calipers, but their mounting method is different. The caliper support has two V-shaped surfaces that are called abutments or ways (**Figure 7-37**). The caliper housing has two matching machined surfaces. The caliper slides onto the caliper support, where the two parts are held together with a caliper support spring, a key, and a key retaining screw. An anti-rattle spring is used to prevent noise from vibration.

When a sliding caliper is replaced, the caliper ways on the caliper support should be inspected closely and caliper movement should be checked to be sure that the replacement caliper slides correctly. It may be necessary to polish the caliper ways with a fine file or emery cloth for proper clearance.

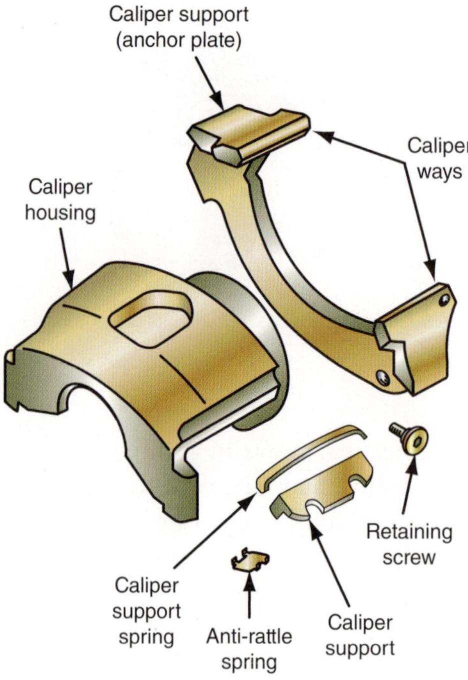

Figure 7-37 A sliding caliper moves on machined ways on the caliper support.

The pins on a floating caliper should be lubricated according to the carmaker's instructions, but lubrication is even more important on a sliding caliper. Both the caliper ways and the mating surfaces on the caliper should be cleaned and then lubricated with high-temperature brake grease, which ensures smooth operation, reduces wear, and prevents corrosion on the sliding surfaces.

REAR-WHEEL DISC BRAKES

Shop Manual
Pages 347

Rear-wheel disc brake calipers may be fixed, floating, or sliding, all of which work just as do the front brake assemblies described previously. The only difference between a front and rear disc brake caliper is the need for a parking brake in the rear.

Many rear disc brakes have a small brake drum built into the center of the rotor (**Figure 7-38**). This is commonly known as a **drum-in-hat** design. Two brake shoes are expanded into the brake drum when the parking brake is applied (**Figure 7-39**). The adjustment of this system is covered in Chapter 9 of the *Shop Manual.* Some rear disc emergency brakes rely on a cable that moves the caliper piston to apply the brake (**Figure 7-40**). The piston is connected to a lever system that can be operated either mechanically by the parking brake lever or pedal or by hydraulic force from the master cylinder. Chapter 9 covers parking brakes in detail.

PERFORMANCE DISC BRAKES

The overall operation of disc brake systems for racing or high-performance vehicles is exactly the same as that for a small subcompact passenger car. The components are made of different materials, however, and are usually a little larger. Many calipers manufactured for racing vehicles are made of aluminum or an aluminum alloy. Pads and possibly rotors could be made of ceramic. Ceramic's main disadvantage is that it does not dissipate heat well. The heat is trapped in the caliper, rotor, pads, and eventually the brake fluid. Carbon fiber may be another material that could be used in racing and eventually on production

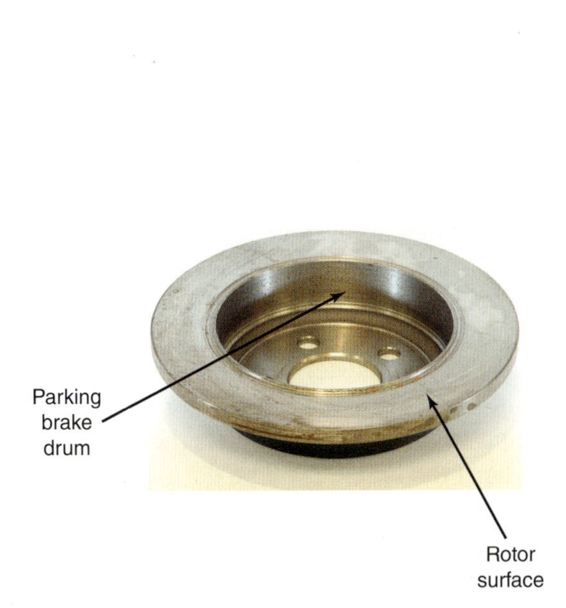

Figure 7-38 This parking brake system has a drum internal to the disc. It is known as a drum-in-hat design.

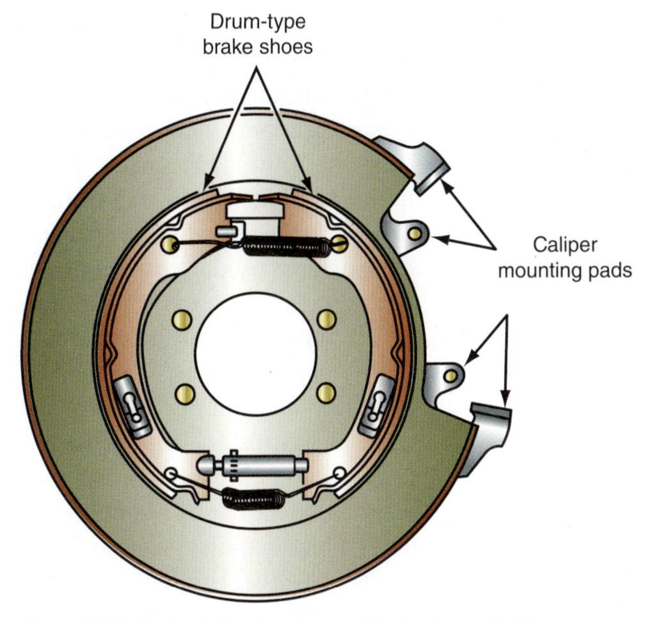

Figure 7-39 Shown is the shoe-type parking brake used with some rear disc brake systems. The actual size is not much larger than the size shown.

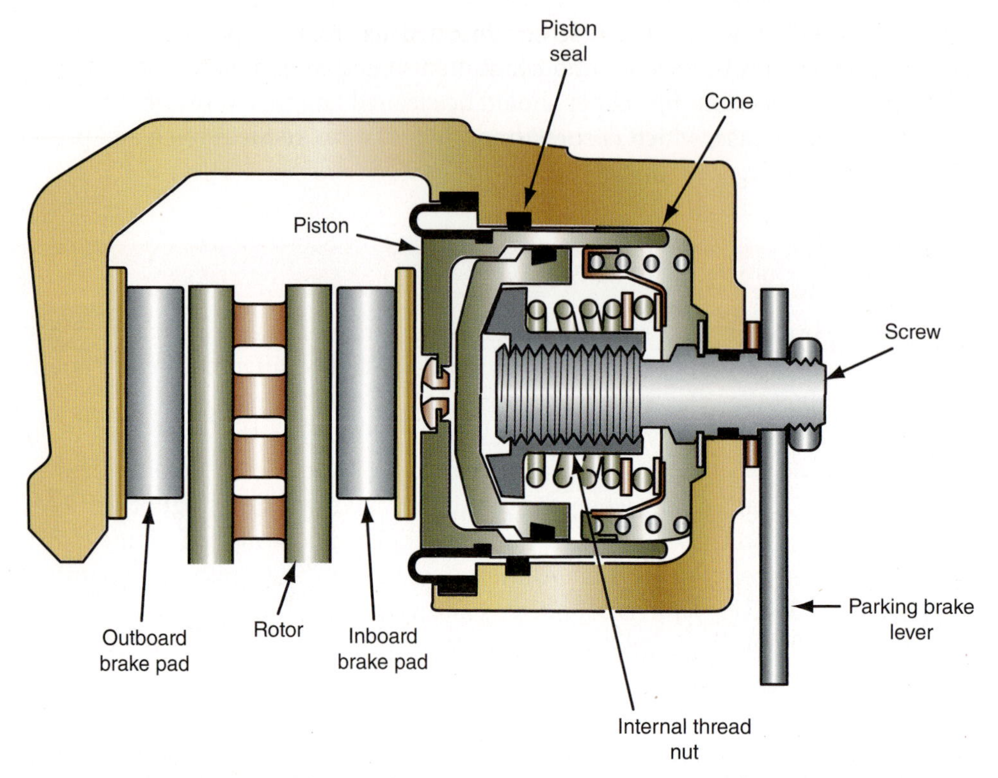

Figure 7-40 Many rear disc brakes have a lever-and-screw mechanism to apply the service brakes for parking.

vehicles. Carbon fiber is light and fairly resistant to heat, but it is expensive. One system used on racing vehicles to dissipate heat that generally has not been installed on production vehicles is the air duct system that moves air from the front bumper or grille and directs that air over the braking mechanism. In general terms, this ductwork is not required on production vehicles.

Some aspects of racing rotors are making it to the production line. The slotted and/or drilled rotors are being installed on many production vehicles. In this case, the industry is finally catching up to motorcycles, which have been using drilled rotors for years. Drilled rotors offer two advantages that take nothing away from braking performance. The first is weight. Drilling a hole removes metal, hence producing a weight reduction. The second advantage is heat dissipation. Passing air can flow through the holes and remove more heat from the rotor. This airflow does nothing to decrease the stability of the vehicle during braking action. The main advantage of slotted rotors is the removal of heat and dispersion of any gas that may form between the pad and rotor. The slots can also help clean the pad. The material used in rotor construction is of a different makeup than the material used for production rotors, so they are more resistant to warping during the hard braking required of a race vehicle. This material is usually cast iron or steel.

Racing calipers may be made of cast aluminum because it allows for weight decrease. However, aluminum does not dissipate heat as well as steel or other heavier metals. In most races, this does not cause a lot of concern because the calipers are supposed to last only for a specific time or distance. Selecting a racing caliper requires the vehicle crew to estimate the amount of braking and how hard the braking will be during an entire race. The caliper manufacturers and the vehicle crews have gotten this estimate down to a near-perfect science, although there were instances when the rotor blew apart during a race, thereby damaging most of the adjacent components. At times the fluid heats up so much in the caliper and attaching hose that vapors increase rapidly and the brakes "go away." Like most other components on a race vehicle, short tracks that require a lot of hard braking require the vehicle to have a larger rotor, pads, and caliper. The long, high-speed-track vehicle has smaller braking components because in a perfect race they need to work only during a pit stop.

The final consideration when selecting a racing caliper and pads is drag. As mentioned earlier, a low-drag caliper where the pads actually separate from the rotors reduces that slight amount of drag inherent in standard disc brakes. Because brake drag reduces the horsepower available to drive the vehicle and increases fuel consumption, low-drag calipers are the best choice for racers. As a reminder, if low-drag calipers are used, then some version of a quick take-up master cylinder must be used or the brakes may engage too late.

Racing pads may be carbon fiber, ceramic, or semi-metallic. Selecting the pads involves several considerations: weight, rotor wear, and durability. As mentioned earlier, the pads are selected based on the use they will endure during a race. The pads are no good, however, if they tend to wear the rotor excessively. The pads and rotors have to match for good braking effect throughout the entire race.

Like most racing equipment, the racing brake components may lead the way to better brakes for production. The manufacturer has two problems adapting racing equipment to production, however: durability and cost. Professional racers with some well-paying sponsors can afford to buy calipers, rotors, and pads that are thrown away after the race is completed. The average consumer is not willing to pay this additional cost without some really great improvements in durability, however.

SUMMARY

- The major parts of a disc brake are the rotor, the hub, the caliper, and the pads.
- The hub contains the wheel bearings and is where the wheel is mounted.
- Rotors come in two types: solid or ventilated.
- The four general kinds of friction pad materials are asbestos, organic, semi-metallic, and metallic.
- The coefficient of friction of the pad materials is indicated by letters and numbers stamped on the edge of the friction material. These letters and numbers are known as the Automotive Friction Material Edge Code.
- Anti-rattle clips and springs are used to keep the pads from making noise when the pads are not in contact with the rotor.
- Brake pad wear indicators alert the driver when the brake linings need servicing.
- Piston seals are used to seal the hydraulic brake fluid and retract the piston and pad from the rotor.
- The two types of calipers are fixed and movable. Movable calipers can be either the floating style or the sliding style.
- Low-drag calipers are movable calipers that move the pads away from the rotor, leaving no contact or drag.
- Rear-wheel disc brakes are similar to front-wheel disc brakes but must operate from hydraulic pressure and from the parking brake cables.

REVIEW QUESTIONS

Short-Answer Essays

1. List the basic parts of a disc brake.
2. Explain the advantages and disadvantages of an aluminum caliper.
3. Why do manufacturers use two-piece rotors on some vehicles?
4. In addition to keeping road splash off the brake assembly, why are splash shields used on disc brakes?
5. Why is asbestos no longer used for brake pad and shoe linings?
6. What are common classifications of brake pad and shoe friction materials?
7. What is the difference between fixed and sliding calipers?
8. How does a low-drag caliper operate?
9. Explain how a floating caliper applies the brake pads.
10. Explain the purpose of a drum-in-hat rotor.

Fill in the Blanks

1. Front calipers are mounted on ______________ ______________ or adapters that may be an integral part of the steering knuckles.
2. When any brake installation reaches its limit of heat dissipation, ______________ ______________ sets in.
3. ______________ rotors are made of different materials, usually cast iron and steel, to reduce weight.
4. ______________ linings usually wear faster than do semi-metallic linings, but they have the benefit of breaking in faster.
5. With disc brakes most of the ______________ area is cooling while only a small area directly contacts the friction materials at any given time.
6. Disc brakes do not suffer from ______________ fade because the rotor does not expand away from the pads.
7. Low-drag calipers require a quick take-up ______________ cylinder.
8. The ______________ ______________ ______________ converts hydraulic pressure from the master cylinder to mechanical force.
9. Many rear disc brakes have a small brake drum built into the center of the ______________.
10. Sliding or floating calipers have a piston only on the______________ side.

Multiple Choice

1. The advantages and disadvantages of disc brake systems are being discussed. Technician A says that disc brakes are more likely to develop brake fade and water fade. Technician B says that reduction of mechanical fade and gas fade is an advantage of disc brakes. Who is correct?

   A. A only
   B. B only
   C. Both A and B
   D. Neither A nor B

2. The operation of disc brake systems is being discussed. Technician A says that composite rotors were developed to save weight. Technician B says that a fixed rotor assembly is part of the hub. Who is correct?

   A. A only
   B. B only
   C. Both A and B
   D. Neither A nor B

3. Disc brake calipers are being discussed. Technician A says that a single caliper may have one, two, or more pistons. Technician B says that the piston seal is known as a square-cut seal. Who is correct?

   A. A only
   B. B only
   C. Both A and B
   D. Neither A nor B

4. Technician A says that solid rotors can usually be machined several times before their minimum thickness limits are reached. Technician B says that composite rotors have cast-iron hubs with steel friction surfaces. Who is correct?

   A. A only
   B. B only
   C. Both A and B
   D. Neither A nor B

5. When discussing disc brake work, Technician A says that organic linings are a composite material made from bonding nonmetallic fibers. Technician B says that semi-metallic linings require higher brake pedal effort but are more fade resistant compared to organic linings. Who is correct?

   A. A only
   B. B only
   C. Both A and B
   D. Neither A nor B

6. Two technicians are discussing a disc brake caliper: Technician A says that calipers use springs to retract the seals after a brake application. Technician B says that sliding calipers typically use one piston on each side of the rotor. Who is correct?

   A. A only
   B. B only
   C. Both A and B
   D. Neither A nor B

7. Technician A says that a return spring is used to retract a caliper piston. Technician B says that the piston seal retracts the caliper piston when hydraulic pressure is released. Who is correct?

   A. A only
   B. B only
   C. Both A and B
   D. Neither A nor B

8. Technician A says that inner and outer pad friction materials may have different coefficients of friction. Technician B says to follow the manufacturer's recommendation for inboard and outboard pad linings when replacing brake pads. Who is correct?

   A. A only
   B. B only
   C. Both A and B
   D. Neither A nor B

9. Technician A says that when the driver hears the audible brake pad wear indicator, he or she should immediately park the car and not use it until the brakes are replaced. Technician B says that the wear pad indicators alert the driver to worn pads that should be replaced soon. Who is correct?

   A. A only
   B. B only
   C. Both A and B
   D. Neither A nor B

10. Technician A says that synthetic brake linings are nonorganic, nonmetallic, and non-asbestos materials. Technician B says that synthetic brake linings are made of aramid or fiberglass. Who is correct?

    A. A only
    B. B only
    C. Both A and B
    D. Neither A nor B

CHAPTER 8
DRUM BRAKES

Upon completion and review of this chapter, you should be able to:

- Describe the basic parts of a drum brake assembly.
- Describe how a drum brake stops a vehicle.
- Describe different types of brake drums.
- Describe the types of friction linings and the three placements of the lining on the shoe.
- Describe the components that make up a wheel cylinder and the purpose of each one.
- Describe the two major drum brake designs and how they differ in operation.
- Describe the different types of self-adjusters used on duo-servo and leading-trailing shoe brake systems and how they work.

Terms To Know

Arcing	Hold-down springs	Secondary shoe
Backing plate	Leading shoe	Self-adjusters
Brake shoes	Leading-trailing brake	Self-energizing operation
Cam-ground lining	Overload spring	Shoe anchor
Composite drum	Pawl	Star wheel
Cup expander	Piston stop	Table
Drum web	Primary shoe	Trailing shoe
Duo-servo brake	Return spring	Web

INTRODUCTION

For more than 80 years, drum brakes at all four wheels were the standard of the automobile industry. Although four-wheel disc brakes have become standard equipment of most cars and light trucks, drum brakes continue in use for a few rear-wheel brakes. No law or regulation requires disc brakes, but the brake performance requirements of FMVSS 105 make front disc brakes virtually mandatory. Nevertheless, drum brakes have certain advantages that will make them an important part of vehicle engineering and service for many more years.

This chapter explains the construction and operation of modern drum brake systems and begins with a summary of their advantages and disadvantages.

The major drum brake advantages are self-energizing and servo action, lack of noise, and efficient parking brake operation without complicated linkage. Drum brake disadvantages include poorer heat dissipation and less fade resistance than disc brakes, lack of self-adjustment without special linkage, and a greater tendency than disc brakes to pull and grab.

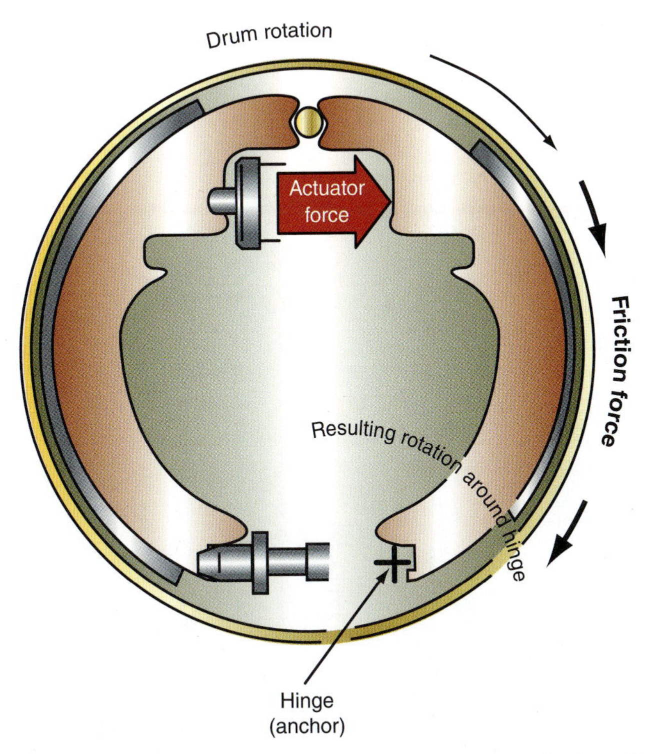

Figure 8-1 Drum rotation adds leverage to the brake shoes as they contact the drum. This process is called self-energizing action.

Drum Brake Self-Energizing and Servo Action

One end of the lining on one shoe of a drum brake contacts the drum before the other end does and becomes a pivot point as friction increases quickly. The brake shoe becomes a self-energizing lever and adds its own mechanical leverage to hydraulic force to help apply the brakes (**Figure 8-1**).

When a component is operated, or applied and that component uses the initial input force to increase that force and transfer it to a matching component, the entire operation is said to be self-energizing. This **self-energizing operation** may be confined to one shoe of a drum brake installation, which is the leading shoe in relation to the direction of wheel rotation. When the self-energizing operation of one shoe applies mechanical force to the other shoe to assist its application, it is called servo action (**Figure 8-2**). Self-energizing operation and servo action can have both operational advantages and disadvantages. Self-energizing drum brakes that use servo operation can always apply more stopping power for a given amount of pedal force than disc brakes can.

Self-energizing operation is the action of a drum brake shoe when drum rotation increases the application force of the shoe by wedging it tightly against the drum.

Drum Brake Pulling and Grabbing

Servo action increases braking force at the wheels, but it must develop smoothly as the brakes are applied. If servo action develops too quickly or unevenly, the result can be brake grabbing or lockup. This is usually experienced as severe pull to one side or the other or reduced vehicle control during braking.

Lack of Noise

Drum brakes are almost noise free. Heavy return springs and hold-down springs hold the brake shoes against the wheel cylinders, anchor pins, backing plates, and away from the drum. Linings are securely bonded or riveted to the brake shoes, and the entire assembly is encased in the brake drum. About the only time that noise is a concern with drum

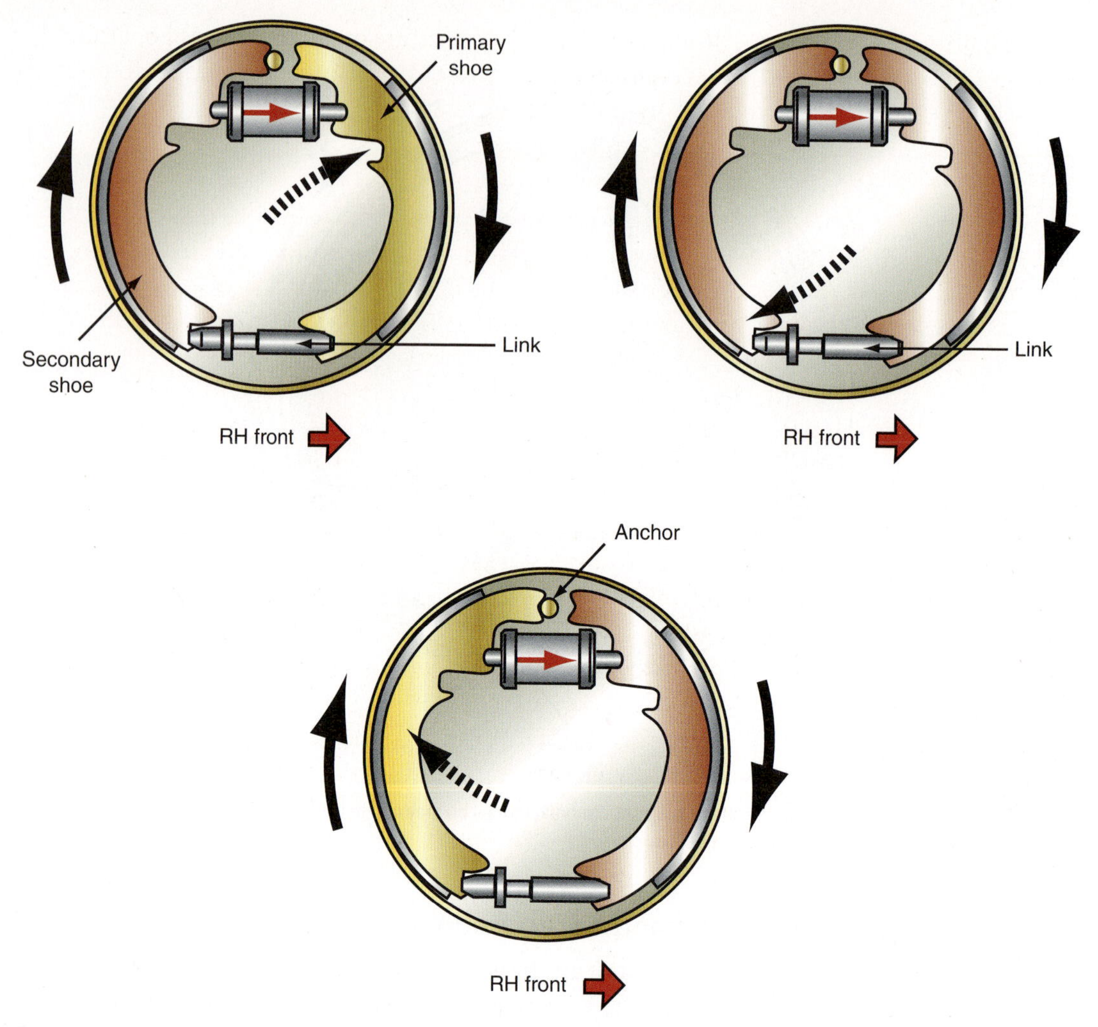

Figure 8-2 Self-energizing action of the primary shoe applies force to the secondary shoe. This process is called servo action.

brakes is when it is the sound of steel brake shoes grinding against the brake drum after the linings have worn away.

Drum Brake Parking Brake Operation

Drum brakes provide a better static coefficient of friction than do disc brakes. The brake linings grab and hold the drums more tightly than disc pads can hold a rotor. In addition, the self-energizing and servo actions of drum brakes contribute to this feature, as does the larger area of brake shoe linings compared to disc brake pads.

For brake installations that have discs at the front and drums at the rear, the parking brake weaknesses of disc brakes are not a problem. Many disc brake applications now use a drum-in-hat style rotor that allows the use of drum style parking brakes. The parking brake shoes are small, but all they have to do is hold the vehicle stationary. Some rear disc brake applications still use a parking brake mechanism that is integral with the disc brake caliper. In most cases, the driver operates a pedal or a lever that pulls on cables attached to the rear brakes. The cables operate levers that mechanically apply the brake shoes. Some vehicles are equipped with electric cable actuators or electric motors that are attached directly to the back of the rear rotors. All mechanical motion is in a straight longitudinal line from the front to the rear of the vehicle.

Drum Brake Self-Adjustment

Brake adjustment is the process of compensating for lining wear and maintaining correct clearance between brake linings and the drum or rotor surfaces. Disc and drum brakes both require adjustment, but self-adjustment is a basic feature of disc brake design. Drum brakes need extra cables, levers, screws, struts, and other mechanical linkage just to provide self-adjustment and proper lining-to-drum clearance.

Fade Resistance

Brake fade is the loss of braking power due to excessive heat that reduces friction between brake linings and the rotors or drums. One factor that contributes to heat dissipation and fade resistance is the swept area of brake drum or rotor, which is the total area that contacts the friction surface of the brake lining. The greater the swept area, the greater the surface available to absorb heat. Although a brake drum can have a relatively large swept area, the entire area is on the inner drum surface. The swept area of a disc brake comprises both sides of the rotor. For any given wheel size, the swept area of a disc brake will always be larger than the swept area of a drum brake (**Figure 8-3**). For example, a 10-inch-diameter rotor will have almost 50 percent more swept area than a 10-inch drum.

Prior to the installation of self-adjusters, periodically the vehicle brakes would have to be adjusted by a technician or an experienced owner.

Mechanical fade is a problem that occurs with drum brakes when the drum becomes very hot and expands outward, away from the brake shoes. As a result, the shoes then must travel farther to contact the drum surface with normal braking force, and the pedal drops lower as the brakes are applied. The increased heat at the braking friction surfaces also reduces the coefficient of friction. The combined result is brake fade.

Lining fade occurs when the linings are overheated and the coefficient of friction drops off severely (**Figure 8-4**). Some heat is needed to bring brake linings to their most efficient working temperature. The coefficient of friction rises, in fact, as brakes warm up. If the temperature rises too high, however, the coefficient of friction decreases rapidly. Because disc brakes are exposed to more cooling airflow than drum brakes, lining fade due to overheating is reduced.

Water fade occurs when water is trapped between the brake linings and the drum or rotor and reduces the coefficient of friction. Gas fade is a condition that occurs under hard braking when hot gases and dust particles are trapped between the brake linings and the drum or rotor and build up pressure that acts against brake force. These hot gases actually lubricate the friction surfaces and reduce the coefficient of friction.

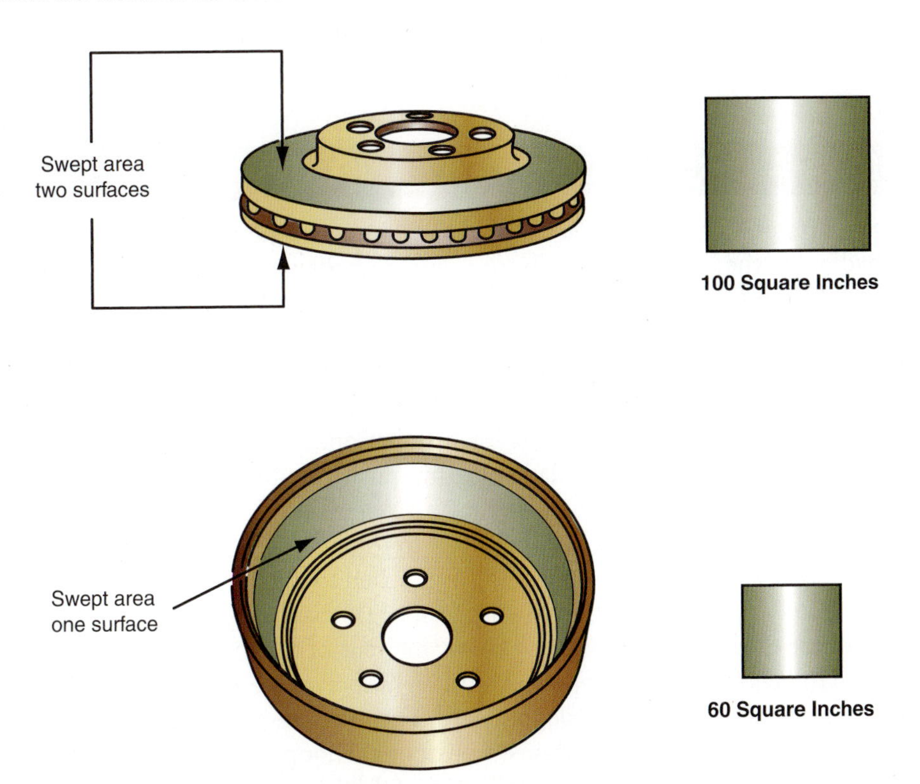

Figure 8-3 For any given size wheel, a disc brake has 35 percent to 50 percent more swept area to dissipate heat.

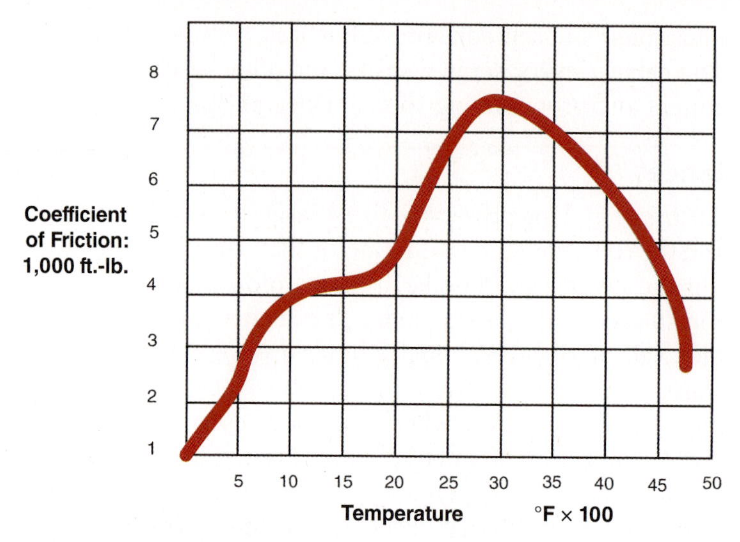

Figure 8-4 Initially, the coefficient of friction increases with heat, but very high temperatures cause it to drop off and cause the brakes to fade.

AUTHOR'S NOTE You can use a piece of sandpaper to demonstrate the effect of heat on friction. Slide the sandpaper slowly over a hard surface, and you can feel the drag or friction. Speed up the sliding action, and shortly you will feel the heat and the reduction of the drag.

Tinnerman nuts are generally discarded the first time the drums are serviced. They are used to hold the drums in place at the assembly plant.

The basic design differences between disc and drum brakes make disc brakes much more self-cleaning to greatly reduce or eliminate water and gas fade. Centrifugal force also works to remove water and gas from the braking surfaces. Additionally, the leading edges of the brake pads help to wipe the rotor surfaces clean. Due to the design of the drum brake, it will trap heat, water, and gases inside the drum. This makes the drum brake more susceptible to brake fade and pull.

DRUM BRAKE CONSTRUCTION AND OPERATION

Shop Manual
page 374

Floating drums are the rule for late-model vehicles that still use drum brakes. One-piece drums were common on older vehicles.

The basic parts of a drum brake are a drum and hub assembly, brake shoes, a backing plate, a hydraulic wheel cylinder, shoe return springs, hold-down springs, and an adjusting mechanism (**Figure 8-5**). Rear brakes also include a parking brake. The drum and shoes provide the friction surfaces for stopping the wheel. The drum has a machined braking surface on its inside circumference. The wheel is mounted to the drum hub by nuts and studs. On later-model vehicles, that drum is usually separate from the hub or axle flange (floating) and fits over the wheel studs for installation. Separate or floating drums are sometimes held to the hub or axle flange by thin push-on nuts commonly referred to as speed nuts or Tinnerman nuts. Screws may also be used to hold drums on. The hub houses the wheel bearings that allow the wheel to rotate.

The **backing plate** provides the mounting platform for the brake assembly and closes the rear of the mounted brake drum.

The hydraulic and friction components are attached to the **backing plate** (**Figure 8-6**), which is mounted on the axle housing or suspension. The backing plate provides the mounting platform for the brake assembly and closes the rear of the mounted drum. The drum encloses and rotates over these components. When the brakes are applied, hydraulic pressure forces the pistons in the wheel cylinder outward. The pressure on the pistons is transmitted to the shoes as the shoes are forced against a pivot or anchor pin and into the rotating drum. Through this action, the shoes tightly wrap up against the drum to provide the stopping action.

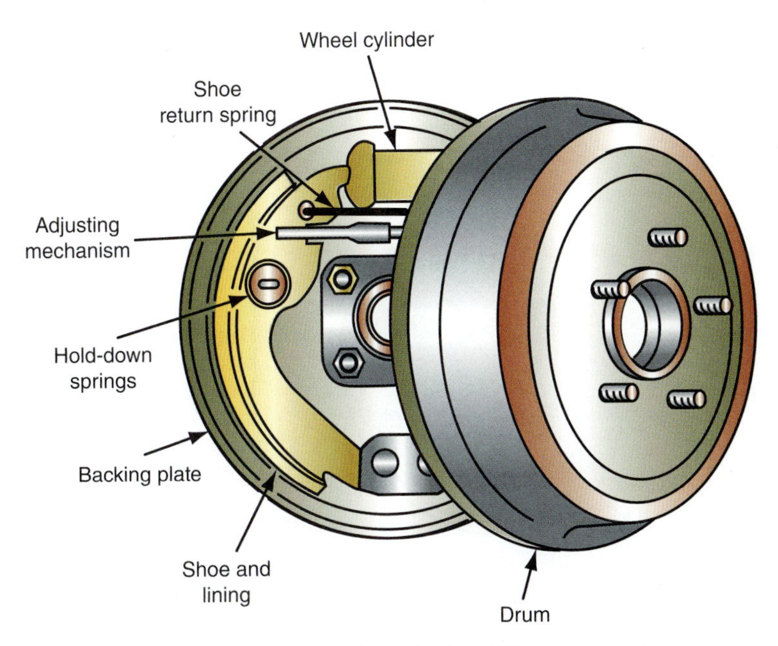

Figure 8-5 Basic parts of all drum brakes. A leading-trailing system is shown.

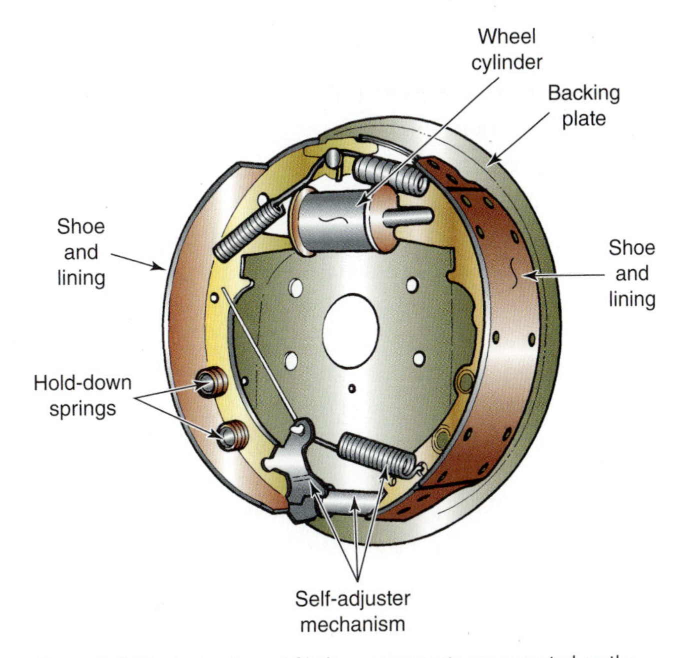

Figure 8-6 The hydraulic and friction components are mounted on the backing plate.

The frictional energy on the drum surface creates heat. This heat is dissipated to the surrounding air as the drum rotates with the wheel. Some drums are finned to help them dissipate heat more efficiently.

Brake Drums and Hubs

The brake drum mounts on the wheel hub or axle and encloses the rest of the brake assembly except the outside of the backing plate. Brake drums are made of cast iron or steel and cast iron. In these variations, iron provides the friction surface because of its excellent combination of wear, friction, and heat-dissipation characteristics.

Shop Manual
page 409

The drum is a bowl-shaped part with a rough-cast or stamped exterior and a machined friction surface on the interior. The open side of the drum fits over the brake shoes and other parts mounted on the backing plate (**Figure 8-7**). It is the side of the drum that is visible when

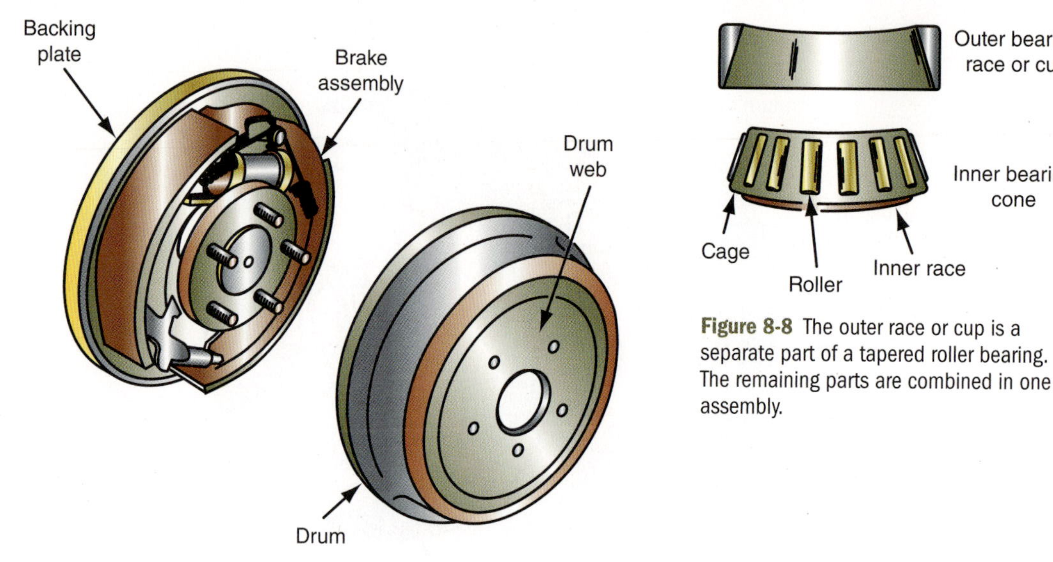

Figure 8-7 The drum fits over the brake shoes and other parts mounted on the backing plate.

Figure 8-8 The outer race or cup is a separate part of a tapered roller bearing. The remaining parts are combined in one assembly.

The **drum web** is the closed side of the drum. The web also holds the wheel bearings in a one-piece drum.

the wheel assembly is removed. The closed side of the drum is called the **drum web** and contains the wheel hub and bearings (one-piece drum) or has mounting holes through which the drum is secured to the axle flange or separate hub (two-piece drum).

Tapered roller bearings, installed in the hubs on drums equipped with hubs, were common on older vehicles; later-model vehicles use sealed wheel bearings on floating drums as the most common bearings used on the rear wheels of FWD cars. The tapered roller bearing has two main parts: the bearing cone and the outer cup (**Figure 8-8**). The bearing cone contains steel tapered rollers that ride on an inner cone and are held together by a cage. The bearing fits into the outer cup, which is pressed into the hub to provide two surfaces, an inner cone and outer cup, for the rollers to ride on. Sealed wheel bearings are used on most front-wheel drive vehicles on the rear wheels with floating drums. Chapter 3 in this *Classroom Manual* contains more detailed information on different kinds of wheel bearings.

On RWD vehicles, the brake assembly is mounted to the backing plate, which is fastened to the rear axle housing (**Figure 8-9**). The axle extends outward past the brake

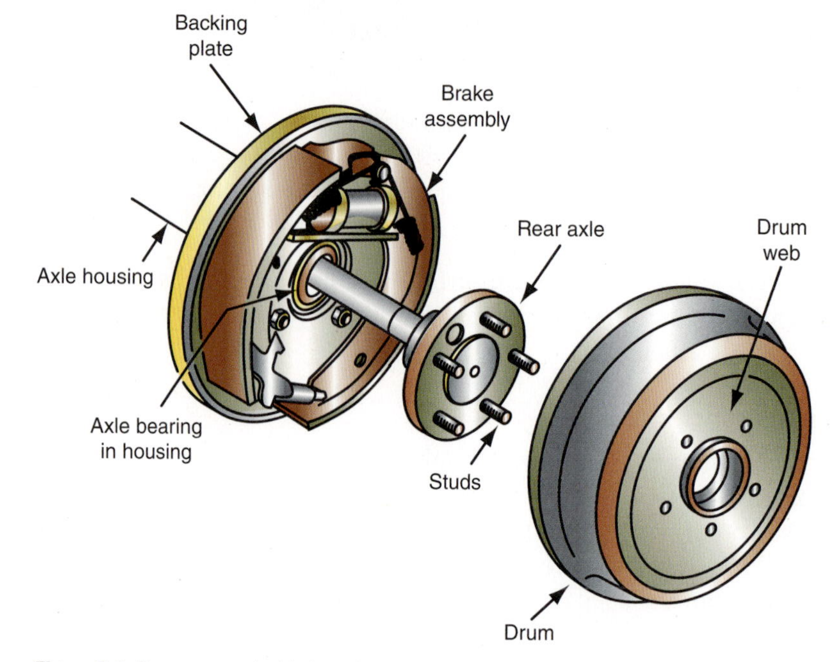

Figure 8-9 Drum removed with the axle partially withdrawn. Shown is the general layout of RWD axle and brake drum.

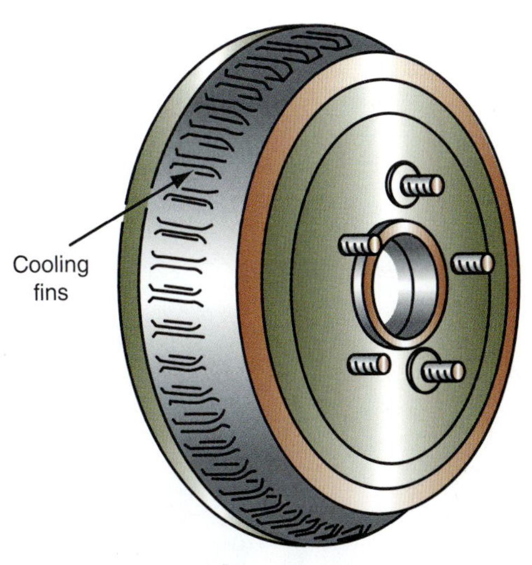

Figure 8-10 This drum has cooling fins cast into its outer circumference.

assembly and has the wheel studs installed into the axle flange. The axle bearing fits into the end of the axle housing. The drum is fitted over the studs and held in place by the placement of the wheel assembly. The outer end of the axle and the brake assembly are enclosed within the drum. Information and service on axle bearings may be found in *Today's Technician Manual Transmissions and Transaxles.*

Some drums have fins cast into the outer circumference to aid in cooling (**Figure 8-10**).

The braking surface area of a drum is determined by the diameter and the depth (width) of the drum. Large cars and trucks, which require more braking energy, have drums that measure 12 inches in diameter or larger. Smaller vehicles use smaller drums. Generally, manufacturers try to keep parts as small and light as possible while still providing efficient braking. Brake drums can be categorized as either solid or composite and by the different ways in which the drums are made.

Solid Cast-Iron Drums. One-piece cast-iron drums are rarely seen on late-model vehicles because of their weight. A solid drum is a one-piece iron casting (**Figure 8-11**). Cast iron has excellent wear characteristics and a coefficient of friction that make it ideal as a braking friction surface. Iron also dissipates heat very well, and it is easy to machine when refinishing is necessary.

Along with these advantages, cast-iron drums have some disadvantages. A large iron casting can become brittle and may crack if overstressed or overheated. Small cracks may be almost invisible to the naked eye, but they can lead to drum failure or heat checking and glazing.

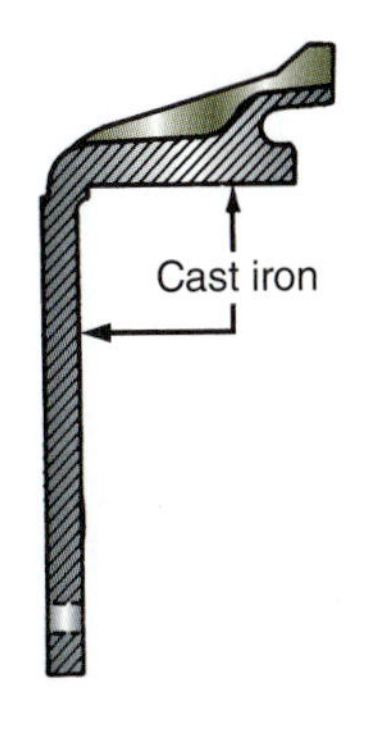

Figure 8-11 Cross section of a cast-iron drum.

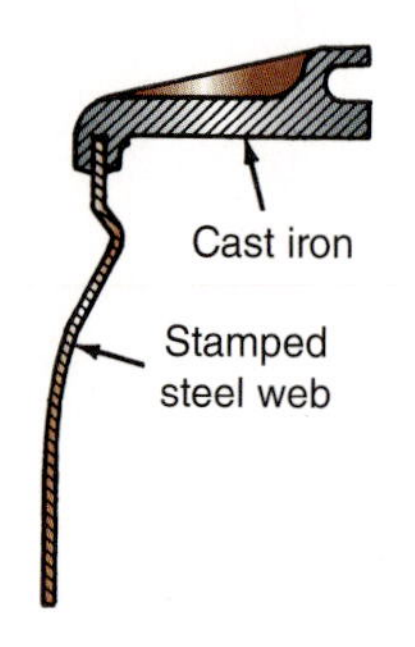

Figure 8-12 Cross section of a composite drum.

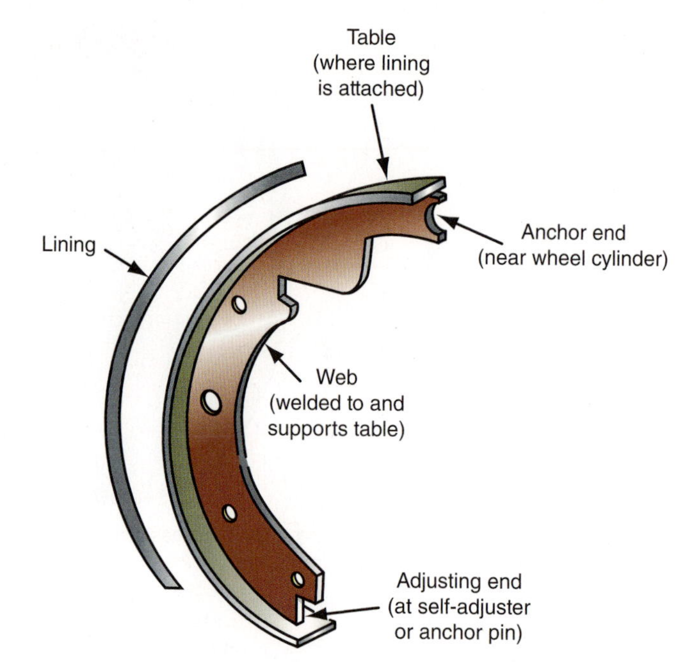

Figure 8-13 A typical brake shoe with the lining detached.

A one-piece cast-iron drum is the heaviest of all drum types, which is both an advantage and disadvantage. The weight and mass of a one-piece drum make it very good at absorbing and dissipating heat. The drum weight, however, adds a lot of weight to the vehicle, and it is all unsprung weight at the wheels. Engineers consider brake drum weight more a disadvantage than an advantage for late-model cars, so most vehicles are being built with composite drums of the following kinds.

composite drums are composed of a stamped steel web mated with the edge of cast-iron drum.

Steel and Iron Drums. Steel and iron **composite drums** are made in two ways. The most common type has a stamped steel web mated with the edge of a cast-iron drum (**Figure 8-12**). The other type of composite drum has a centrifugally cast iron liner inside a stamped steel drum. To make this kind of drum, a stamped steel drum is rotated at high speed while molten iron is poured into it. Centrifugal force causes the molten iron to flow outward and bond tightly to the inner circumference of the steel drum.

Brake shoes are the components that mount the friction material to the backing plate.

Steel and iron composite drums are lighter and cheaper to make than one-piece cast-iron drums, but they are less able to absorb and dissipate heat and resist fade. They work well on the rear drum brakes of compact cars, however.

The brake shoe **table** is the metal part of the brake shoes to which the linings are attached.

Brake Shoes and Linings

Brake shoes are the components that mount the friction material or linings that contact the drums. The shoes also provide the area to mount the shoe to the backing plate. Brake shoes are usually made from welded steel, although some drum brakes use aluminum shoes. The outer part is called the **table** (or sometimes the rim) and is curved to match the curvature of the drum (**Figure 8-13**). The brake lining is riveted or bonded to the brake shoe table (**Figure 8-14**). Many brake shoes have small notches or nibs along the edge of the table that bear against the backing plate and help to keep the shoe aligned in the drum. The **web** is the inner part of the shoe that is perpendicular to the table and to which all of the springs and other linkage parts attach.

Some aftermarket brake shoe sets are fitted with equal amounts of lining on each shoe. This allows one brake shoe set to replace several types of original manufacturer shoe.

Duo-servo and leading-trailing non-servo brakes are described in detail later in this chapter, but all drum brakes have a pair of shoes at each wheel. The front shoe on duo-servo brakes is called the primary shoe, and the rear shoe is called the secondary shoe. The front and rear shoes on leading-trailing brakes are called just that: the leading shoe and the trailing shoe, respectively.

Brake linings react against the brake drum with friction that is dissipated as heat.

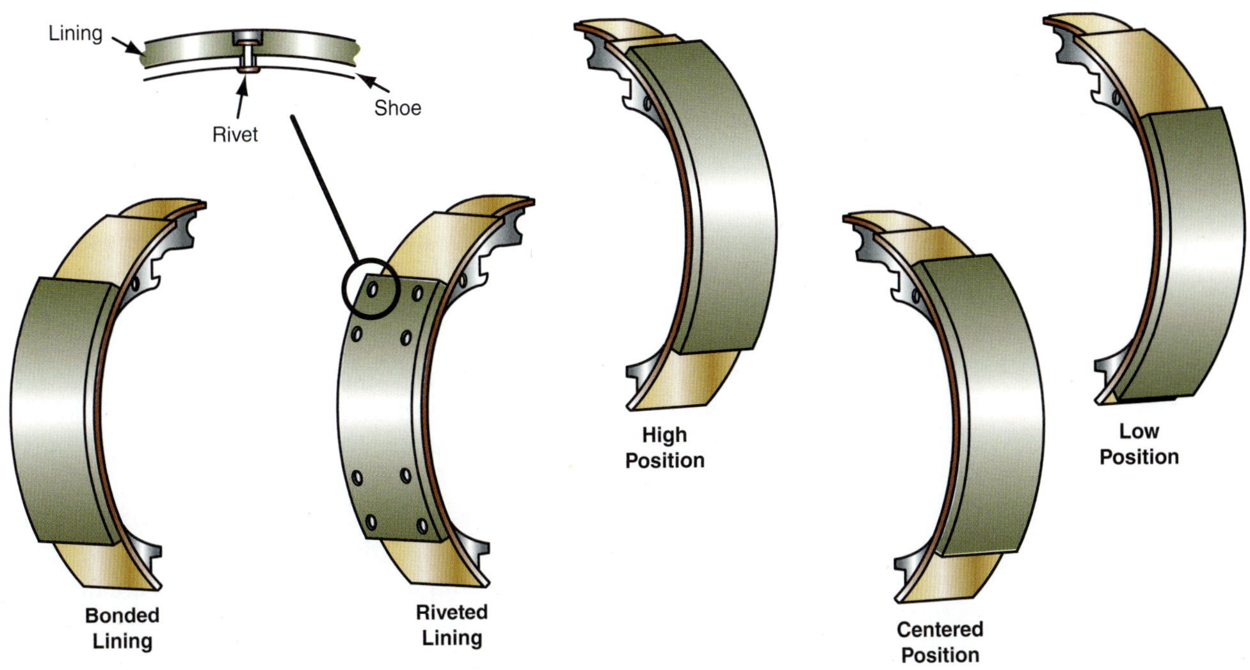

Figure 8-14 Linings may be bonded or riveted to the shoe.

Figure 8-15 Lining can be attached to the shoe in different positions, depending on the desired stopping characteristics.

On leading-trailing brakes, the lining on the leading and trailing shoes are usually the same length and positioned in the same location on each shoe. On a duo-servo brake, however, the primary shoe lining is shorter than the secondary lining and may even have a different coefficient of friction. As explained later in this chapter, the primary shoe of a duo-servo brake is self-energized by drum rotation and then applies servo action to the secondary shoe. The secondary shoe applies most of the stopping force and thus requires a larger lining. The coefficient of friction for the secondary lining is often higher than for the primary lining to provide good stopping power. However, the lower coefficient of friction for the primary lining helps to keep the brakes from applying too harshly and locking.

Lining for the secondary shoe is almost always centered from top to bottom on the shoe table. Primary lining may be centered from top to bottom on the table, or it may be offset toward either the top or the bottom (**Figure 8-15**), depending on the operating characteristics of a specific brake installation.

Brake Friction Materials. Both drum and disc brake friction materials are covered in Chapter 7 of this *Classroom Manual*.

AUTHOR'S NOTE On a few older light-duty trucks and large passenger sedans, the shoes were marked "L" or "R" and were to be mounted on either the left (L) or right (R) side of the vehicle. This was necessary because the shoe web had a reinforced area at the top outside of the web. If installed on the wrong side, this reinforced area would drag on the backing plate and cause grabbing. The reinforcement enabled the manufacturer to use an older model shoe web on heavier vehicles. The original web design could not withstand the braking forces of the heavier vehicle. In theory, this was supposed to save manufacturing costs, but it was basically a failure and was discontinued after a few model years.

Lining-to-Drum Fit. When disc brakes are applied, a flat pad contacts a flat rotor, so the surface fit of the pad to the rotor is usually not a concern. With drum brakes, however, a semicircular lining contacts a circular drum, and the lining does not move in a straight line toward the drum. The complete lining surface does not contact the drum all at the same time, so the surface fit of the lining to the drum is an important operating consideration.

The leverage that operates in drum brakes and the motion of brake shoes as they are applied work against full lining contact with the drum. Therefore, lining shape must be adjusted to overcome these natural conditions and work toward providing full lining-to-drum contact.

The hydraulic force of the wheel cylinder pistons and the rigid positions of the shoe anchors tend to force the upper ends of the shoes into contact with the drum before the center of the shoes. Most drum brake noise complaints are caused by binding at the ends of the shoes as they contact the drum before the center does. More important, if the full surface of new brake linings does not contact the drum as the linings are broken in, the linings can overheat and become glazed, which reduces braking effectiveness.

Some brake linings are fitted to provide more clearance at the ends than the center. **Arcing** used to be done as part of the brake job, but today most linings are manufactured with a cam ground lining at the factory.

To compensate for these natural characteristics of drum brakes, linings are fitted to shoes to provide more clearance at the ends than at the center. This process is known as **arcing** the shoes, and it forms either cam-ground or undersized linings. Arcing or cam-grounding of the lining is sometimes very visible on some brake shoes. On others, the arc or ground is not noticeable without close inspection. A **cam-ground lining** is thinner at the ends than at the center (**Figure 8-16**), and the lining surface is not a portion of a circle with a constant radius. An undersized lining has a uniform thickness and a constant radius (**Figure 8-17**), but it has a smaller outside diameter than the inside diameter of the drum. A fixed-anchor arc is a variation of an undersized lining in which the overall arc is offset slightly so that one end of the lining is thicker than the other (**Figure 8-18**).

Arc grinding, or arcing, of new brake linings used to be a standard part of brake service. Today, however, brake shoe linings are seldom arced as part of the brake installation job.

Lined shoes are sold with the lining surface already contoured for proper drum fit. Several factors led to the decline of arc grinding, and concerns about airborne asbestos and brake dust in general were the leading cause. In addition, rear drum brakes used with front disc brakes provide a smaller percentage of overall braking for late-model FWD cars.

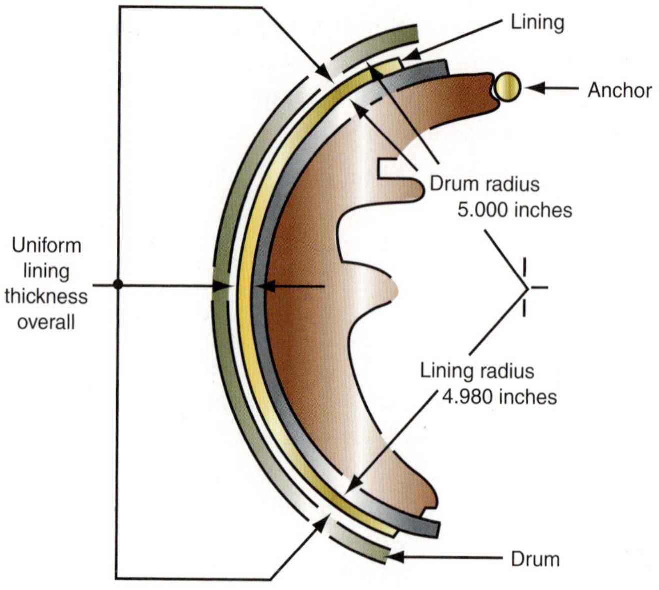

Figure 8-16 A cam-ground lining is thinner at the ends than in the center (clearances exaggerated).

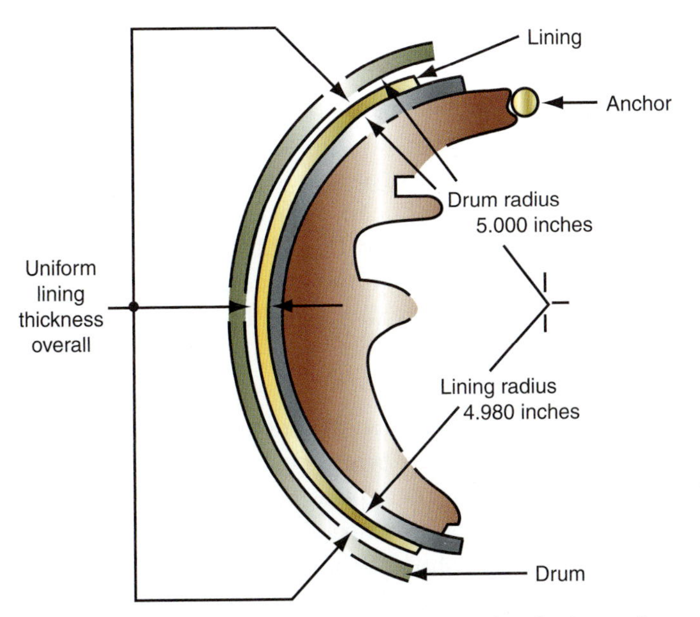

Figure 8-17 An undersized lining has a shorter radius than the drum radius (clearances exaggerated).

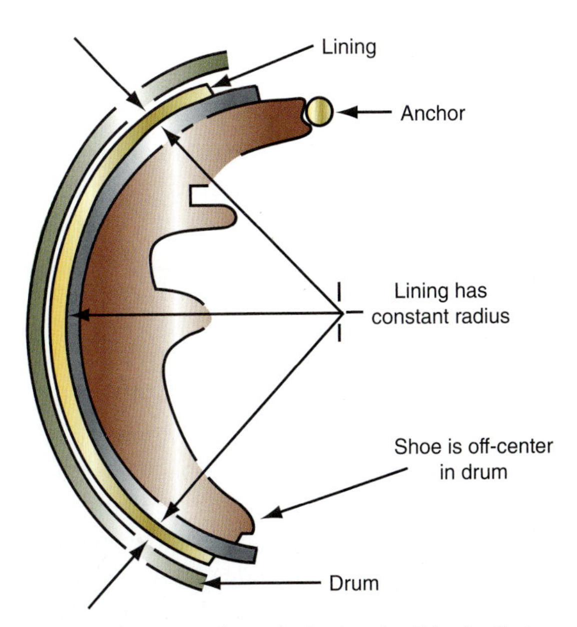

Figure 8-18 The overall arc of a fixed-anchor lining is offset slightly, and one end of the lining is thicker than the other (clearances exaggerated).

Brake linings and drums are less prone to overheating, and lining wear is not as severe as it was in the past. Therefore, arcing requirements are not as great, and precontoured linings are more easily provided by manufacturers.

Backing Plate

The backing plate is bolted either to the steering knuckle on the front suspension or to the axle flange or hub at the rear (**Figure 8-19**). The backing plate is the mounting surface for all other brake parts except the drum. The circumference of the backing plate is curved to form a lip that fits inside the drum circumference and helps to keep dirt, water, and road debris out of the brake assembly.

Brake **shoe anchors** are attached to the backing plate to support the shoes and keep them from rotating with the drum. Most modern brakes have a single anchor that is either

Brake **shoe anchors** hold the brake shoes on the backing plate.

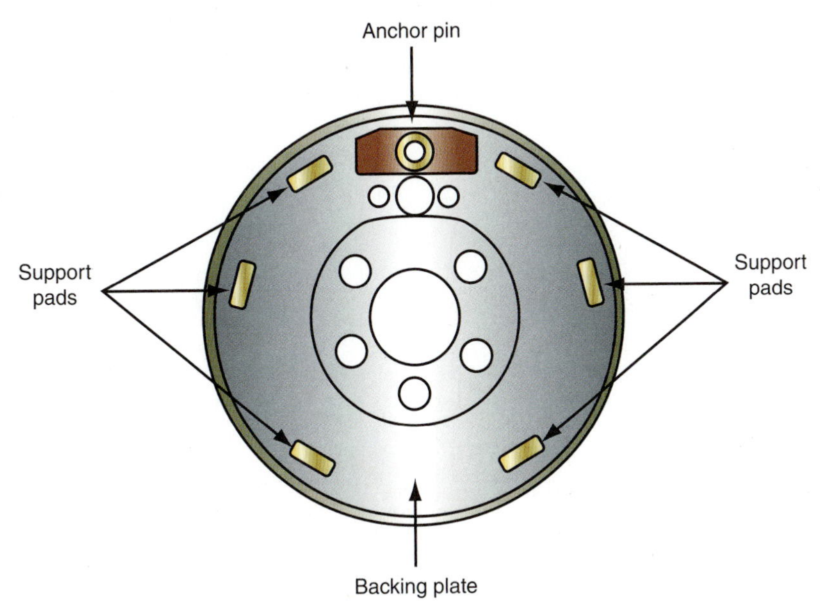

Figure 8-19 The backing plate holds the complete drum brake assembly. The shoe slides across the lubricated shoe contact pads as they are applied and released.

Piston stops keep the wheel cylinder pistons from coming out of the wheel cylinder.

Shop Manual
page 395

a round post or a wedge-shaped block. Some older drum brakes and a few late-model versions have separate anchors for each shoe.

All backing plates have some form of shoe-support pads stamped into their surfaces. The edges of the shoes slide against these pads as the brakes are applied and released. The pads thus keep the shoes aligned with the drum and other parts of the assembly. A light coat of brake lubricant should be applied to the pads to aid shoe movement and reduce noise.

Many backing plates have **piston stops**, which are steel tabs at the ends of the wheel cylinder. These stops keep the pistons from coming out of the cylinder during service. The wheel cylinder must be removed from a backing plate with piston stops for disassembly.

Wheel Cylinders

Wheel cylinders convert hydraulic pressure from the master cylinder to mechanical force that applies the brake shoes. Historically, many kinds of wheel cylinders have been used on different kinds of drum brakes. These included single-piston cylinders and stepped-bore cylinders with pistons of different diameters. Some wheel cylinders were installed to slide on the backing plate as they applied the brake shoes. Wheel cylinders for late-model drum brakes, however, are almost all two-piston, straight-bore cylinders that are mounted rigidly to the backing plate. The basic parts of all cylinders are the body, the pistons, the cups (seals), the spring and cup expanders, and two dust boots (**Figure 8-20**). Some cylinders include two shoe links, or pushrods, to transfer piston movement to the shoes. In other designs, the shoe webs bear directly against the cylinder pistons.

AUTHOR'S NOTE Some drums have a slot extending around the outer edge of the drum. The lip of the backing plate fits into this groove, providing better protection for the brake components.

The wheel cylinder body is usually cast iron, but aluminum cylinder bodies are becoming more common. The cylinder bore is finished to provide a long-wearing, corrosion-resistant surface. A fitting for a hydraulic line is provided at the center of the cylinder, between the two pistons. The bleeder screw also is tapped into the cylinder body at its highest point.

A piston is installed in each end of the wheel cylinder, and the inside end of each piston is sealed with a cup seal. Hydraulic pressure against the seal forces the seal lip to

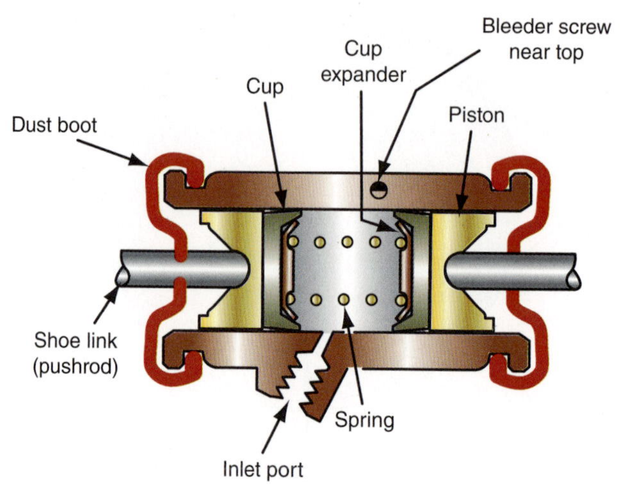

Figure 8-20 Almost all modern drum brakes use straight-bore, two-piston wheel cylinders such as the one shown in this cross section.

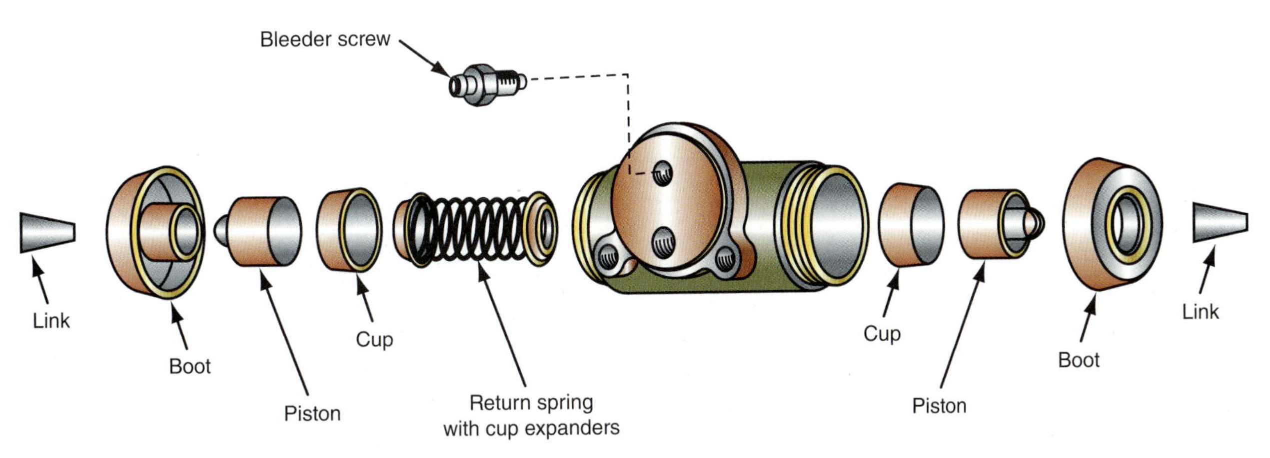

Figure 8-21 An exploded view of a typical wheel assembly.

expand against the cylinder bore and form a leak-free seal. Most wheel cylinders also have spring-loaded metal **cup expanders** that bear against the inner sides of the cups (**Figure 8-21**). Piston cup expanders were made to keep the lips of the seal in place when the brakes are released. The spring takes up any slack between the pistons and their pushrods and helps to center the pistons in the cylinder bore.

piston **cup expanders** hold the piston seals against the wheel cylinder walls when the brakes are released.

AUTHOR'S NOTE The wheel cylinder springs are ***not*** return springs. The shoe return springs push the pistons back in place.

Each end of the wheel cylinder is sealed with a dust boot to keep dirt and moisture out of the cylinder. The dust boots also prevent minor fluid seepage from getting onto the brake linings.

Return and Hold-Down Springs

Strong **return springs** retract the shoes when the brakes are released and hold them against their anchors and the wheel cylinder pushrod. Most return springs are tightly wound coil springs in which the coils touch each other when retracted (**Figure 8-22**).

return springs retract the brake shoes when the brakes are released.

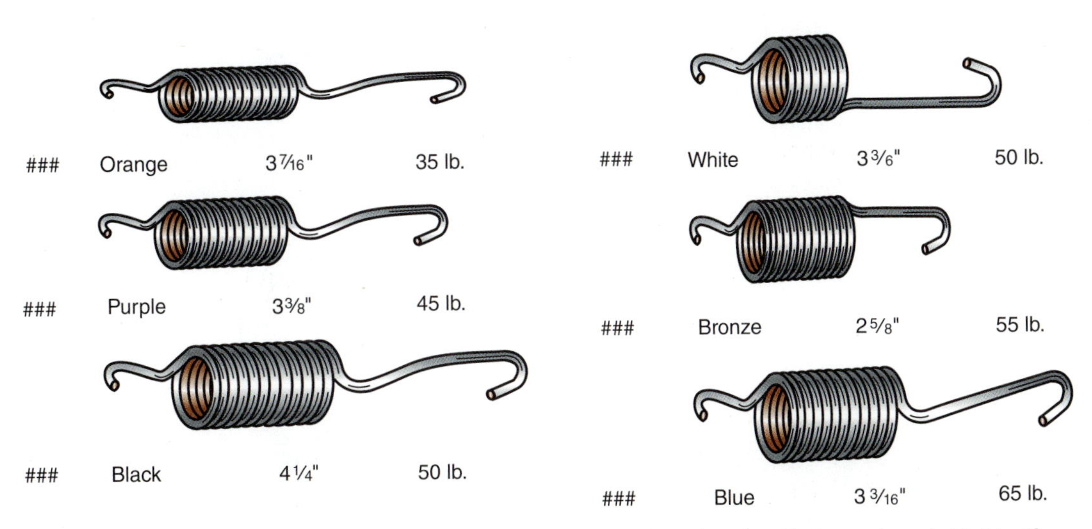

Figure 8-22 Return springs are identified by their part number, free length, and tension. Most are color coded to identify tension differences among similar-looking springs.

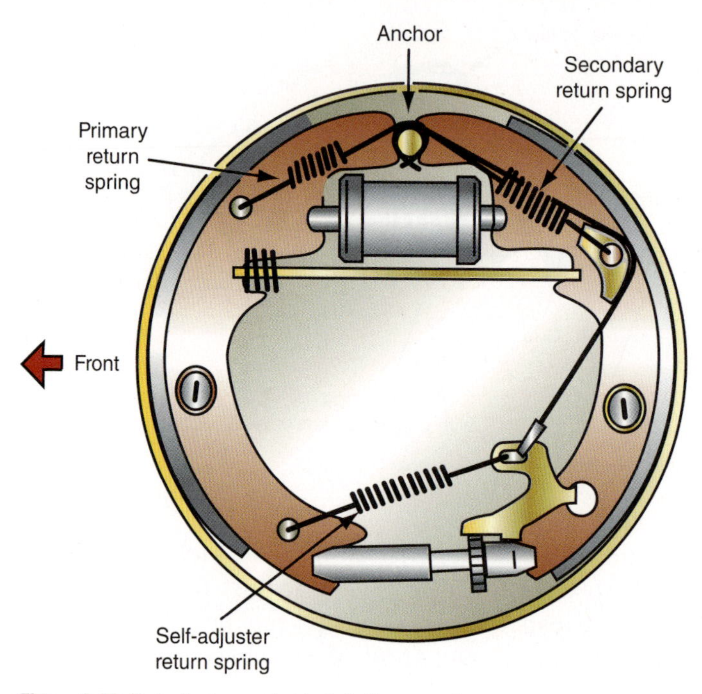

Figure 8-23 Typical return spring installation on a duo-servo system.

Shop Manual
page 398

Hold down springs are part of the brake shoe anchor.

Return springs often are color coded to indicate different tension values among springs of the same size and shape.

The type, location, and number of springs vary, but return springs are installed either from shoe to shoe or from each shoe to an anchor post (**Figure 8-23**).

Hold-down springs, clips, and pins also come in various shapes and sizes, but all have the same purpose of holding the shoes in alignment with the backing plate. Hold-downs must hold the shoes in position when providing flexibility for their application and release. **Figure 8-24** shows some of the varieties of hold-down springs and clips.

Some GM cars and some imported vehicles use a single large horseshoe-shaped spring that acts as both a shoe return spring and a hold-down spring (**Figure 8-25**).

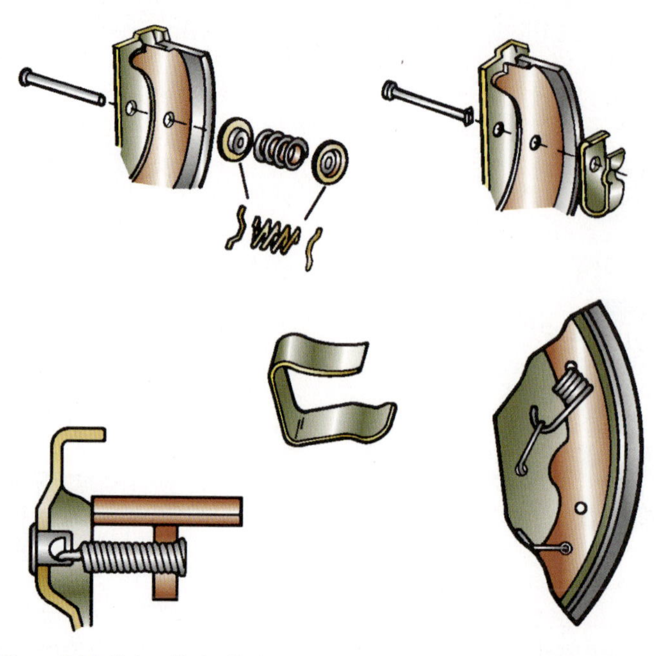

Figure 8-24 Various kinds of hold-down springs are used on different drum brakes.

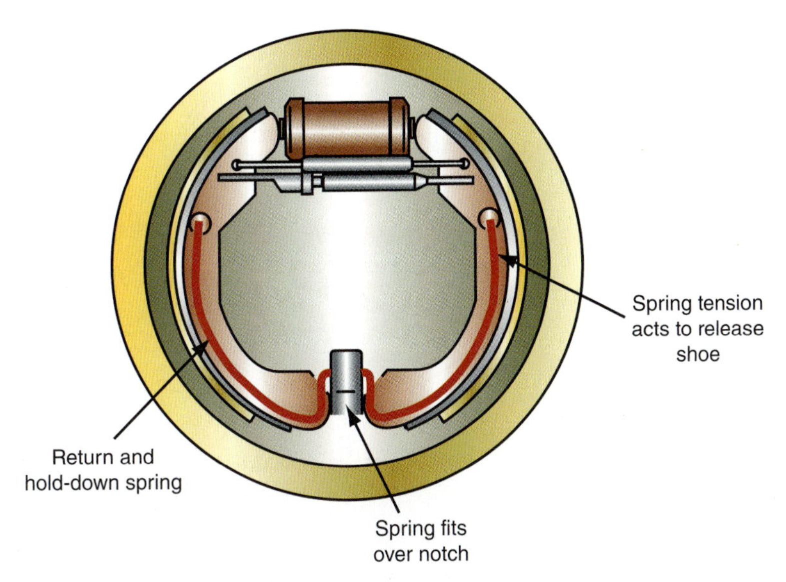

Figure 8-25 The big horseshoe-shaped spring serves as both a shoe return spring and a shoe hold-down on some late-model GM rear brakes.

Self-Adjusters

As the brakes are applied, the lining wears and becomes thinner. As the distance between the shoe and the drum becomes greater, the shoes must move farther before the lining contacts the drum. Therefore, the pistons of the wheel cylinder also must move farther, and more brake fluid must come from the master cylinder to apply the brakes. To provide more brake fluid, brake pedal travel will increase.

Shop Manual
page 403

Brake adjustment compensates for lining wear and maintains correct clearance between brake linings and the drum. Disc and drum brakes both require adjustment, but self-adjustment is a basic feature of disc brake design. Drum brakes, however, need extra cables, levers, screws, struts, and other linkage to provide self-adjustment and proper lining-to-drum clearance. Many different kinds of manual and automatic adjustment devices have been used on drum brakes over the years, but today automatic **self-adjusters** that operate star wheel or ratchet adjustment mechanisms are the most common. Automatic self-adjusters use the movement of the brake shoes to maintain proper shoe/drum clearance.

self-adjusters maintain proper brake shoe to drum clearance automatically.

Duo-Servo Star Wheel Adjusters. The bottoms of the shoes in duo-servo brakes are connected to each other by a link, and the shoes are held against the link by a strong spring (**Figure 8-26**). The link keeps the shoes aligned with each other and transfers the servo action of one shoe to the other during braking. It is not attached to the backing plate, but it moves back and forth with the shoes.

One half of the link is internally threaded, and the other is externally threaded. The externally threaded part has a **star wheel** near one end. When the two halves of the link are assembled, rotating the star wheel screws them together or apart. In this way, the adjustment link is lengthened or shortened to adjust the lining-to-drum clearance.

Even with self-adjusters, the initial adjustment of a duo-servo brake is made this way after new shoes are installed and the drum is resurfaced. Further adjustment is done automatically, however, by the self-adjustment mechanism for the life of the linings.

Duo-servo brakes commonly have self-adjusters operated by the secondary shoe. A cable, a heavy wire link, or a lever is attached to the secondary shoe. The cable, link, or lever is attached to a smaller lever or **pawl** that engages the star wheel (**Figure 8-27**). During braking in reverse, the secondary shoe moves away from the anchor post. If the lining is worn far enough and the shoe can move far enough, it will pull the cable, link, or lever to move the pawl to the next notch on the star wheel. When the brakes are released,

A **pawl** is a hinged or pivoted component that engages a toothed wheel or rod to provide rotation or movement in one direction while preventing it in the opposite direction.

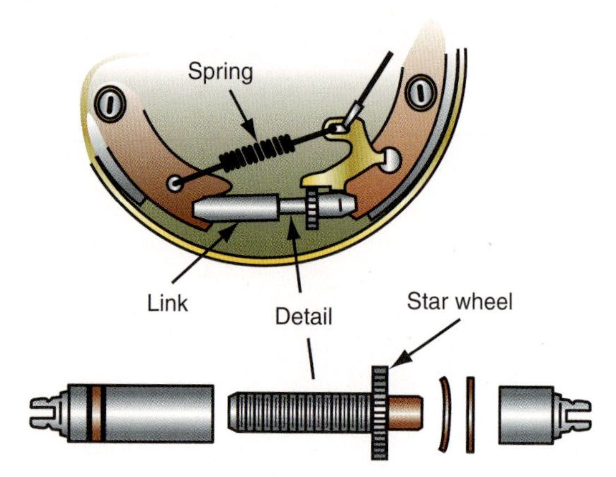

Figure 8-26 The link between the shoes of a duo-servo brake contains the adjustment star wheel.

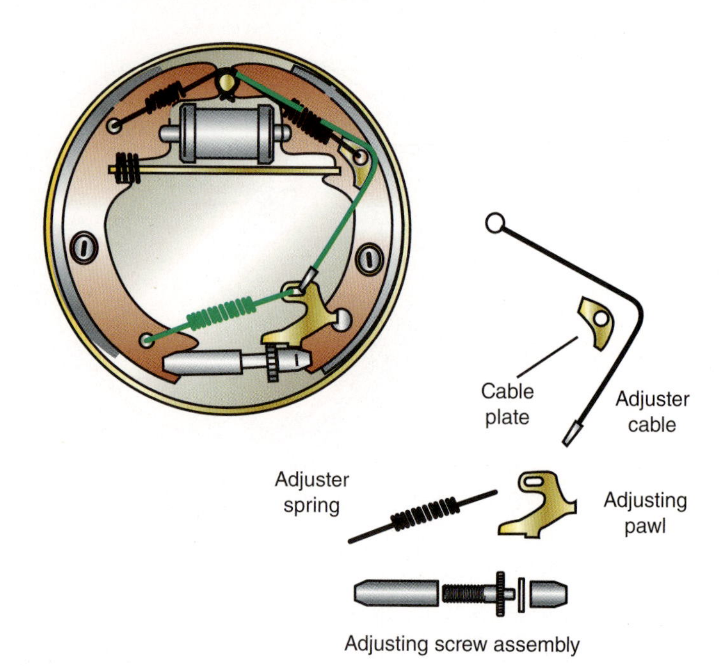

Figure 8-27 A cable-operated self-adjuster on a duo-servo brake moves the lever (pawl) upward when the secondary shoe moves off its anchor during reverse braking.

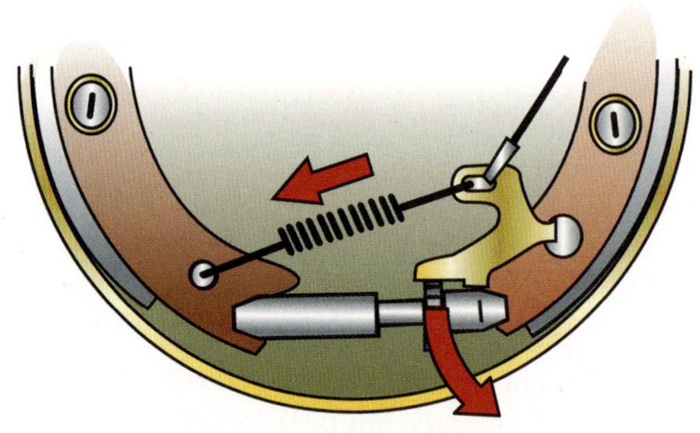

Figure 8-28 As the adjuster spring pulls the lever (pawl) downward, it turns the star wheel one notch to expand the adjuster link.

the pawl spring moves the pawl to rotate the star wheel (**Figure 8-28**). The rotation of the star wheel then expands the link one notch to take up the clearance between the linings and the drum.

A BIT OF HISTORY

Self-adjusters on drum brakes became common and quickly progressed to standard equipment on cars of the late 1950s and early 1960s. Mercury and Edsel models from Ford are credited with leading the industry to self-adjusting brakes with their 1958 models.

Ford was not the first carmaker to make self-adjusting brakes standard equipment, however. A decade earlier, Studebaker models of 1947 featured self-adjusting brakes, and they remained a Studebaker exclusive through 1954. Public disinterest (and technician distrust) caused Studebaker to drop this feature until the early 1960s when the rest of the industry finally recognized the benefits.

Automatic adjusters for drum brake systems were first used in 1957. Automatic adjusters have been used on all domestic cars and some import cars since 1963. A few years later even trucks had automatic brake adjusters installed.

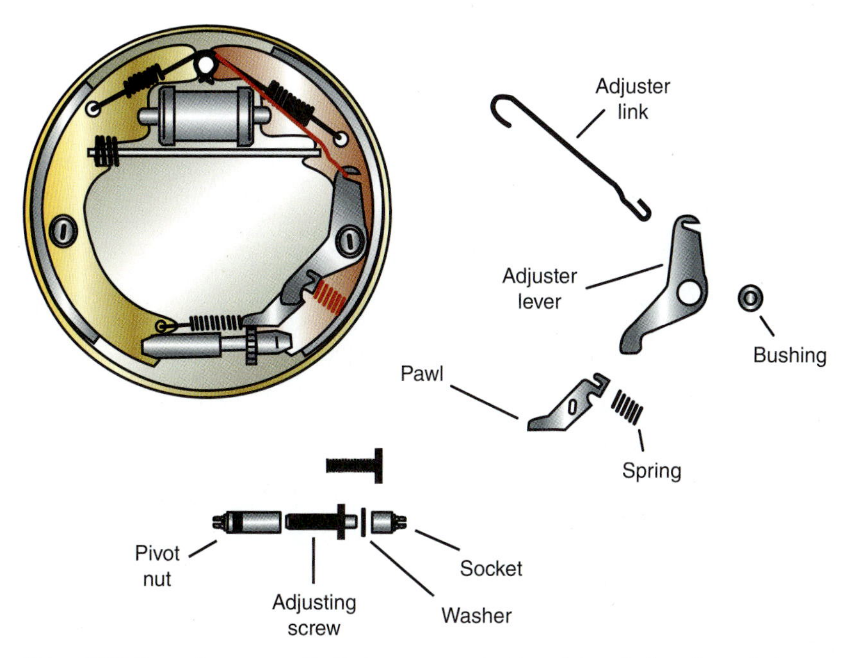

Figure 8-29 Some duo-servo self-adjusters are operated by a link and a lever instead of a cable.

Quite a bit of variety exists in the specific linkage used on different duo-servo brakes. Cable- and link-operated self-adjusters usually move the pawl to engage the next notch in the star wheel as the brakes are applied in reverse. Lever-operated self-adjusters (**Figure 8-29**) usually move the pawl to rotate the star wheel as the brakes are applied. Some cable-operated self-adjusters move a pawl mounted under the star wheel to adjust the brakes during application, not release. These cable-operated adjusters usually have an **overload spring** in the end of the cable that lets the cable move without breaking if the pawl or star wheel is jammed.

Self-adjuster **overload springs** let the cable move without breaking if the pawl or star wheel is damaged.

The overload spring also prevents overadjustment during very hard braking when the drum may distort and let the shoes move farther out than normal. To prevent this, the overload spring stretches with the cable and does not let the pawl actuate. Overadjustment will not occur during hard braking when the brakes are in normal adjustment.

Left- and right-hand threaded star wheel adjusters are used on the opposite sides of the car. Therefore, parts must be kept separated and not intermixed. If a star wheel adjuster is installed on the wrong side, the adjuster will not adjust at all and may unadjust, causing excessive shoe/drum clearance.

AUTHOR'S NOTE Duo-servo brakes only self-adjust in reverse, but leading-trailing brakes will self-adjust as necessary going in either direction, or even when the parking brake is applied.

Leading-Trailing-Shoe Star Wheel Adjusters. Even more variety exists in the linkage used for star wheel adjusters on leading-trailing-shoe brakes. The star wheel self-adjusters can be operated by either the leading or the trailing shoe and can work whenever the brakes are operated in either forward or reverse.

The star wheel is usually part of the parking brake strut that is mounted between the two shoes. A pawl can be operated by either the leading or the trailing shoe to engage and rotate the star wheel as the brakes are applied or released.

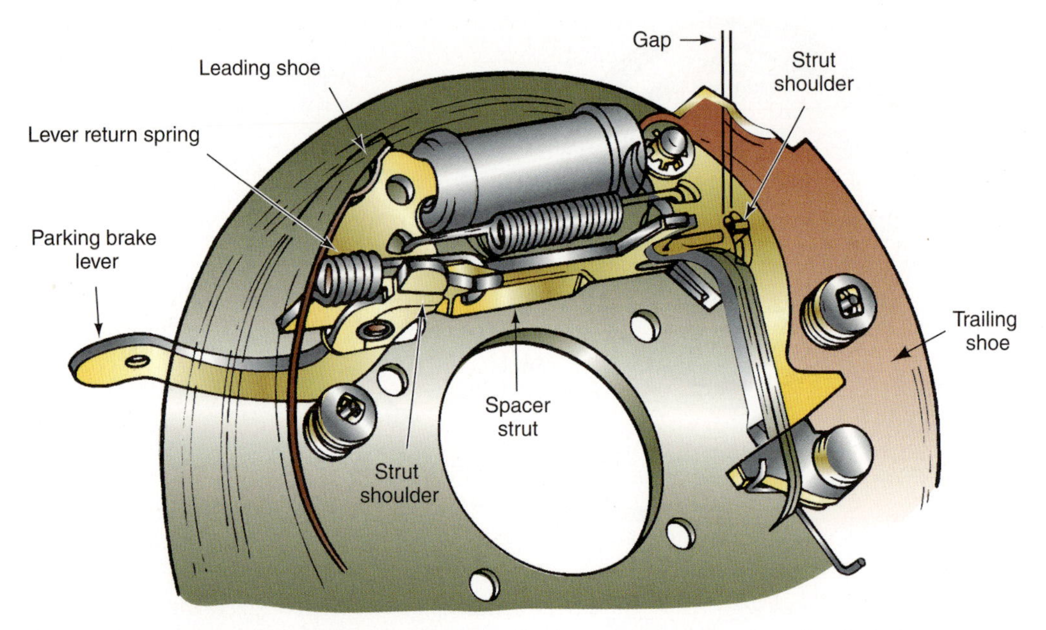

Figure 8-30 The ratchet-type self-adjuster for leading-trailing brakes is operated by the parking brake lever.

Leading-Trailing-Shoe Ratchet Adjusters. A lever-latch ratchet adjuster has a large lever and small latch attached to the leading shoe. Teeth on the lever and the latch form a ratchet mechanism that moves the shoes outward to take up excess clearance as the brakes are applied.

Another kind of ratchet adjuster consists of a pair of large- and small-toothed ratchets and a spacer strut mounted between the shoes (**Figure 8-30**). The strut is connected to the leading shoe through the hand brake lever and the inner edge of a large ratchet. As the gap between the shoes and the drums becomes greater, the strut and the leading shoe move together to close the gap. More movement will cause the large ratchet on the trailing shoe to rotate inward against a small spring-loaded ratchet and reach a new adjustment position.

Other similar systems are actuated by the parking brake and adjust when the parking brake is applied and released. An adjustment strut is attached to the parking brake lever (**Figure 8-31**). As the lining wears, application of the parking brake will restore the proper lining clearance.

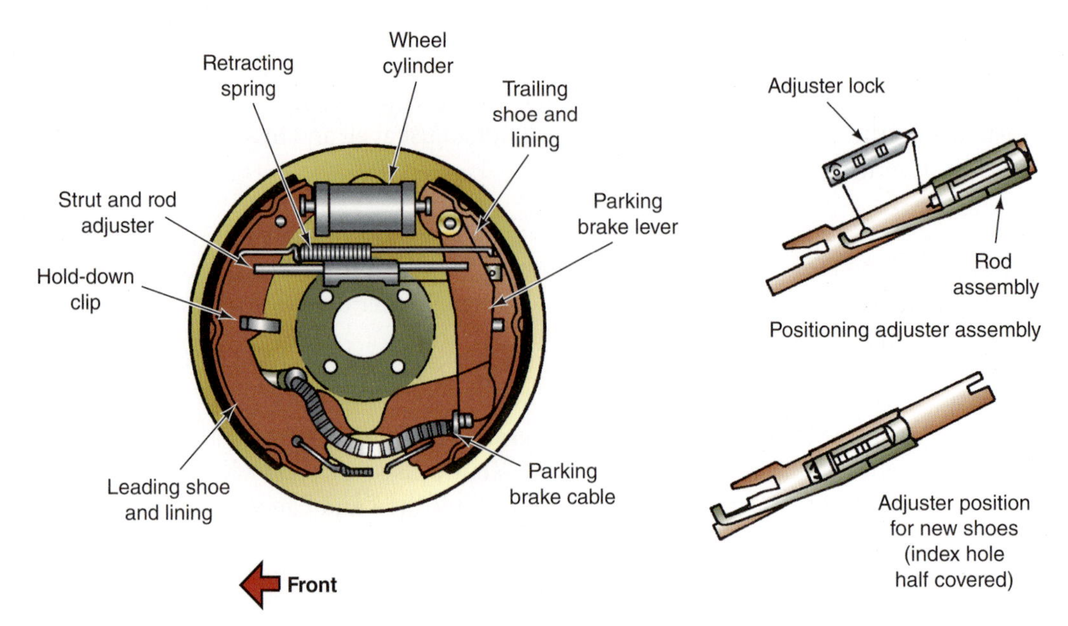

Figure 8-31 Semiautomatic self-adjuster for leading-trailing brakes.

The strut-quadrant self-adjuster is a variation of the strut-rod adjuster shown in Figure 8-31. The strut-quadrant adjuster has a rotating semicircular quadrant as half of the engaging ratchet mechanism.

AUTHOR'S NOTE Installation of these types of self-adjusters can be a real nightmare at times. Nothing seems to fit and the springs are strong and awkward to grip. The service manual instructions usually leave a lot to the imagination. The best method I have found is to look for a technician with experience on these systems and pick his or her brain for the best and quicker way to assemble and mount the different parts.

Leading-Trailing-Shoe Cam Adjusters. Cam-type adjusters use cams with an adjuster pin that fits in a slot on the shoes (**Figure 8-32**). As the brake shoes move outward, the pin in the slot moves the cam to a new position if adjustment is needed. Shoe retraction and proper lining clearance are always maintained because the pin diameter is smaller than the width of the slot on the shoe. These brakes can be adjusted even while the vehicle is at a standstill because brake pedal application is all that is needed to move the cams into proper adjustment. Because the brakes can be adjusted completely with one application of the brake pedal, this adjuster is sometimes called the one-shot adjuster.

Self-Adjuster Precautions. The self-adjuster mechanisms described in the preceding paragraphs are just a few common examples of such devices used on late-model drum brakes. When servicing these brakes, examine the self-adjuster parts closely and pay attention to how they are assembled. One previous paragraph mentioned that left- and right-hand parts exist for the left and right wheels of most systems. If parts are interchanged from side to side, the self-adjusters will not work.

Examine self-adjuster parts closely for wear and damage. Teeth wear off and adjusters freeze up. All threaded parts must move freely. Cable guides wear at the points where cables slide back and forth. Spring anchors can bend, stretch, or wear. Holes and slots wear where parts pivot or springs are anchored. Cables stretch with age and use. Many shop owners make it standard practice to replace self-adjuster cables whenever brakes are serviced and other parts when they show any sign of wear. Finally, if unfamiliar with a

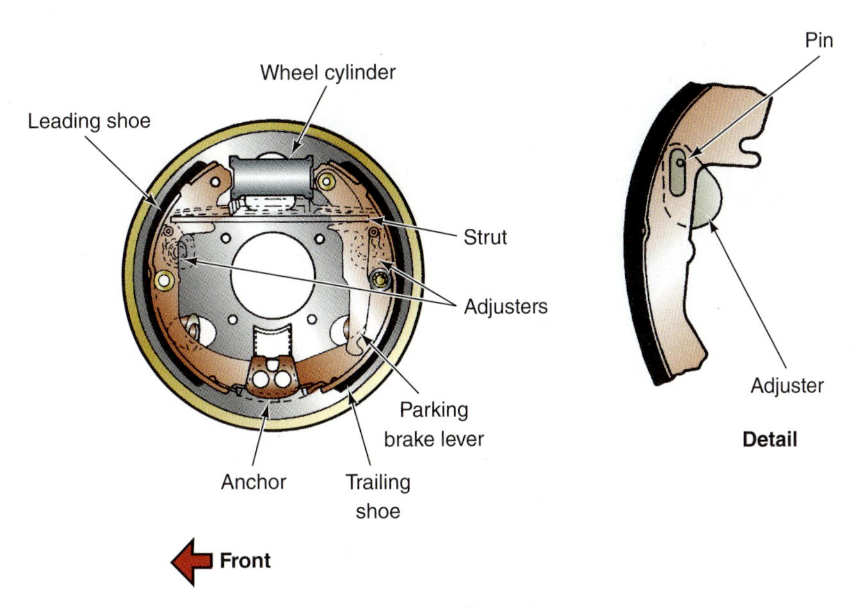

Figure 8-32 Cam-type self-adjuster for leading-trailing brakes.

In most cases of current drum leading-trailing brake systems, it is assumed that the leading shoe is always the shoe toward the front of the vehicle. This is assumed because the vehicle is normally braked while moving forward, and the wheel cylinder is at the top of the assembly.

particular self-adjuster installation, service one wheel at a time and use the opposite brake assembly as a reference for parts installation.

Parking Brake Linkage

Almost all rear drum brake installations include mechanical parking brake linkage. The linkage basically consists of a cable, a lever, and a strut. The lever and strut spread the shoes against the drum when the cable pulls on the lever. The parking brake strut may contain part of the self-adjuster as explained in previous paragraphs. Chapter 9 covers parking brakes in detail.

DRUM BRAKE DESIGNS

Dozens of drum brake designs have been used over the decades of automobile manufacturing. Full-servo, partial-servo, non-servo, two-leading-shoe, two-trailing-shoe, and centerplane are just a few of the brake designs that are part of automobile history. Today, two designs account for almost all of the drum brakes installed on late-model vehicles:

1. **Leading-trailing brakes** (**Figure 8-33**), also called partial-servo or non-servo brakes
2. **Duo-servo brakes** (**Figure 8-34**), also called dual-servo or full-servo brakes

The following sections describe these brake designs and explain their operations.

Self-Energizing and Servo Actions

Two terms used to describe drum brake operation are "self-energizing" and "servo." They refer to the leverage developed on the brake shoe as it contacts the drum and the action of one shoe to help apply the other.

Brake shoes are described as leading or trailing and primary or secondary. Leading and trailing shoes are components of leading-trailing (servo) brakes. The assumption is that the **leading shoe** is the one that moves from the wheel cylinder's end that points in the direction of drum rotation. To identify the leading brake shoe, put a hand at the

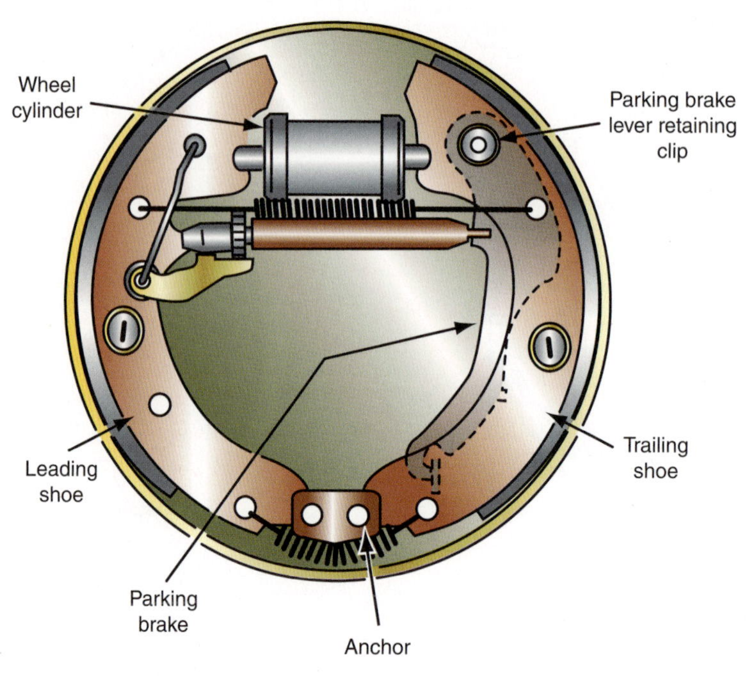

Figure 8-33 A typical leading-trailing brake.

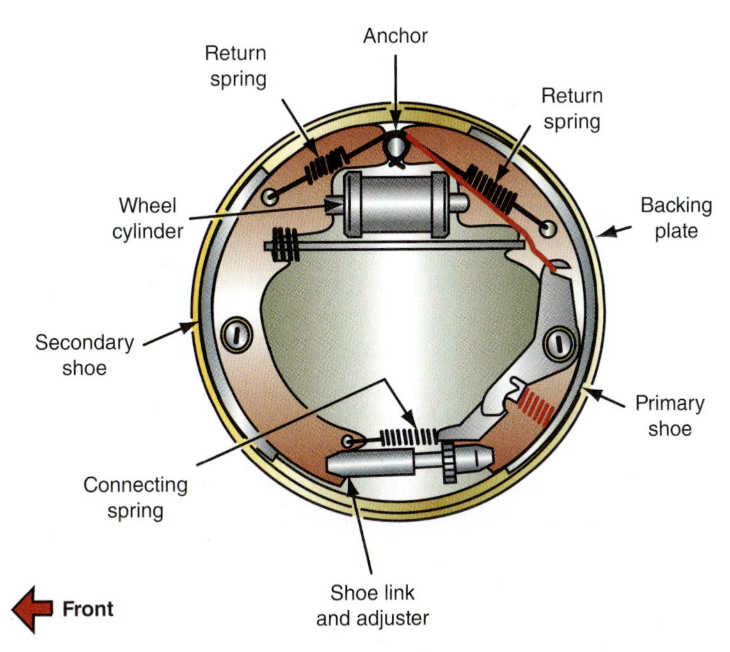

Figure 8-34 Typical duo-servo brake.

position of the wheel cylinder, and then point in the direction of drum rotation: clockwise or counterclockwise. The shoe that is pointed to first is the leading shoe. Think about this for a moment, and it is obvious that the leading shoe can be the front or the rear shoe, depending on whether the drum is rotating forward or in reverse.

When the brakes operate, the wheel cylinder forces one end of the leading shoe outward against the drum. The other end of the shoe is forced back solidly against its anchor post or anchor block. The cylinder end of the leading shoe contacts the drum first and develops friction against the rotating drum. The drum friction actually pulls the shoe into tighter contact with the drum (**Figure 8-35**) and aids the hydraulic force of the cylinder

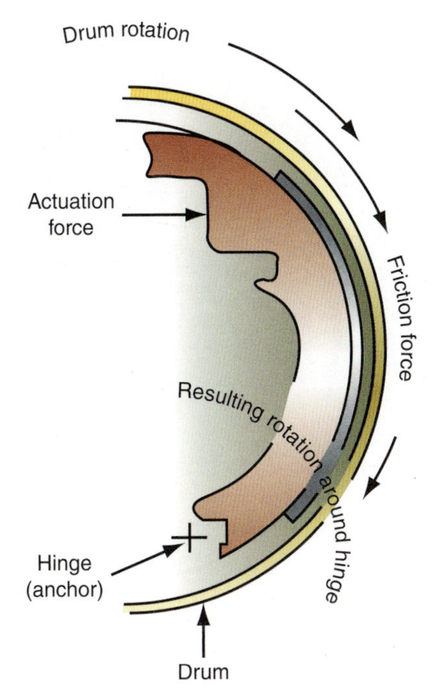

Figure 8-35 Self-energizing operation of a brake shoe develops as friction against the rotating drum pulls the shoe into tighter contact with the drum.

to apply the brake shoe. As drum-to-lining contact increases, the rest of the lining is forced by the cylinder and pulled by friction against the drum to stop rotation. This is self-energizing action by the leading shoe.

In a leading-trailing brake, the reaction of the **trailing shoe** to drum rotation is opposite to the reaction of the leading shoe. As the wheel cylinder tries to force the trailing shoe outward against the drum, rotation tries to force the shoe back against the cylinder. Eventually, hydraulic force overcomes drum rotation as the wheel slows, and the trailing shoe contributes to braking action. The trailing shoe, however, is said to be non-self-energizing (**Figure 8-36**).

In another brake design, the self-energizing action of one brake shoe can be used to help apply the other shoe. In a duo-servo brake, the **primary shoe** has the position of the leading shoe in a leading-trailing brake. The primary shoe is the shoe toward the front of the vehicle and many times has a shorter lining than the secondary shoe. The ends of both shoes opposite the wheel cylinder are not mounted on a rigid anchor attached to the backing plate as in a leading-trailing brake. The ends of the two shoes are linked to each other through the star wheel adjuster, and the ends (normally by the bottom) of the shoes float in the drum.

As the primary shoe is applied by the wheel cylinder, it develops self-energizing action as in a leading-trailing brake. Instead of being forced against an anchor, however, the primary shoe is forced against the **secondary shoe** and applies leverage to force the secondary shoe against the drum. This is duo-servo action (**Figure 8-37**), or the action of mechanically multiplying force. The secondary shoe is then forced against the drum by the wheel cylinder force at one end and the servo action of the primary shoe at the other.

Self-energizing action and servo operation are forces that naturally occur in internal-expanding drum brakes but do not exist in disc brakes. Engineers put these actions to work in different ways in different brake designs that are described and illustrated in the following sections.

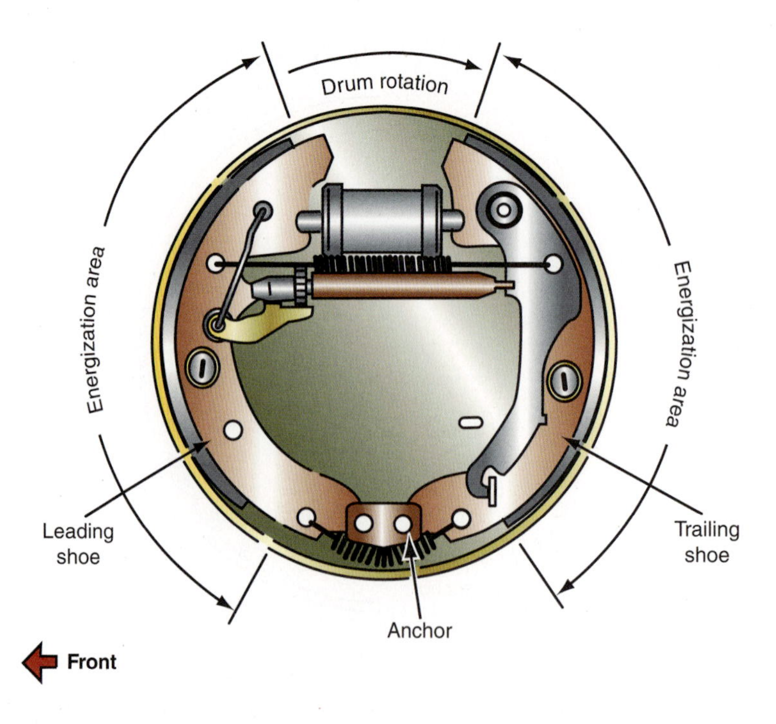

Figure 8-36 The leading shoe is self-energizing, but the trailing shoe is non-self-energizing in a leading-trailing brake.

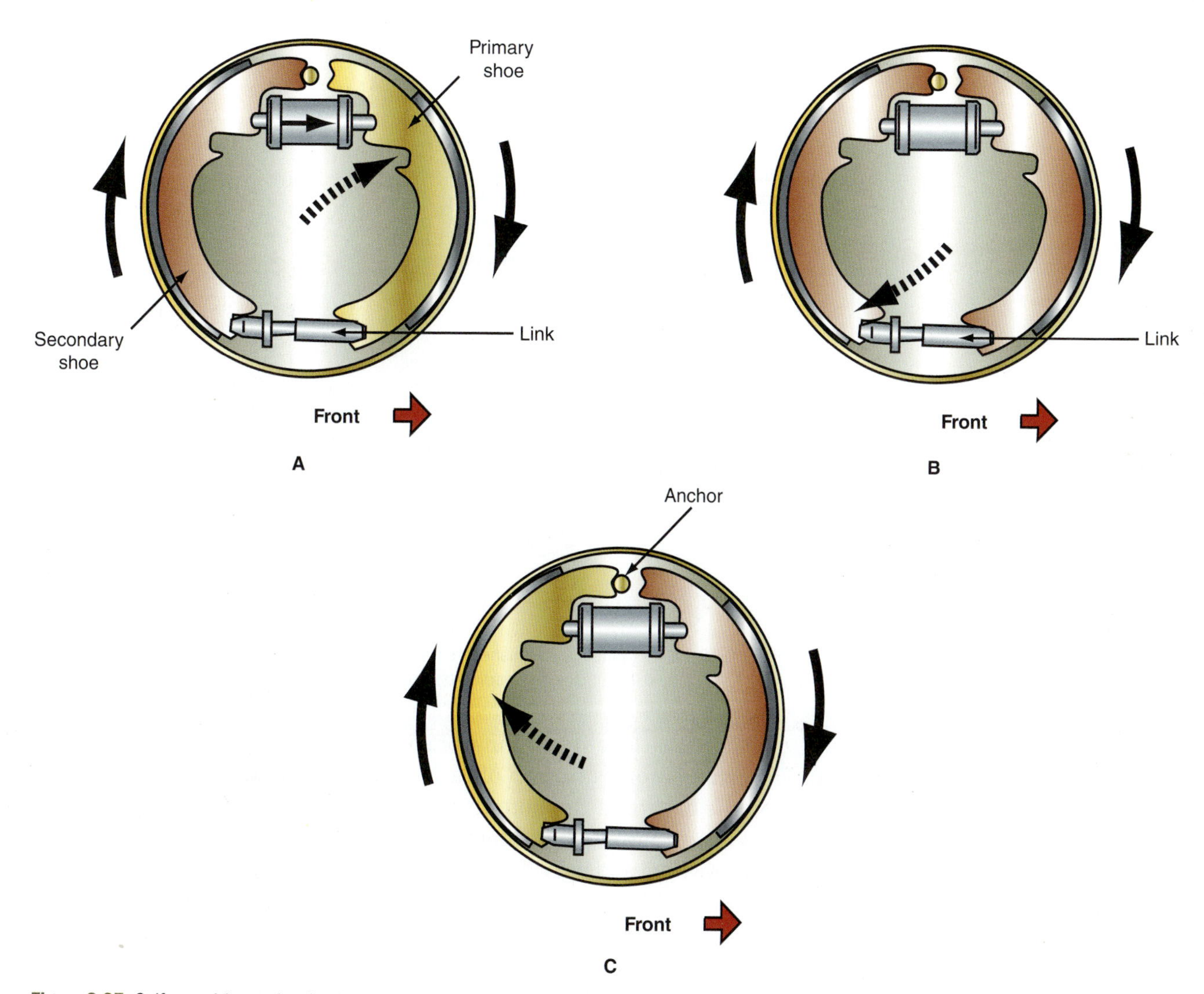

Figure 8-37 Self-energizing action for the primary shoe begins as the wheel cylinder forces it against the rotating drum (A). Reaction force of the primary shoe is transferred through the link to the secondary shoe (B). This transfer of force is servo action. The servo action of the primary shoe against the secondary shoe causes it also to become self-energizing (C). The top of the secondary shoe is forced against the anchor to complete the brake application.

Leading-Trailing Brakes Operation

Leading-trailing or non-servo brake design was common in four-wheel drum brake systems up to the late 1960s but was steadily replaced by duo-servo four-wheel drum brakes as cars got heavier and faster. Duo-servo brakes are a more powerful drum brake design, and leading-trailing brakes almost vanished from the industry in the 1960s. They began to reemerge, however, as rear brakes on smaller and lighter cars with front disc brakes in the 1970s. Today, leading-trailing brakes are used as rear drum brakes on most FWD automobiles, as well as on some light trucks with front discs.

In a typical leading-trailing brake installation, the wheel cylinder is mounted at the top of the backing plate, and the cylinder pushrod or shoe links bear against the upper ends of the shoe webs. The lower end of each shoe bears against, or is held onto, an anchor block or anchor post toward the bottom of the backing plate. One or two strong return springs usually hold the lower ends of the shoes against the anchor, and another return spring usually is installed between the upper ends of the shoes to hold them together and hold them against the wheel cylinder pushrod. Various kinds of self-adjusters and parking brake linkage described earlier in this chapter also are installed.

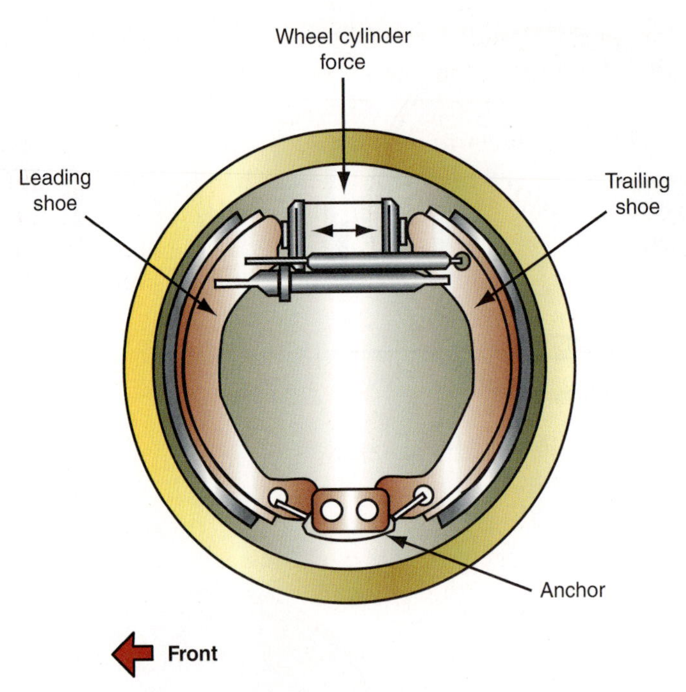

Figure 8-38 Wheel cylinder force pushes both the leading and trailing shoes against the fixed anchor in a leading-trailing brake.

The wheel cylinder of a leading-trailing brake acts equally on each brake shoe (**Figure 8-38**). The cylinder forces the top of each shoe outward toward the drum, and each shoe pivots on the anchor at the bottom. Each shoe thus operates separately and independently from the other. The leading shoe develops self-energizing action as described previously, however, and provides most of the braking force. The force of drum rotation works against the wheel cylinder force on the trailing shoe, so the trailing shoe is not self-energizing. The trailing shoe often is said to be nonenergized even though it is technically energized by the wheel cylinder. Because of this operation, the leading shoe wears much faster than the trailing shoe, so during a brake inspection, this situation is not unusual.

The front shoe is the leading shoe for forward braking in a leading-trailing brake as described here. For reverse braking, however, the roles are reversed, and the rear shoe becomes the self-energized leading shoe.

When comparing a leading-trailing brake to a duo-servo brake described in the next section, it will be understood that a duo-servo brake provides greater braking force. One might then ask why engineers would install a "weaker" brake on an advanced, late-model automobile. Part of the reason is that FWD automobiles generate up to 80 percent of their braking force on the front wheels. Overly powerful duo-servo brakes at the rear could easily cause the rear wheels to lock as weight shifts forward and the rear end becomes very light during braking. Ironically, a more powerful drum brake could reduce overall braking performance.

Equally important, antilock brakes make overall brake balance at all four wheels a critical factor. To make brake lockup easier to control while maintaining the best total braking efficiency, engineers have actually had to "detune" the rear drum brakes on some FWD cars. For the sake of brake balance and overall efficiency, leading-trailing brakes will be used on the rear wheels of many vehicles for a long time to come.

Duo-Servo Brakes Operation

Just as do leading-trailing brakes, duo-servo brakes have a single, two-piston wheel cylinder mounted toward the top of the backing plate. A return spring holds each shoe against the cylinder pushrod. A duo-servo brake is distinguished from a leading-trailing brake by a single anchor post at the top of the backing plate. The top of each shoe web has a semicircular notch, and the return springs hold the notches tightly against the anchor post. The cylinder pushrod acts on the shoes a couple of inches below the anchor. The bottoms of the shoes are not anchored to the backing plate but are joined to each other by the adjuster link. Another spring holds the shoes tightly together and against the adjuster link. Think of the shoes of a duo-servo brake as hanging from the anchor post and floating within the drum at the bottom, and that is the key to their operation.

The forward shoe of a duo-servo brake is called the primary shoe, and its lining is shorter than the lining on the rear shoe, which is called the secondary shoe. As the wheel cylinder applies force against both brake shoes during forward braking, the cylinder force pushes the top of the primary shoe away from the anchor. The top edge or leading edge of the primary shoe contacts the rotating drum first and develops self-energizing action just as does the leading shoe of a leading-trailing brake. The self-energized primary shoe is both drawn and forced against the drum, but its lower end does not bear against a fixed anchor. Instead, the lower end of the primary shoe applies servo force to the bottom of the secondary shoe.

The secondary shoe then becomes self-energized, with the self-energizing action beginning at the bottom of the shoe. This combined servo and self-energized action actually works against the wheel cylinder, but servo force applied by the primary shoe to the secondary shoe is greater than the hydraulic force of the wheel cylinder alone. Servo force and the self-energizing action of the secondary shoe force its upper end against the anchor post and wrap the shoe tightly into the drum. The secondary shoe—with its longer lining—thus applies most of the braking force for forward braking.

During reverse braking, the roles of primary and secondary shoes are reversed. The forward shoe with its smaller lining becomes the secondary shoe and receives servo action from the rear shoe. Even though the forward shoe with smaller lining must apply most of the braking force in reverse, it is not a problem because vehicle speed is usually much lower and the vehicle travels a shorter distance in reverse. Braking efficiency remains within safe limits.

Duo-servo brakes can apply a lot of braking power very efficiently, but excessive servo action can lead to brake grabbing and locking. Engineers, therefore, must balance several factors to take full advantage of servo power but maintain smooth braking. These factors include lining size on both shoes and placement of the lining on the primary shoe, which can be centered or mounted high or low on the shoe. Primary and secondary linings also can have different coefficients of friction to achieve smooth brake application.

AUTHOR'S NOTE Duo-servo drum brakes can develop a lot of braking force through combined hydraulic and mechanical actions. From the 1950s through the mid-1960s, duo-servo brakes with large diameter drums, wide linings, and cooling fins to help heat dissipation provided powerful braking for powerful cars. Disc brakes, however, provide even more powerful and efficient braking. By the mid-1960s, disc brakes began to replace drum brakes on front wheels and on all four wheels of some high-performance cars. The 1976 revision to FMVSS 105 made disc brakes the most practical designs to achieve the brake system performance requirements. However, duo-servo drum brakes are still used on the rear wheels of older vehicles, particularly RWD cars, trucks, and sport utility vehicles. Today, four-wheel disc brakes are found even on trucks.

SUMMARY

- The basic parts of a drum brake are a drum, a hub, brake shoes, a backing plate, a hydraulic wheel cylinder, shoe return springs, hold-down springs, an adjusting mechanism, and a parking brake.
- Rear drums mount on a flange attached to the axle on RWD vehicles.
- The wheel cylinder has pressurized brake fluid applied to the pistons inside the cylinder. These pistons act on the brake shoes to move them into contact with the brake drum.
- The wheel cylinder and brake shoes are mounted on a solid backing plate, which is attached to the axle housing or steering knuckle.
- The brake shoes are a primary (or a leading) shoe and a secondary (or trailing) shoe.
- Brake linings are primarily made of semi-metallic or synthetic materials that have replaced asbestos materials. Lining materials are attached to the shoes by either bonding or riveting.
- Brake shoes are held against their anchors and adjusters by springs.
- Servo brakes are systems in which one brake shoe increases the braking force of the other brake shoe.
- In a self-energizing brake, the rotation of the drum increases the braking force of the lining against the drum. The force of the brake drum is added to the force supplied by the wheel cylinder.
- A duo-servo brake is a servo brake in which servo action takes place whether the vehicle is moving forward or backward.
- A non-servo brake is a brake in which one shoe does not apply a force to the other shoe.
- In a leading-trailing brake, the leading shoe is self-energizing but the other is not self-energizing.
- Primary shoes and leading shoes face to the front of the vehicle.
- Secondary shoes and trailing shoes face to the rear of the vehicle.
- Self-adjusting brakes automatically adjust lining-to-drum clearances.

REVIEW QUESTIONS

Essay

1. List the components of a typical drum brake assembly.
2. Describe the function of a wheel cylinder.
3. Describe brake gas fade.
4. What are the major advantages to using drum brakes?
5. Since duo-servo brakes supply more braking force than leading-trailing drum brakes, why would engineers use the leading-trailing design?
6. What determines the braking surface area of a drum brake?
7. What is the function of a self-adjuster?
8. What is a self-energizing brake?
9. What is the difference between the braking action of leading-trailing and duo-servo brakes?
10. How are the brake shoes returned to their released position when the brake pedal is released?

Fill in the Blanks

1. _______________ are the components that mount the friction material or linings that contact the drums.
2. The hydraulic and friction components are attached to the _______________.
3. _______________ operation is the action of a drum brake shoe when drum rotation increases the application force of the shoe by wedging it tightly against the drum.
4. The _______________is usually part of the parking brake strut that is mounted between the two shoes of leading-trailing brakes.
5. A duo-servo brake is distinguished from a leading-trailing brake by a single anchor post at the _______________ of the backing plate.
6. _______________brakes, also called partial-servo or non-servo brakes.
7. Duo-servo brakes commonly have self-adjusters operated by the _______________ shoe.
8. Cam-type adjusters use cams with an _______________ pin that fits in a slot on the shoes.
9. Wheel cylinders convert _______________ pressure from the master cylinder to _______________ force that applies the brake shoes.
10. The forward shoe of a duo-servo brake is called the _______________shoe.

Multiple Choice

1. *Technician A* says that for the same brake pedal force, duo-servo brakes apply a greater braking force than leading-trailing-shoe brakes. *Technician B* says that the servo action of the duo-servo brakes takes place only when the car is moving forward. Who is correct?

   A. A only
   B. B only
   C. Both A and B
   D. Neither A nor B

2. *Technician A* says that the leading shoe lining receives the greatest wear on a leading-trailing shoe brake. *Technician B* says that leading-trailing brakes are non-servo brakes. Who is correct?

   A. A only
   B. B only
   C. Both A and B
   D. Neither A nor B

3. *Technician A* says that some leading-trailing-shoe brakes with an adjustable parking brake self-adjust when the service brakes are applied. *Technician B* says that some brakes with an adjustable parking brake strut self-adjust when the parking brake is applied. Who is correct?

   A. A only
   B. B only
   C. Both A and B
   D. Neither A nor B

4. *Technician A* says that on a duo-servo brake, the force applied by the drum to the primary shoe aids the force applied by the wheel cylinder. *Technician B* says that on a leading-trailing brake, the force applied by the drum to the primary shoe opposes the force applied by the wheel cylinder. Who is correct?

   A. A only
   B. B only
   C. Both A and B
   D. Neither A nor B

5. Two technicians are discussing automatic brake adjusters. *Technician A* says that self-adjusting brakes were first used in 1947 by Studebaker. *Technician B* says that self-adjusting brakes were used on all domestic drum brakes in 1963. Who is correct?

   A. A only
   B. B only
   C. Both A and B
   D. Neither A nor B

6. *Technician A* says that rear brake drums are usually separate assemblies from the rear hubs. *Technician B* says that Tinnerman nuts at the wheel studs hold the drum onto many rear hubs. Who is correct?

   A. A only
   B. B only
   C. Both A and B
   D. Neither A nor B

7. While discussing a typical leading-trailing brake installation, *Technician A* says the wheel cylinder is mounted at the top of the backing plate, and the cylinder pushrod or shoe links bear against the upper ends of the shoe webs. *Technician B* says the lower end of each shoe bears against, or is held onto, an anchor block or anchor post toward the bottom of the backing plate. Who is correct?

   A. A only
   B. B only
   C. Both A and B
   D. Neither A nor B

8. *Technician A* says that most self-adjusters are the same for both sides of the vehicle. *Technician B* says that some technicians service one side of the vehicle at a time and use the other side as a reference while assembling drum brake assemblies. Who is correct?

   A. A only
   B. B only
   C. Both A and B
   D. Neither A nor B

9. *Technician A* says on leading-trailing brakes, the lining on the leading and trailing shoes are usually the same length and positioned in the same location on each shoe. *Technician B* says on duo-servo brake, however, the primary shoe lining is shorter than the secondary lining and may even have a different coefficient of friction. Who is correct?

   A. A only
   B. B only
   C. Both A and B
   D. Neither A nor B

10. *Technician A* says that self-adjusters perform lining-to-drum adjustments without driver intervention. *Technician B* says that self-adjusting duo-servo brakes occur when the vehicle is moving forward and the brakes are applied. Who is correct?

    A. A only
    B. B only
    C. Both A and B
    D. Neither A nor B

CHAPTER 9

PARKING BRAKES

Upon completion and review of this chapter, you should be able to:

- Explain the function of parking brakes.
- Identify the basic types of parking brake systems.
- Identify types of parking brake controls.
- Identify the types of cables used to operate the parking brakes.
- Identify and explain the operation of disc brake and drum brake parking brakes.
- Explain the operation of electric parking brakes.

Terms To Know

Ball-and-ramp	Equalizer	Parking brake control
Conduit	Intermediate lever	Screw-and-nut
Eccentric		

INTRODUCTION

After the service brakes stop the moving car, the parking brakes hold it stationary. Parking brakes are often mistakenly called "emergency" brakes, but parking brakes are not intended to be used as an alternative to the service brakes to stop vehicles. The stopping power available from parking brakes is much less than from service brakes. Because the parking brakes work only on two wheels or on the driveline, much less friction surface is available for braking energy. Also, since the hydraulic system has two circuits, total hydraulic failure should be a very rare occurrence.

PARKING BRAKE OPERATION

The parking brake system is not a part of the hydraulic braking system. It is either mechanically operated by cables and levers to apply the rear brakes, or it can be operated mechanically or by its own hydraulic system to activate a drum brake on the transmission or drive shaft.

Many vehicles have a "press-to-release" feature on the parking brakes. To release the parking brakes, pressure (force) is applied to the parking brake pedal. This releases the locking ratchet and allows the pedal to return to the off or up position.

Most parking brake systems use the service brake shoes or disc pads. Some heavy-duty trucks use a separate set of shoes or pads, such as transmission or drive shaft parking brakes, which are called independent parking brakes. These brakes actually lock the drive shaft in place to keep the vehicle from rolling.

Parking brake actuators may be operated either by hand, foot, or electronically. Many small- and medium-size vehicles use a hand-operated parking brake lever mounted in the console between the front seats (**Figure 9-1**). When the lever is pulled up, the parking brakes are applied. A ratchet-and-pawl mechanism acts to keep the brake lever applied. To release the lever and the brakes, a button on the lever is pressed and the lever is moved

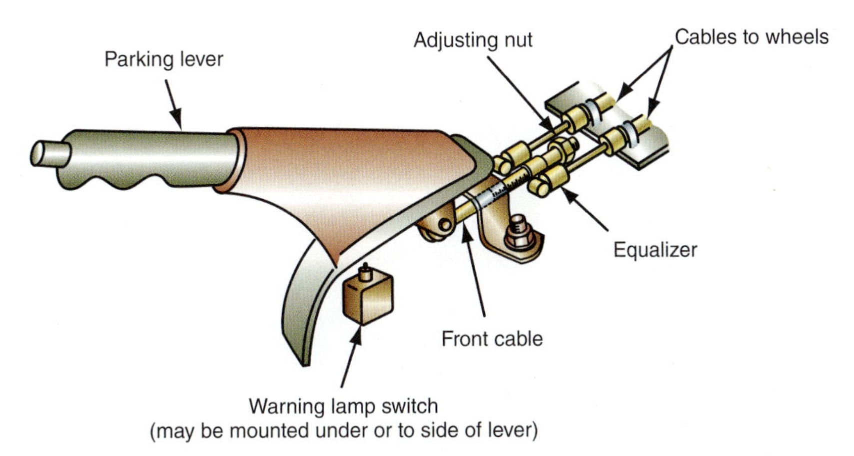

Figure 9-1 A typical lever-operated parking brake control.

to unlock the ratchet. Many late-model vehicles use an electronically actuated parking brake system, where an electric motor is used to apply the cables with the push of a button. The electrically applied parking brake can also be released at the touch of a button or when the vehicle is taken out of the park mode. Some medium trucks and mobile construction equipment use the hydraulic service brakes as the parking brakes. With the vehicle/equipment stopped and the service brakes applied, an electric solenoid is activated. The solenoid closes the hydraulic lines between the wheels and master cylinders, effectively locking all wheels. The service brake pedal can be released until it is time to unlock the wheels.

Figure 9-2 shows a typical foot-operated pedal with a ratchet and pawl. Stepping on the pedal applies the brakes and engages the ratchet and pawl. A release handle and rod or cable is attached to the ratchet release lever. When the release handle is pulled, the pawl is lifted off the ratchet to release the brakes.

Some vehicles engage and disengage the parking brakes with a switch (**Figure 9-3**). These parking brakes are applied by an electric motor (**Figure 9-4**) or the parking brakes

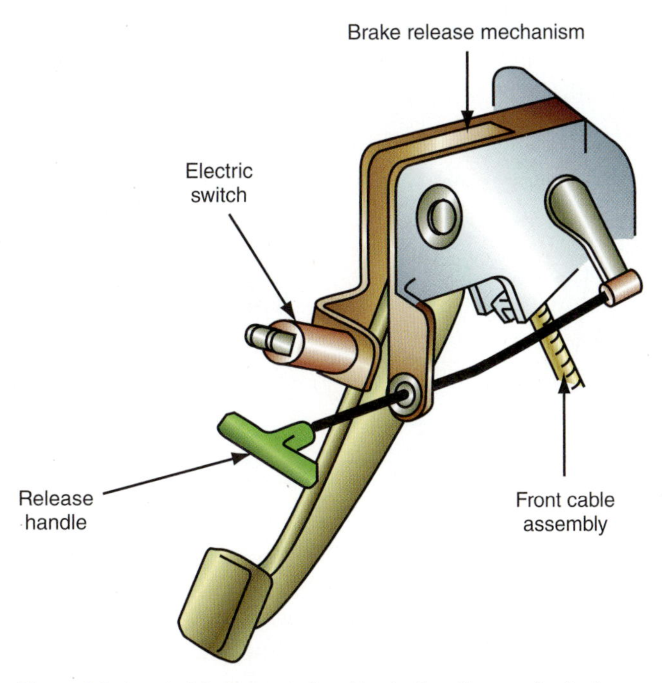

Figure 9-2 A typical foot-operated parking brake with a mechanical release handle.

Figure 9-3 Parking brakes are applied and released electronically on many vehicles.

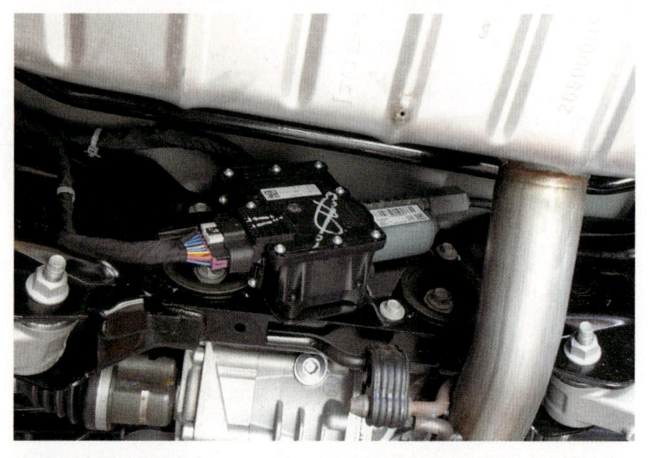

Figure 9-4 An electric parking brake actuator can be used to apply the parking brake cable.

disengage whenever the vehicle starts to move. Some electronic parking brakes can be utilized as a "hill hold" feature for use in vehicles with a manual transmission, or some are equipped with an autostop feature that applies and holds the vehicle whenever it comes to a complete stop and releases as soon as the accelerator is reapplied.

This chapter explains the most common types of parking brake levers, handles, cables, and other linkage parts as well as warning lamps and switches. The final sections of this chapter describe typical drum, disc, and drive shaft parking brake assemblies and their operation.

PARKING BRAKE CONTROLS—LEVERS AND PEDALS

Shop Manual
page 435

The parking brake on many late-model vehicles is applied and released electrically by a parking brake actuator (Figure 9-4). Many older and a few current vehicles have a handle under the instrument panel that is pulled to release the parking brake (**Figure 9-5**). Aside from the operation of the **parking brake control** the linkage for most parking brakes remains the same.

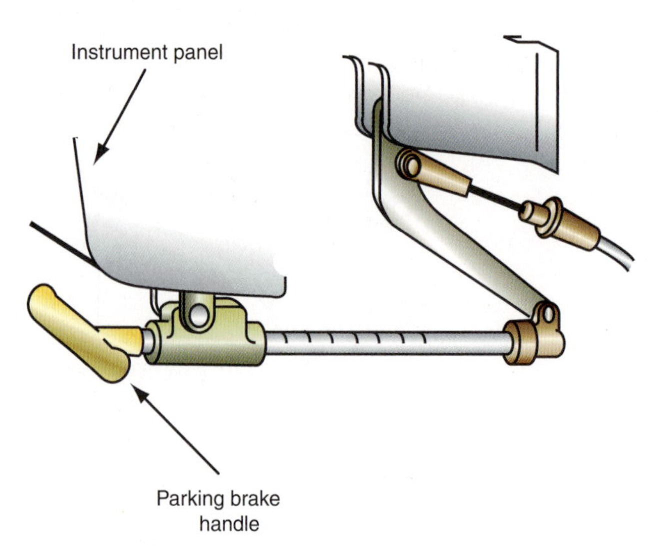

Figure 9-5 Many vehicles still use a manual release for the parking brake.

Most parking brakes use the service brake shoes or pads to lock the rear wheels after the vehicle is stationary. The parking brakes can be set most securely if the service brake pedal is pressed and held when the parking brake control lever or pedal is applied. The hydraulic system applies greater force to the shoes or pads than the parking brake mechanical linkage can apply. When the hydraulic system is used to set the brakes, the parking brake linkage simply takes up slack in the system and holds the shoes or pads tightly against the drums or rotors.

Levers

The control lever for lever-operated parking brakes usually is installed between the two front seats. As the lever is pulled upward, the ratchet mechanism engages to keep tension on the cables and hold the brakes applied. To release the brakes, the spring-loaded button in the end of the lever is pressed and held while the lever is lowered to the floor.

The lever-operated parking brakes on some Chevrolet Corvettes are examples of a design in which the lever drops back to the floor after the brakes are applied. The cables and linkage hold the brakes applied, but the lever returns to the released position. To actually release the parking brakes, pull up on the lever until some resistance is felt; then press and hold the button in the end of the lever while moving the lever back to the released position. The parking brake control lever on these Corvettes is located between the driver's seat and the door sill. If the control lever stayed in the upward position with the brakes applied, it would be difficult to climb in and out of the car.

AUTHOR'S NOTE For those of you who are lucky enough to operate or maintain a right-handed vehicle (steering wheel on right), the parking brake lever will be to the left of the driver and a braking brake pedal under the right side of the dash. Other than that, all other aspects of the parking brake system are the same.

Pedals

In a pedal-operated parking brake system, the pedal and its release mechanism are mounted on a bracket under the left end of the instrument panel. As the pedal is pushed downward by the driver's foot, the ratchet mechanism engages to keep tension on the cables and hold the brakes applied (**Figure 9-6**). A spring-loaded handle or lever is pulled or the pedal is pressed to release the brakes. A return spring moves the pedal to the released position. A rubber bumper is used to absorb the shock of the released parking brake pedal. If this bumper is missing, the pedal will break the warning light switch after a few operations.

FMVSS 105 requires that parking brakes must hold the vehicle stationary for 5 minutes on a 30 percent grade in both the forward and reverse directions (**Figure 9-7**). FMVSS 105 also specifies that the force needed to apply the parking brakes cannot exceed 125 pounds for foot-operated brakes or 90 pounds for hand-operated brakes. Some heavy full-size cars built in the late 1970s and early 1980s had trouble meeting the brake-holding requirements without exceeding the allowed maximum application force.

Manufacturers solved the problem with a pedal that had a very high leverage ratio but required two or three applications with the foot to set the brakes completely. The first pedal stroke partially applied the brakes, and the ratchet mechanism held the linkage in this position when the pedal was released. The second or third pedal stroke applied the brakes completely. A single pull on the release handle released the brakes.

Figure 9-6 The pedal ratchet is part of the parking brake pedal assembly.

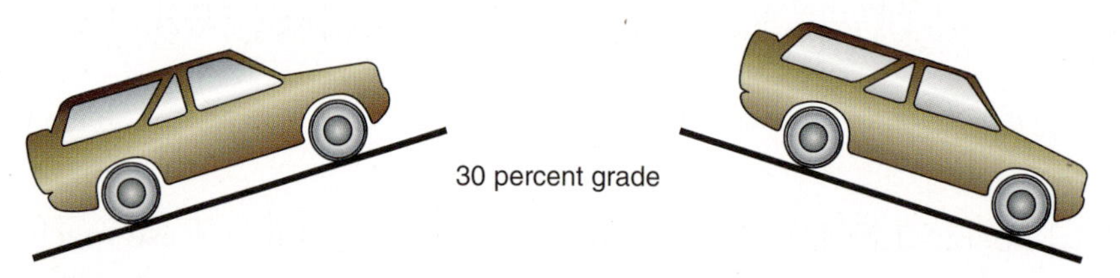

Figure 9-7 The parking brake must hold the vehicle on a 30 percent grade for 5 minutes in both the forward and reverse directions.

Automatic Parking Brake Release

An automatic parking brake release mechanism is a convenience feature offered by many carmakers. Electric parking brake systems can be released by turning the key on and pressing on the service brake pedal and turning the parking brake off, or by simply putting the vehicle in gear and hitting the accelerator pedal.

WARNING LAMPS

Shop Manual
page 435

On many vehicles, the same warning lamp used to indicate a low fluid level in the master cylinder will also indicate that the parking brake is applied. A normally open switch on the control linkage closes as the pedal is pressed or the lever is pulled. The lamp will not light, however, unless the ignition is on. Parking brake lamp switches are adjusted so that the lamp stays lit until the brake is released completely. In most vehicles, the red brake warning light will come on during key on, engine off, even if the fluid is correct and the parking brake is off. This is part of the "lamp check" or "prove out mode" to check the operation of the various warning light systems. The lamp should switch off when the engine is running if everything is correct. In electronic parking brake systems, the parking brake module controls the warning lamp function by providing a message to the instrument panel cluster that the brake has been released.

PARKING BRAKE LINKAGE

Parking brake linkage transmits force equally from the control pedal or lever to the shoes or pads at the rear wheels. There are as many different linkage designs as there are different vehicles, but all have the same job and work in basically the same way. The following paragraphs explain the cables, rods, levers, and equalizers or adjusters used in a typical parking brake linkage.

Shop Manual
page 438

Cables

Most parking brakes use cables to connect the control lever or pedal to the service brakes (**Figure 9-8**). The cables are constructed of high-strength strands of steel wire that are tightly twisted together. In vehicles with electronic parking brake actuators, an electric motor controls the application of the parking brake cables and, in some cases, the parking brake cable tension. Parking brake cables must transmit hundreds of pounds of force without jamming, breaking, or stretching. The ends of the cables have different kinds of connectors that attach to other parts of the linkage. Some cables have threaded rods or clevises at their ends. Others have ball or thimble-shaped connectors that fit into holes and slots on other parts of the linkage.

Parking brake cables are made of high-strength strands of steel wire that are tightly twisted together.

The front cable connects the parking brake lever or pedal to the **equalizer**, which provides balanced braking force to each wheel. The equalizer is only a lever mounted on a pivot or a U-shaped grooved guide. Pulling the front cable moves the lever or guide. The lever or guide transmits the force equally to the two wheel cables (**Figure 9-9**).

The **equalizer** may be referred to as the parking brake adjuster. In some cases, it is the point at which the parking brake is, in fact, adjusted.

Some vehicles have a three-part cable installation, which includes an intermediate cable that passes through the equalizer. Typically, one of the two rear cables is attached directly to this intermediate cable with a connector (**Figure 9-10**). The other rear cable is attached to the intermediate cable with some kind of cable adjuster, usually a turn buckle.

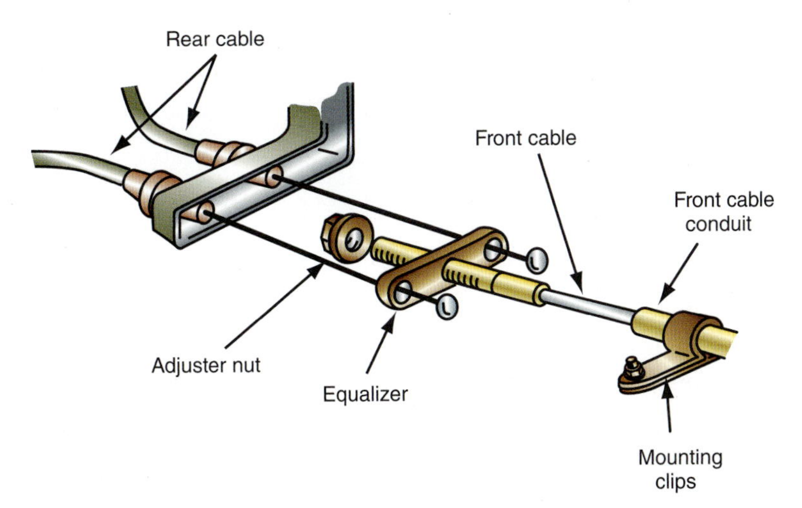

Figure 9-8 Typical parking brake cable installation showing an adjusting mechanism.

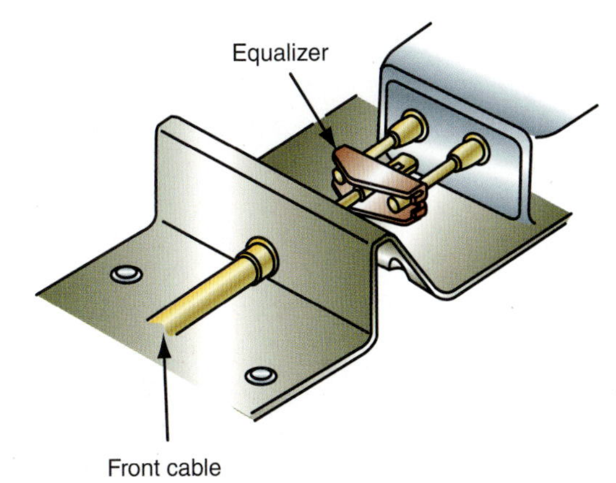

Figure 9-9 This parking brake equalizer is under the center console.

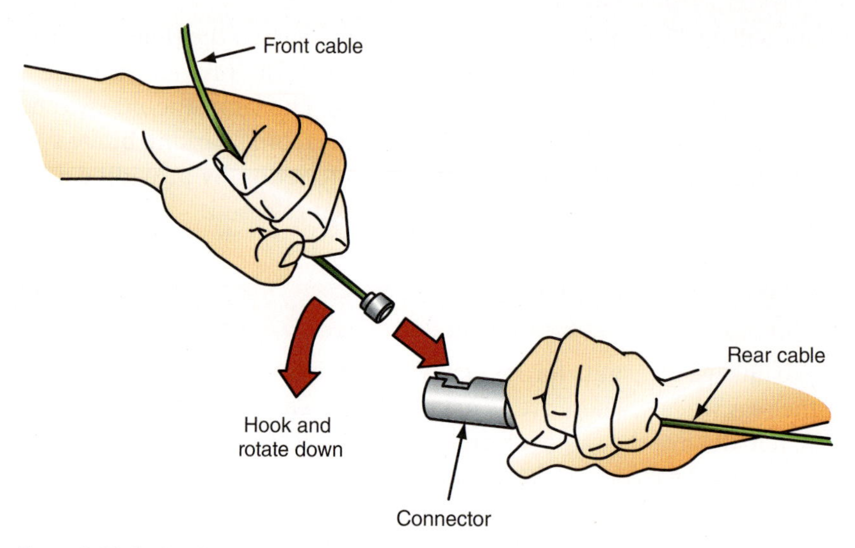

Figure 9-10 Parking brake cable connector.

Some lever-operated parking brakes have a separate cable for each rear wheel attached to the control lever. Each cable is adjusted separately, and an equalizer is not necessary. The wheel cables run from each wheel all the way to the pedal or lever mechanism within the passenger compartment. Adjustment of the cables may be made at the pedal or lever.

Cable retainers and hooks maintain cable position on the rear axle, frame, and underbody of the vehicle (**Figure 9-11**). These retainers allow the cable to flex and move at their point of body attachment and help the equalizer to provide its equalizing action.

A **conduit** is a flexible metal housing or jacket that houses the parking brake cables to protect them from dirt, rust abrasion, and other damage.

Most control cables and rear brake cables are partially covered with a flexible metal **conduit** or housing (**Figure 9-12**). The cable slides inside the conduit and is protected from chaffing or rubbing against the underside of the vehicle. One end of the cable conduit is fastened to a bracket on the underside of the vehicle with some type of retaining clip, and the other end is attached to the brake backing plate (**Figure 9-13**). Many cables are coated with nylon or plastic, which allows them to slide more easily through the conduit. The coatings help to reduce corrosion and contamination and make parking brake application easier.

AUTHOR'S NOTE Conduits are common in other systems as well. Check the area around your house or shop, and note the conduits that protect the electrical wiring. Parking brake conduits perform the same protective function for the brake cables.

Rods

The most common use of solid steel rods in parking brake linkage is in lever-operated systems to span a short distance in a straight line to an equalizer or intermediate lever. The linkage rod is usually attached to the control lever by a pin. The other end of the rod is often threaded to provide linkage adjustment. This control system has been dropped from typical passenger cars and light trucks.

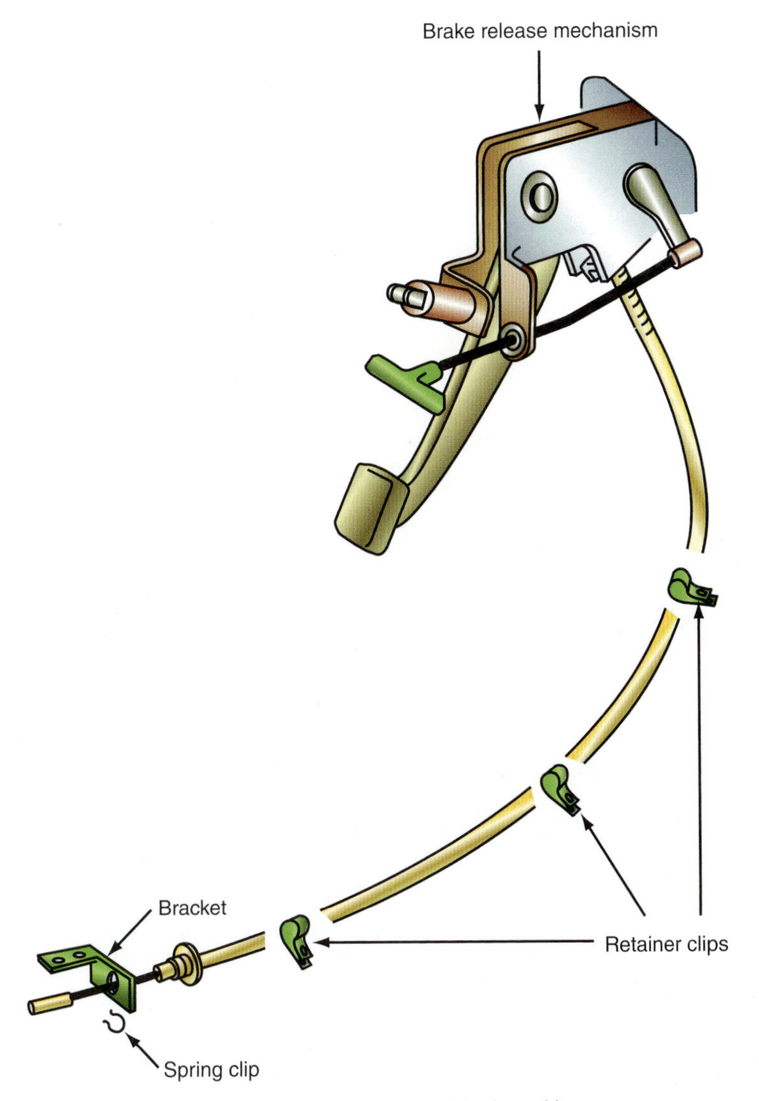

Figure 9-11 Cable retainers and clips hold the cables in position.

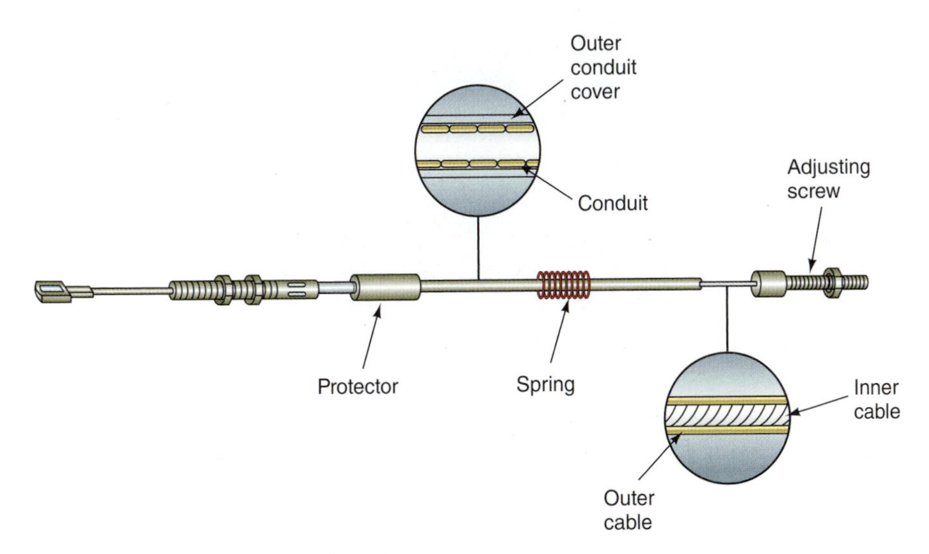

Figure 9-12 Typical cable and conduit.

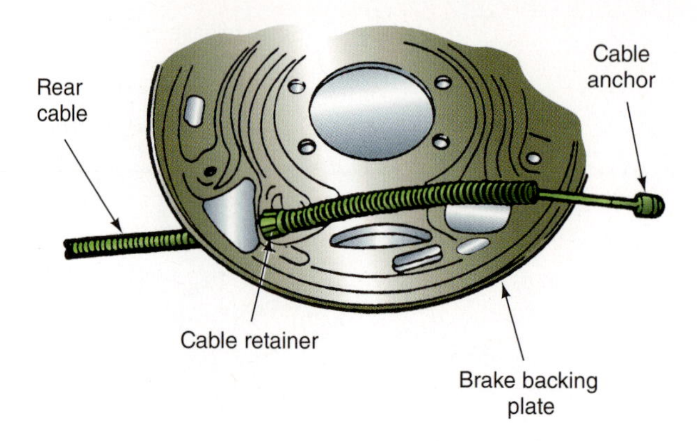

Figure 9-13 A cable retainer secures the conduit to the backing plate.

A BIT OF HISTORY

A historic cause of anxiety for novice drivers learning to cope with manual transmissions is learning how to start from a complete stop, going up hill. Many student drivers panic while trying to hold the brake pedal, release the clutch, and apply the throttle at the same time. If only they had a third foot, life would be so much easier. In 1936, the Wagner Electric company patented the *NoRol* hill holder, which first appeared on Studebaker models of that year. The *NoRol* hill holder consisted of linkage from the clutch pedal to the brake master cylinder that maintained hydraulic pressure after the brake pedal was released and until the clutch was engaged. Pontiac and Graham also offered versions of the hill holder as optional equipment, and it continued as standard on Studebakers with manual transmissions through 1964. Several later Subaru models had similar automatic hill-holding devices. Now some vehicles with electronic parking brakes have a feature that applies the brakes called auto-hold. Whenever the vehicle comes to a complete stop auto-hold (**Figure 9-14**) can maintain the vehicle on a hill until the accelerator pedal is applied.

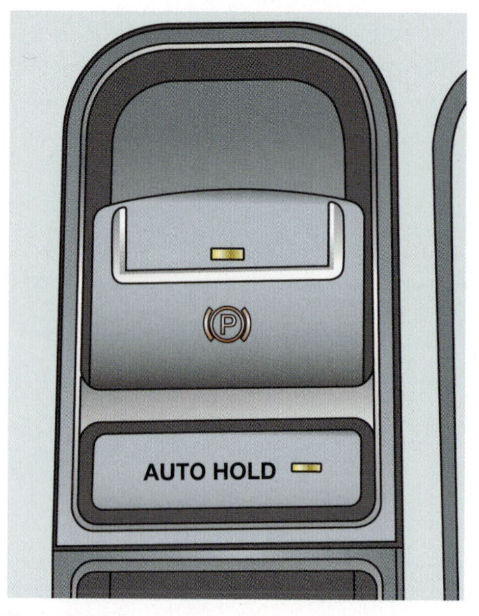

Figure 9-14 An electronic parking brake with an auto-hold feature.

Levers

Chapter 2 of this *Classroom Manual* explains how levers are used to multiply force in mechanical linkage. Mechanical leverage is necessary in parking brake linkage to make brake application easy for the driver. The parking brake pedals and control levers multiply the force applied by the driver. Many parking brake installations also have an **intermediate lever** under the vehicle body to increase the application force even more. The intermediate lever also is designed to work with the equalizer to ensure that force is applied equally to both rear wheels. Intermediate levers are most common on large cars and trucks that need greater force to apply the parking brakes.

Equalizers and Adjusters

Parking brakes that use the service brakes—either shoes or pads—to lock the wheels must apply equal force to each wheel. If application force is unequal, the parking brakes may not hold the vehicle safely. To meet this requirement, most parking brake linkage installations include an equalizer mechanism. The equalizer also usually contains the linkage adjustment point.

The simplest example of an equalizer is a U-shaped cable guide attached to a threaded rod. The rear cable (or an intermediate cable) slides back and forth on the guide to balance the force applied to each wheel. In some installations, the equalizer guide is attached to a lever to increase application force.

Another kind of equalizer is installed in a long cable that runs from the driver's position to one rear wheel. A shorter cable runs from the equalizer to the other wheel. When the parking brakes are applied, the long cable applies its brake directly and then continues to move forward after the shoes or pads lock the wheel. The continued forward motion pulls the equalizer and the shorter cable to lock the brake at the other wheel.

Rear Drum Parking Brakes

Rear drum parking brakes that use the rear service brakes to lock the wheels are a common kind of parking brake system. Mechanical linkage that works with drum brake shoes is a relatively simple and economical design, and the self-energizing action of the brake shoes provides excellent holding power.

Figure 9-15 shows a typical parking brake installation with rear drum brakes. The brake cable runs through a conduit that goes through the backing plate. The cable end is attached to the lower end of the parking brake lever. The parking brake lever is hinged at the top of the web of the secondary or trailing shoe and connects to the primary or leading

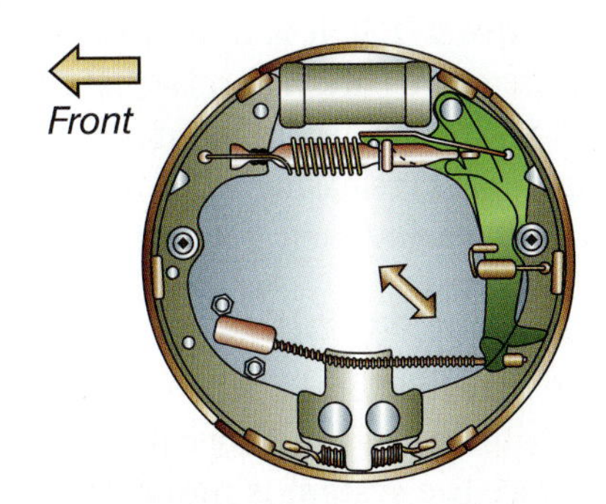

Figure 9-15 A parking brake is applied by a lever working on the trailing or secondary shoe, depending on the system.

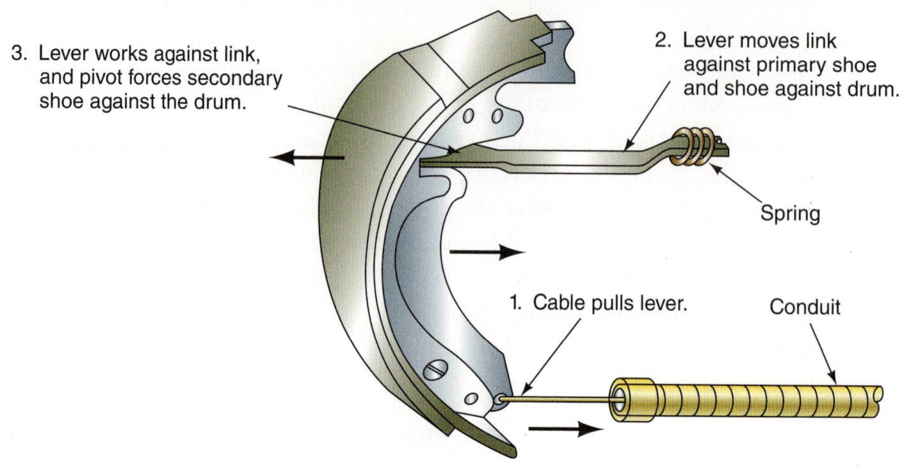

Figure 9-16 Parking brake lever and strut operation.

The parking brake strut connects the primary shoe to the secondary shoe.

shoe through a strut. When activated, the lever and strut move the shoes away from both the anchor points and into contact with the brake drum (**Figure 9-16**). When tension on the cable is released, the return springs move the shoes back to their unapplied positions.

The details of parking brake parts vary with different brake designs, but all work in the same way basically. Many parking brakes include various springs and clips to prevent rattles and to hold the parts in alignment.

ELECTRICAL PARKING BRAKE SYSTEMS

As discussed earlier, true electric parking brakes are being introduced today. The first is an extension of the ABS and active brake systems. When parking, the driver can switch on the parking brakes, and the hydraulic modulator applies fluid pressure to at least the rear wheels. The switch can also be activated by the movement of the transmission shift mechanism into park.

Some of the most recent electric parking brakes are available from Continental Teves North American and Siemens. Both manufacturers offer electric parking brake systems that can be found on many top-line luxury and midsize models. Both systems dispense with lever/pedal controls and both will provide some form of emergency braking should the service brakes completely fail.

The Continental Teves system uses an electrical motor or solenoid that moves cables running to the rear wheels (**Figure 9-17**). It is designed for drum-in-hat or disc caliper parking brakes. The system is an active parking brake that can be integrated into an electronic stability program (ESP) with appropriate interfaces. The system offers the advantages of automatically locking in park, releasing when engaging a drive gear, reducing or preventing rollback on slopes, increasing theft deterrent, and assisting in parking if integrated with a distance sensor system.

One version of the parking brake system has an electric mechanism at each rear wheel caliper that applies the rear disc brake when activated (**Figure 9-18**). The two mechanisms are mounted just inside of the rear calipers (**Figure 9-19**). The manual switch to activate and deactivate the system is mounted near the transmission's console shift lever or at a central point on the dash. An electrical signal can be mounted to sense transmission shift lever movement and engage/disengage the parking brake accordingly.

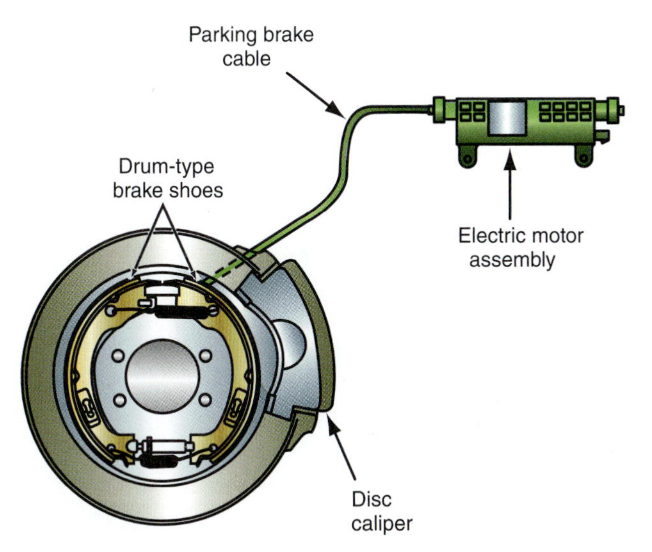

Figure 9-17 This electric motor assembly pulls on the cables to apply the parking brakes.

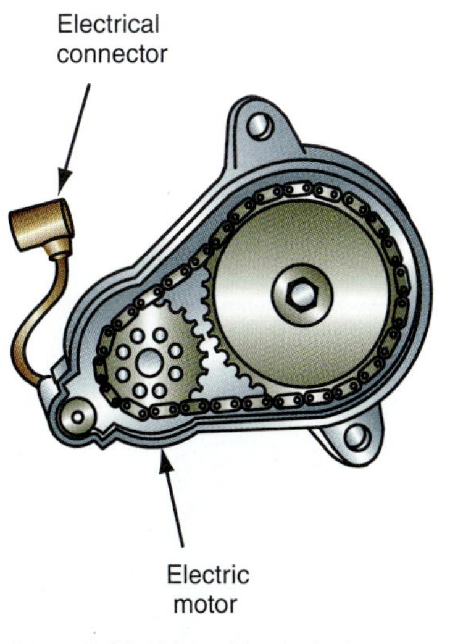

Figure 9-18 This parking brake is applied by an electric motor and cam operating directly on the caliper.

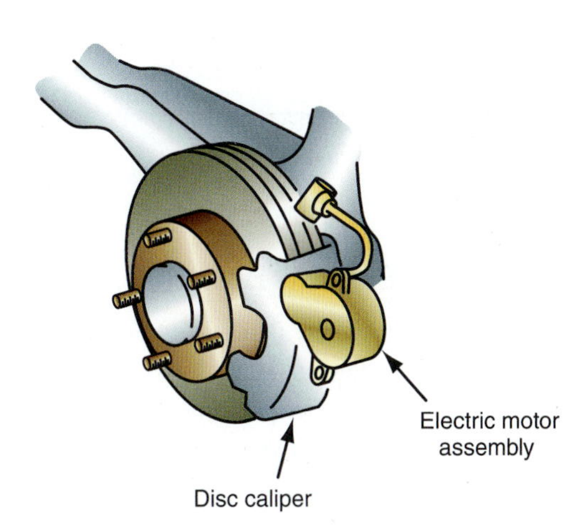

Figure 9-19 The electric motors are connected to the caliper and its adapter.

REAR DISC PARKING BRAKES

Shop Manual
pages 443

Two different types of parking brakes are used with rear disc brakes: auxiliary drum parking brakes and integral caliper-actuated parking brakes. Both are more complicated designs than parking brakes that are part of rear drum service brakes.

Auxiliary Drum Parking Brakes

Many sliding caliper rear disc brakes have a small drum cast into each rotor (**Figure 9-20**). A pair of small brake shoes is mounted on a backing plate that is bolted to the axle housing or the hub carrier. These parking brake shoes operate independently from the service brakes. They are applied by linkage and cables from the control pedal or lever. The cable at each wheel operates a lever and strut that apply the shoes in the same way that standard rear drum parking brakes work. These auxiliary drum parking brakes must be adjusted manually with star wheels that are accessible through the backing plate or through the outboard surface of the drum. They do not have self-adjusters.

These types of auxiliary parking brakes are sometimes called drum-in-hat parking brakes.

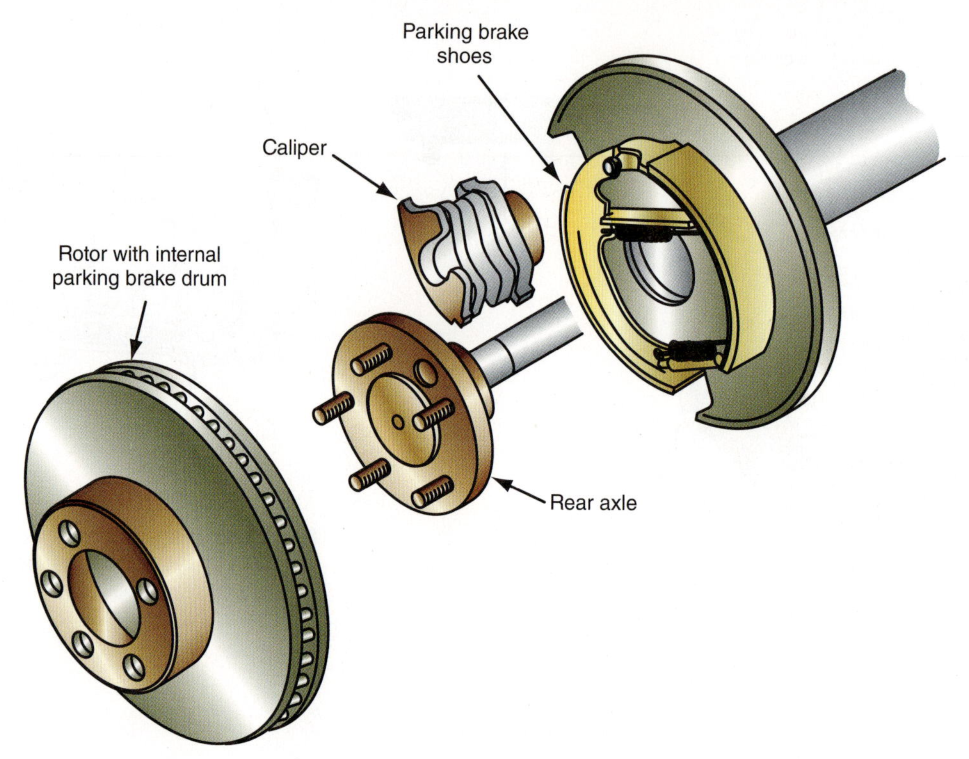

Figure 9-20 An auxiliary parking brake installation for rear disc brakes.

Assuming that the driver does not drive with the parking brakes applied, the parking brake shoes within the drum will last the life of the car. However, they may get corroded over a period and should be checked when servicing the rear service brakes.

Screw-and-nut and **ball-and-ramp** are two different calipers with integral parking brake mechanisms. See figures 9-21 and 9-22 on the page 223.

Eccentric means not round or concentric.

Integral (Caliper-Actuated) Parking Brakes

Some floating or sliding caliper rear disc brakes have components that mechanically apply the caliper piston to lock the pads against the rotors for parking. These parking brake mechanisms are termed integral because the parking brake mechanism is part of the caliper. All caliper-actuated parking brakes have a lever that protrudes from the inboard side of the caliper. The caliper levers are operated by linkage and cables from the control pedal or lever. As with most brake assemblies and subassemblies, detail differences exist from one brake design to another. The two most common kinds of caliper-actuated parking brakes are the **screw-and-nut** type and the **ball-and-ramp** type. A few imported cars have a third kind that uses an **eccentric** shaft and a rod to apply the caliper piston. This type is not as common as the first two, however. An eccentric acts like a cam. One portion of the shaft is oval shaped. As the shaft turns, the high part of the oval pushes the operating rod out of applying the brakes.

Screw-and-Nut Operation. General Motors' floating caliper rear disc brakes are the most common example of the screw-and-nut parking brake mechanism (**Figure 9-21**). The caliper lever is attached to an actuator screw inside the caliper that is threaded into a large nut. The nut, in turn, is splined to the inside of a large cone that fits inside the caliper piston. When the parking brake is applied, the caliper lever rotates the actuator screw. Because the nut is splined to the inside of the cone, it cannot rotate so it forces the cone outward against the inside of the piston. Movement of the nut and cone forces the piston outward. Similarly, the piston cannot rotate because it is keyed to the brake pad, which is fixed in the caliper. The piston then applies the inboard brake pad, and the caliper slides as it does for service brake operation and forces the outboard pad against the rotor.

An adjuster spring inside the nut and cone rotates the nut outward when the parking brakes are released to provide self-adjustment. Rotation of the nut takes up clearance as the brake pads wear.

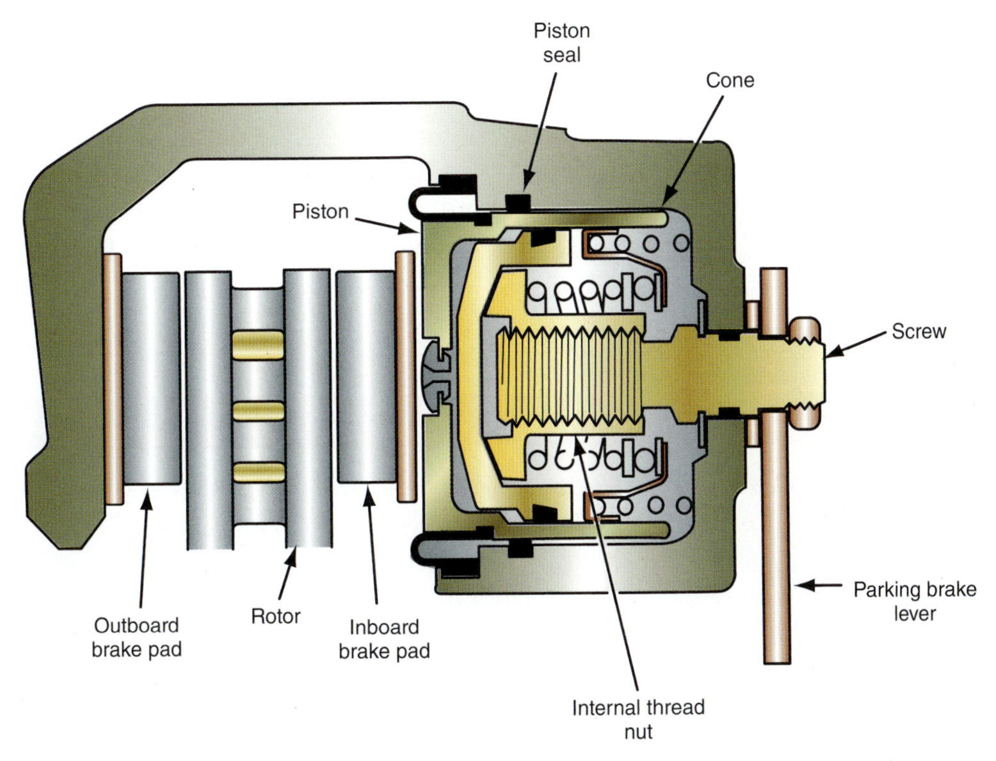

Figure 9-21 A GM screw-and-nut parking brake mechanism for rear disc brakes.

Ball-and-Ramp Operation. Ford's floating caliper rear disc brakes are the most common example of the ball-and-ramp parking brake mechanism (**Figure 9-22**). The caliper lever is attached to a shaft inside the caliper that has a small plate on the other end. Another plate is attached to a thrust screw inside the caliper piston. The two plates face each other, and three steel balls separate them. When the parking brake is applied, the caliper lever rotates the shaft and plate. Ramps in the surface of the plate force the balls outward against similar ramps in the other plate. As the plates move farther apart, the thrust screw forces the piston outward. The thrust screw cannot

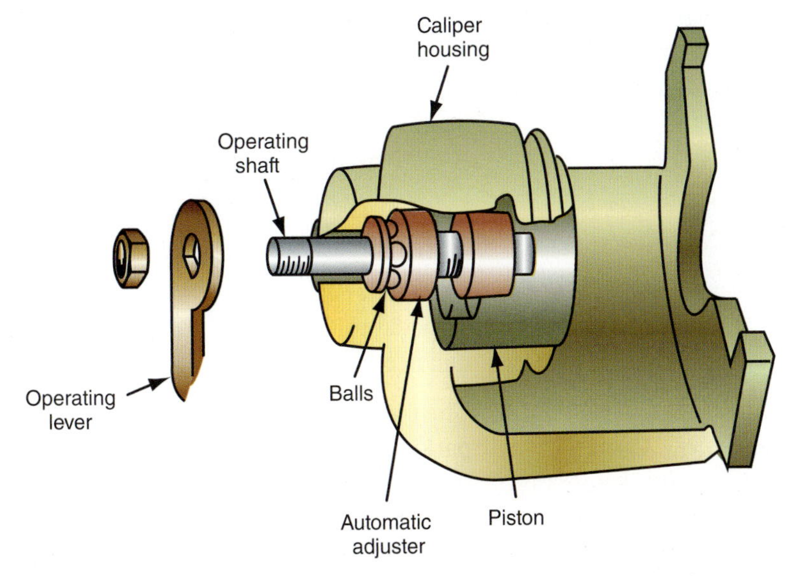

Figure 9-22 A Ford ball-and-ramp parking brake mechanism for rear disc brakes.

rotate because it is keyed to the caliper. The piston then applies the inboard brake pad, and the caliper slides as it does for service brake operation and forces the outboard pad against the rotor.

When the caliper piston moves away from the thrust screw, an adjuster nut inside the piston rotates on the screw to take up clearance and provide self-adjustment. A drive ring on the nut keeps it from rotating backward. It should be noted that each of these systems are still in use by other vehicle manufacturers in addition to General Motors and Ford.

SUMMARY

- The parking brake prevents the vehicle from moving when parked.
- Parking brakes may be operated by electronically controlled electric motor or either hand lever or foot pedal. The release mechanism may be either a manual release or an automatic release using an electrical switch or when the vehicle begins moving on electronic parking brakes.
- Electronically operated parking brakes are released when the vehicle is taken out of park.
- Equalizers are used to balance the forces applied to the parking brakes during application.
- Equalizer levers are used to multiply the effort of the driver applying the parking brakes.
- Rear drum brakes use a lever and strut to move the shoes into contact with the drum.
- Electric parking brakes may operate conventional parking cables or directly apply the rear disc brakes.
- Integral disc parking brakes use the normal disc calipers as parking brakes to hold the vehicle while it is parked.
- Auxiliary drum parking brakes are contained inside the rotor of some rear disc brakes.

REVIEW QUESTIONS

Short-Answer Essays

1. Explain the purpose of a parking brake.
2. Describe why a parking brake should not be used as an "emergency" brake.
3. How do the levers connected to equalizers on some brake systems multiply the driver's application effort?
4. What is the function of an equalizer in a parking brake system?
5. What is the construction of a parking brake cable?
6. Why are some parking brake cables plastic coated?
7. How are integral disc parking brakes applied?
8. Describe how the rear drum brakes are applied to hold the vehicle when it is parked.
9. Describe how the internal-expanding shoe transmission-type brake works.
10. How are rods used in some parking brake assemblies?

Fill in the Blanks

1. The parking brake system is not a part of the _______________ braking system.
2. Parking brake actuators may be operated either by _______________, _______________, or _______________.
3. The front cable connects the parking brake lever or pedal to the _______________, which provides balanced braking force to each wheel.
4. The two most common kinds of caliper-actuated parking brakes are the _______________ type and the _______________ type.

5. The parking brake ______________ is hinged at the top of the web of the secondary or trailing shoe.

6. The parking brake______________ connects the primary shoe to the secondary shoe.

7. Ford's floating caliper rear disc brakes are the most common example of the ______________ and ______________parking brake mechanism.

8. General Motors' floating caliper rear disc brakes are the most common example of the ______________and ______________ parking brake mechanism.

9. Two different types of parking brakes are used with rear disc brakes: auxiliary ______________ parking brakes and integral ______________ ______________ parking brakes.

Multiple Choice

1. *Technician A* says that the parking brakes are provided with equal cable tension by the cable adjusters. *Technician B* says that the parking brakes are provided with equal cable tension by the equalizer. Who is correct?

   A. A only
   B. B only
   C. Both A and B
   D. Neither A nor B

2. *Technician A* says that the parking brakes are mechanically operated because mechanical brakes are much more effective than hydraulic brakes. *Technician B* says that parking brakes are mechanical because the parking brakes must operate separately from the service brakes. Who is correct?

   A. A only
   B. B only
   C. Both A and B
   D. Neither A nor B

3. *Technician A* says that the usual device that releases the parking brakes whenever the transmission gear selector is in drive or reverse is an electronic parking brake controller. *Technician B* says that the device that releases the parking brakes can be a lever controlled by the driver. Who is correct?

   A. A only
   B. B only
   C. Both A and B
   D. Neither A nor B

4. While discussing FMVSS 105, *Technician A* says it requires that parking brakes hold the vehicle stationary for 15 minutes on a 30 percent grade in both the forward and reverse directions. *Technician B* says FMVSS 105 also specifies that the force needed to apply the parking brakes cannot exceed 200 pounds for foot-operated brakes or 190 pounds for hand-operated brakes. Who is correct?

   A. A only
   B. B only
   C. Both A and B
   D. Neither A nor B

5. *Technician A* says that some electric parking brakes use a special feature called auto-hold. *Technician B* says that some electric parking brakes may be applied by moving the vehicle. Who is correct?

   A. A only
   B. B only
   C. Both A and B
   D. Neither A nor B

6. *Technician A* says that a nut-and-screw type caliper assembly forces the disc pads against the rotor in a rear disc parking brake system. *Technician B* says that a ball-and-ramp type caliper assembly forces the disc pads against the rotor in a rear disc parking brake system. Who is correct?

   A. A only
   B. B only
   C. Both A and B
   D. Neither A nor B

7. Which of the following applies the rear caliper by cables for parking?

   A. Continental Teves
   B. Pedal control
   C. Lever control
   D. All of the above

8. *Technician A* says that another name for a rear (drum-in-hat) parking brake is auxiliary drum brake. *Technician B* says that another name for a rear (drum-in-hat) parking brake is integral brake. Who is correct?

   A. A only
   B. B only
   C. Both A and B
   D. Neither A nor B

9. *Technician A* says that it is important to remember that the parking brake is not part of the vehicle hydraulic brake system. *Technician B* says that it is important to remember that the parking brake is applied mechanically. Who is correct?

   A. A only
   B. B only
   C. Both A and B
   D. Neither A nor B

10. While discussing console mounted emergency brakes, *Technician A* says some lever-operated parking brakes have a separate cable for each rear wheel attached to the control lever. *Technician B* says each cable is adjusted separately, and an equalizer is not necessary. Who is correct?

    A. A only
    B. B only
    C. Both A and B
    D. Neither A nor B

CHAPTER 10
ELECTRICAL BRAKING SYSTEMS

Upon completion of this chapter, you should be able to:

- Define and understand the electronic terms commonly associated with electrical braking systems.
- Identify the components of a typical ABS.
- Explain the operation of a typical ABS.
- Describe the differences between integrated and non-integrated ABS systems.
- Discuss the general operation of Bosch 9.0 ABS.

Terms To Know

CAN protocol
Class 2 protocol
Controller antilock brake (CAB) module
Dynamic rear proportioning (DRP)
Electrohydraulic unit
Electronic brake control module (EBCM)
Electronic brake traction control module (EBTCM)
Enhanced traction system (ETS)
Flux lines
Isolation/dump valve
Negative wheel slip
Permanent magnet (PM) generator
Reluctor

INTRODUCTION

Although 1987 was not the first year when antilock brake systems (ABSs) were used, it was the first year when they were installed on all light trucks and small vans. The systems were first used to reduce rear-wheel lockup during hard braking. The types of vehicles noted were the ones that were most likely to lock the rear wheels during braking. Since then rear-wheel antilock systems have evolved to today's initial offering of full electrical braking systems. Certain technical terms and components from ABSs have been incorporated into the newer systems and are discussed in the first section.

COMMON COMPONENTS AND TERMS

Sensors

Sensors are electrical devices that measure some physical action and change that action to an electrical signal (**Figure 10-1**). They also are used to measure electrical voltage or frequency. Common actions measured by sensors are mechanical movement, temperature, pressure, and speed. Most sensors are simple to operate although the engineering design may be complex. The simplest are on–off switches, whereas others are actually small electrical generators, or digital signal generators.

Figure 10-1 The throttle position sensor (TPS) is mounted to the side of the throttle body. As the throttle valve is moved by the accelerator cable, a wiper arm inside the TPS moves over a variable resister wire. This changes the input 5-volt signal to a return signal between 0.5 volt and 4.5 volts.

A BIT OF HISTORY

Ford Motor Company was the first U.S. carmaker to experiment with ABSs on a production car when it offered an antiskid option on the 1954 Lincoln Continental Mark II. It worked—sort of—but added too much weight to the car and cost too much, so it wound up in the technological trash can.

Wheel Speed Signals

Signals refer to the electrical signal sent from the sensor to the controller. They may be analog or digital (**Figure 10-2**). Digital signals are more consistent because they have the same amplitude (voltage level) regardless of wheel speed. With digital signals, only the frequency changes, and the digital wheel speed sensors are accurate all the way down to a stopped wheel. An analog voltage signal is infinitely variable; in other words, it flows smoothly as it changes voltage levels. The wheel speed signal is compared to the ones from different wheels, and the electronic brake control module determines which wheel is about to lock up or skid; the slower the wheel, the lower the frequency produced. The frequency of the signal is still being measured because the signal is changed from analog to digital by the EBCM (electronic brake control module), but the voltage level of the signal (amplitude) varies greatly based on the wheel speed. At low speeds, the signal may not be readable.

Shop Manual
pages 479

The digital signal could be on or off, high or low, or yes or no. For simplicity in this discussion, the signal is either on or off. Digital signals can differ in the amount of on or off time and the rapidity of repeating signals over time, known as frequency. (**Figure 10-3**). A slow wheel as monitored by the wheel speed sensor will produce a low-frequency signal, whereas a faster wheel will cause a comparatively higher frequency, In this case, the computer can compare wheel speeds between individual wheels and determine if one wheel or wheels is slowing down faster than the others, compared to vehicle speed. The hydraulic modulator can then determine the proper command, if any, to be issued to the hydraulic modulator. Digital wheel speed sensors (WSSs) are now used on most vehicles.

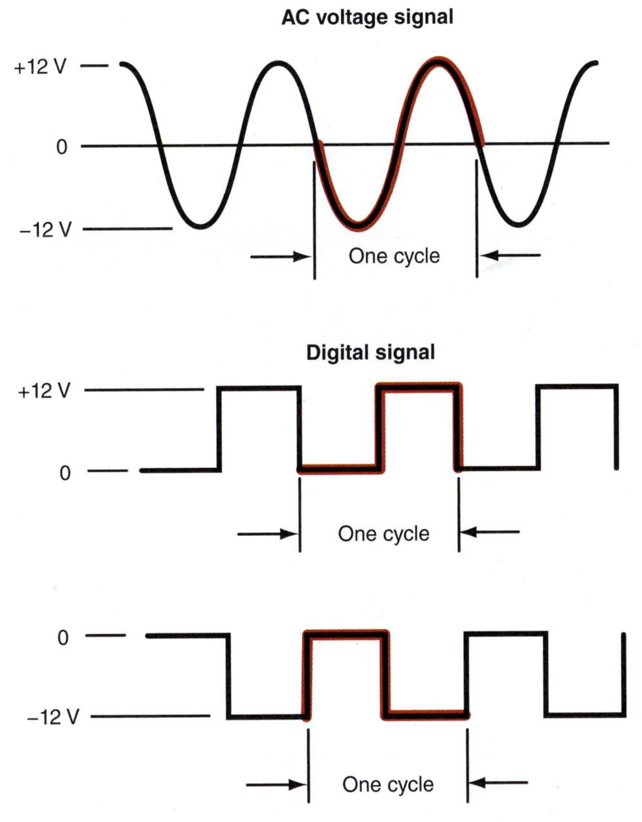

Figure 10-2 An analog signal switches from positive to negative to positive. A digital signal is usually positive or negative, but not both.

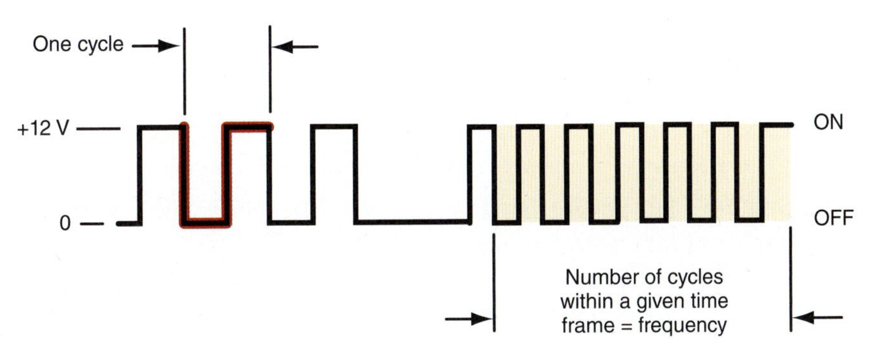

Figure 10-3 This signal indicates the amount of time on versus time off.

Actuators

Actuators are the workhorse of electrical/electronic systems. Actuators convert electrical energy into mechanical movement. They perform some action based on commands from their controllers, and many are based on duty cycle. Different levels of activity (duty-cycle) can be based on the percentage of "on" time verses "off" time. Actuators are usually motors or solenoids. Some of the most common actuators are cooling fans, fuel injectors, and transmission shift solenoids (**Figure 10-4**). In most cases, the controller can determine if the actuator is electrically and mechanically active by measuring the current flowing through that circuit. If the current is too high or too low, the controller may disable the actuator and/or illuminate a light in the dash.

Figure 10-4 A fuel injector is a solenoid that will allow a specific amount of fuel to flow when commanded to open.

Controller

A controller is a computer programmed to perform certain decisions based on sensor signals and to issue electrical commands to actuators (see Figure 10-4). Controllers are simple computers compared to the ones at work or school that can use multiple programs to perform different tasks. The automotive computer works within its own programmed parameters to control the engine, transmission, climate control, and other systems. Data are shared with other onboard computers to reduce the number of sensors and actuators needed to control multiple vehicle systems.

Automotive Networking

Consider networking to be a small Internet on the vehicle (**Figure 10-5**). Each computer has access to all data, but an individual computer will use only the data pertinent to its programming. The most common and most powerful controllers are the engine control module (ECM), the body control module (BCM), and the transmission control module (TCM). They control many smaller computer modules that have an even more specific function, such as adjusting the blower speed on an automatic climate control. There may be over 50 different controllers on a single vehicle, all networked.

AUTHOR'S NOTE The advantage of networking is the elimination of redundant sensors and the accompanying wiring. An example of this is that several years ago, vehicles had a temperature sensor to control the engine cooling fan, a temperature sensor for the warning lamp in the instrument cluster, and finally, a temperature sensor for the ECM. Now, the engine coolant temperature (ECT) for the ECM takes care of all three. The coolant temperature sensor information is shared among modules as necessary. The ECM controls the cooling fan directly, and the instrument cluster is provided temperature readings directly over the vehicle network.

Hydraulic Modulator

The hydraulic modulator is the electrical/hydraulic unit used on ABS, traction control system (TCS), and other braking systems to control hydraulic braking pressure to one or more wheels (**Figure 10-6**). The hydraulic modulator can hold or release pressure for ABS action, or increase, hold, or decrease pressure for traction and stability control. The hydraulic modulator is sometimes called the electrohydraulic modulator because the EBCM is often attached to the unit.

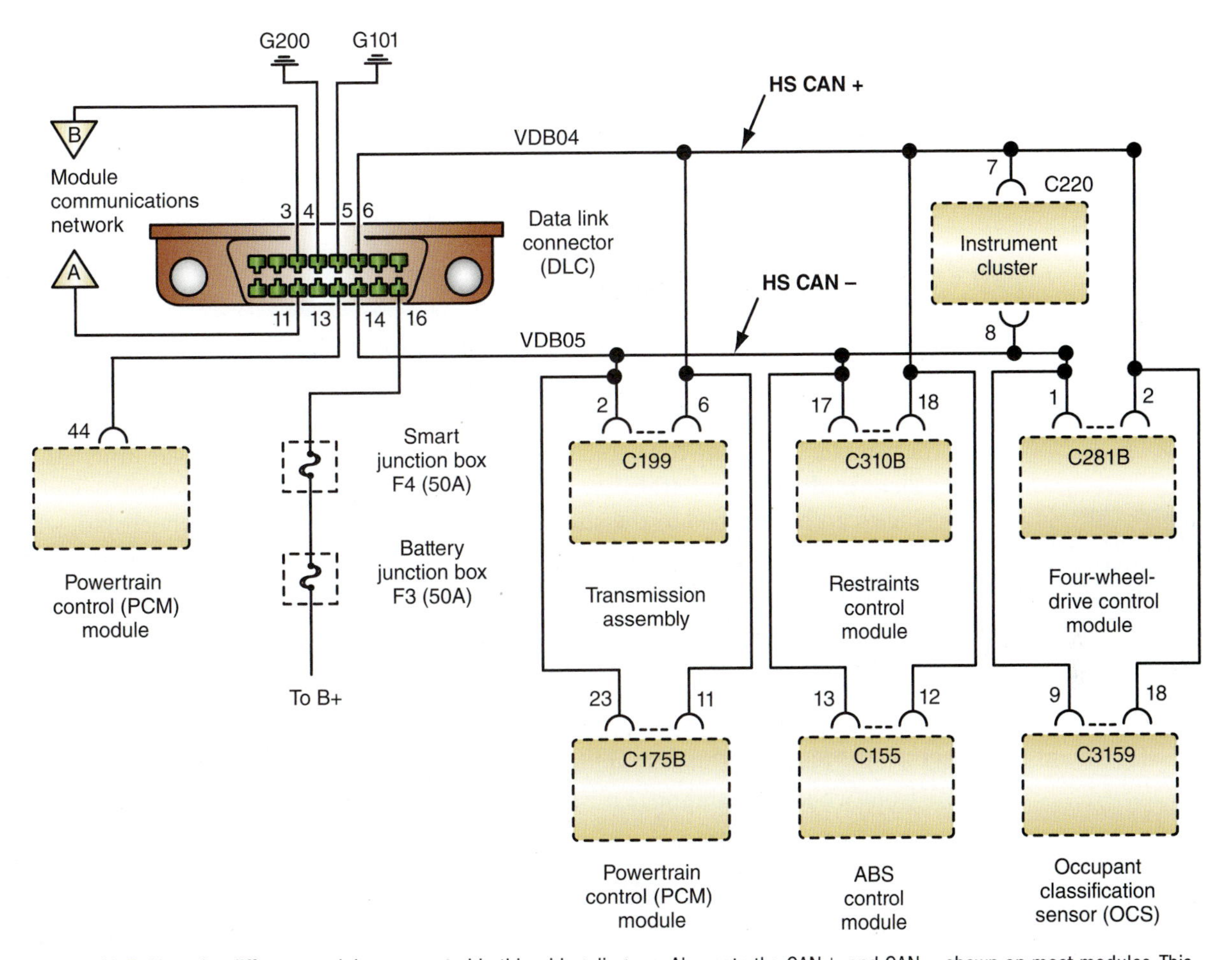

Figure 10-5 Note the different modules connected in this wiring diagram. Also note the CAN+ and CAN− shown on most modules. This is a controlled area network (CAN).

Figure 10-6 A typical ABS hydraulic control modulator.

Commands

Commands are nothing more than an electrical signal or output signal to an actuator. A command from a module can be performed by completing a power circuit, a ground circuit or both power and ground to an actuator. When the circuit is complete, the actuator can then function.

ANTILOCK BRAKE SYSTEM AND VEHICLE CONTROL

Negative wheel slip means the wheel is locked up and skidding.

Skidding wheels have lost traction and can not be steered, so **directional stability** and control have been lost.

The ABS is not the end-all to safe braking. The system was designed to provide some means of controlling wheel skid or **negative wheel slip**. If the driver overdrives the road condition or the capability of the vehicle, however, the ABS will not assist much in controlling the vehicle. Assuming the vehicle is operating within road and vehicle limits, the ABS is excellent in **directional stability** and **directional control**. Stability is achieved when the vehicle can be stopped in the shortest possible distance without wheel skid, whereas control refers to the fact that the driver can steer the vehicle during a panic or ABS stop. A locked wheel (which is sliding) has little, if any, traction and will not grip the road surface sufficiently for braking or steering. It should also be noted that the ABS will not necessarily stop the vehicle quicker than standard brake systems.

AUTHOR'S NOTE One thing I have learned is that you cannot steer a vehicle if the front wheels are sliding. I missed a turn one day, many years ago, and slammed on the brakes of a 1969 VW Fastback. The front wheels locked, and I steered to the right. The car continued to go straight ahead. I realized I was going to miss the turn anyway so I released the brakes. The wheels stopped sliding and the car immediately went into a hard right turn—I missed the turn and aged 10 years in less than 5 seconds. This experience has stayed with me for over 40 years: Do not try to steer the vehicle if the front wheels are locked and sliding. The ABS reduces this possible problem to almost zero.

ABS TYPES AND GENERAL OPERATIONS

Shop Manual page 490

Basically, the ABS controller is monitoring all four wheel speeds while driving. When the ABS senses a wheel slowing faster than the other wheels, the hydraulic modulator (**Figure 10-7**) goes into "isolation" mode. In isolation mode, brake pressure is isolated from the master cylinder even if the driver applies more pressure to the affected wheel. If the wheel continues to slow down faster than the other wheels, and the ABS goes into a "pressure release" or "dump" and the fluid is returned back to the reservoir, or in some systems a low-pressure reservoir. If the wheel speed gets back into line with the speeds of the other wheels, then the pressure is allowed build up again from the master cylinder. This process can happen several times per second. This action also causes the pedal to click and drop when the ABS is active and is entirely normal. The commands are based on the signals from the WSSs, and the commands are to the hydraulic modulator. With all newer vehicles since the advent of stability control (and a few before) the ABS will be able to control all four wheels individually. This wasn't always the case, as many of the earlier systems controlled either the rear wheels only as an axle set, meaning if one wheel locked up, both wheels were treated the same way. Later, we had three-channel systems that controlled the rear wheels as an axle set but the front wheels individually. Finally, we saw four-channel systems, in which all four wheels were controlled individually. We really needed to have four channels to implement vehicle stability control.

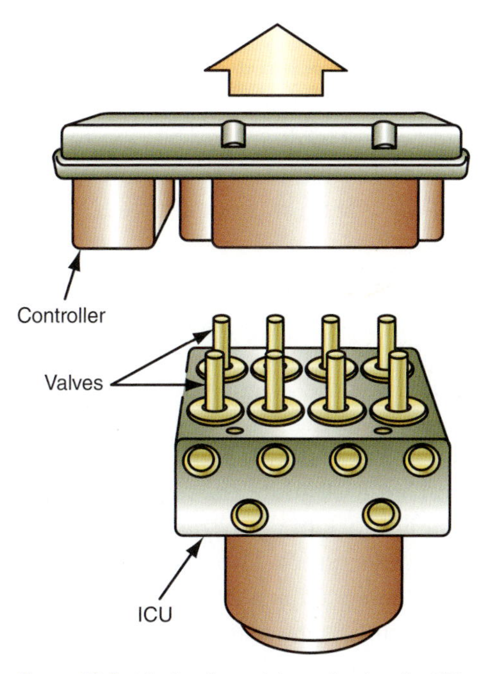

Figure 10-7 A hydraulic modulator showing the ABS control valves.

Integrated and Nonintegrated ABS

There were two major types of earlier ABS braking systems: integrated and nonintegrated. The integrated systems have a hydraulic brake booster, a master cylinder, and an ABS computer referred to as an EBCM all built into one unit. For the most part, these units had very few serviceable parts (**Figure 10-8**). This led to some problems with the earlier versions. The earlier ABS modulators went through some growing pains that are typical of new systems, but these were particularly painful to some vehicle owners. Some modulator/master cylinder units cost as much as $3,000. Because this is an integrated unit with few, if any, parts serviced separately, any failure required the owner to spend a large amount of money to fix the vehicle properly.

The nonintegrated system was also called an "add-on" system, because it used a conventional vacuum booster and basic brake system with a separate hydraulic modulator. The nonintegrated ABS system was a benefit to all concerned, from the manufacturer to

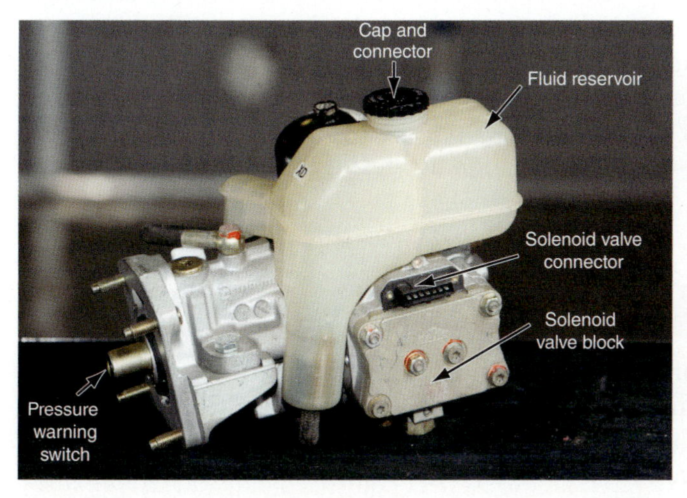

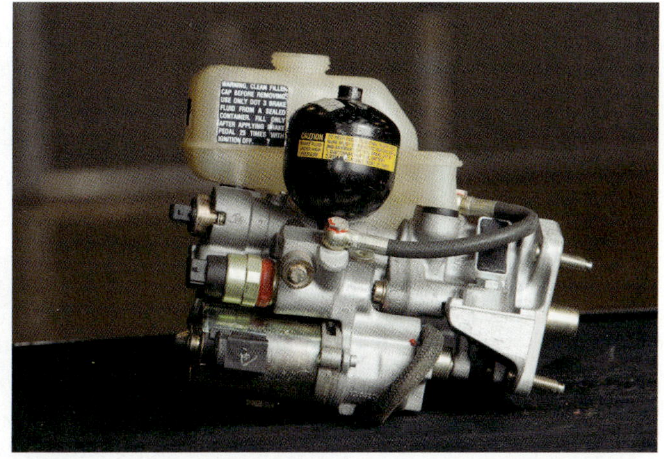

Figure 10-8 An integrated ABS combines the master cylinder, brake booster, and ABS components in a single unit.

the technician and especially the vehicle owner. The hydraulic modulator was separated from the master cylinder and usually was installed just below the master cylinder or somewhere on the left side of the engine compartment. This left the master cylinder as a separate unit that could be replaced alone. The cost of separating the two was nominal because the biggest cost was in engineering and a few extra feet of brake tubing and wires. The overall efficiency of the system was not affected.

Rear-Wheel ABS (RWAL or RABS)

One of the most difficult vehicles to design a braking system for was the light truck. A manufacturer had no way of knowing how or even if the vehicle would be loaded. The first attempts at a solution was the height-sensing proportioning valve, followed by rear wheel antilock braking. This was the first true mass-production ABS installed on vehicles. The **RWAL/RABS** system was first installed on minivans and pickup trucks in late 1987. *RWAL* means rear-wheel antilock and is the most common term used by most manufacturers. *RABS* means rear antilock brake system and is the term commonly used by older Fords and some other manufacturers. The system essentially controls wheel skid only on the rear wheels. Usually the speed sensor was located on the differential housing and measured the speed of the differential ring gear.

Four-Wheel ABS

After the rear-wheel ABS was developed, it was quickly followed by four-wheel ABS. It can control the braking effect at each wheel through one of two systems: three-channel and four-channel. The three-channel was the first four-wheel ABS system. In the three-channel system, the rear-wheel brakes are controlled as one, but the two front wheels were controlled independently of each other. For example, if the right rear wheel started to skid, the ABS would reduce the brake pressure to both rear wheels. To do this requires three WSSs and a larger, more sophisticated hydraulic modulator. Each front wheel has a speed sensor, whereas another is placed to measure the rear wheels. The hydraulic modulator basically has three "isolation/dump" valves in a single unit (see **Figure 10-9**). The basic principle of operation remained the same, but the different components improved drastically in design and serviceability. The controller's programming was increased to handle the additional workload and to effectively control the hydraulic module. The three-channel system works best with front disc and rear drum brake systems. This is mainly because the speed of release or application of drum brakes is a little slower than disc brakes.

The four-channel ABS works along the same lines as the three-channel, but all four wheels have a speed sensor with dedicated valves for each wheel. The hydraulic modulator is a full-fledged hydraulic unit housing four or more valves commanded by a stronger controller and usually with the controller integrated into the modulator. The time span for development between the first RWAL/RABS and the four-channel system allowed for better programming, better manufacturing, better materials, and a better fit into the service brake's hydraulic system. The four-channel works best with four-wheel disc but will work satisfactorily with a disc/drum combination. Its best feature is its adaptability to TCSs and braking systems.

ABS BRANDS

ABS components are not manufactured by automotive manufacturers. They are designed and built for a specific vehicle line, but all function in much the same manner. Two of the different brands are discussed later. The primary manufacturers of ABSs are Bosch (and Delco-Bosch), Continental Teves, Kelsey-Hayes, Delphi Chassis, Bendix, Nippondenso, Nissan, and Sumitomo.

Figure 10-9 This early rear wheel antilock is an example of a separate isolation/dump valve.

ABS COMPONENTS

ABS components are additions to the standard braking hydraulic/mechanical components covered in earlier chapters. Common components are the controller, hydraulic modulator, WSSs, and brake switch.

Shop Manual
page 493

Controllers

The antilock controller or microprocessor is the computer for the ABS. It is commonly referred to as the **electronic brake control module (EBCM)** although different manufacturers use different names. **Controller antilock brakes (CAB) module** is another common term for ABS controllers. If the controller is built into or attached to the hydraulic modulator, then the whole assembly is known as the **integrated control unit (ICU), or an electrohydraulic modulator** or electrohydraulic unit. All controllers do basically the same thing: receive input signals, process the signals, store the signals for possible use later, and issue output commands. Most automotive computers, including the EBCM, are programmed with an adaptive memory capability or learning mode. This seems to mean that the computer can "learn," but that is not exactly true because computers do not have the ability to learn. Computers can process signals and determine exactly how this particular sensor or actuator, including the vehicle driver, responds to different conditions, however. This "learning" can be done only within the program parameters. For instance, if a new WSS is installed, the computer will determine the limits of the sensor. The new sensor's high/low signals may be a little (tenths of a volt) different from the old one. If the new signals are within set limits, the computer can accept them as the norm for that sensor. If not, then the computer activates the ABS warning light on the dash and disables the ABS entirely.

Anti-lock brake controllers are known as **electronic brake control module (EBCM)** or **controller anti-lock brake (CAB)**.

If the hydraulic modulator has the controller attached, then the unit may be called **integrated control unit (ICU) or electrohydraulic modulator**.

Incorporating the EBCM with the hydraulic modulator helps to eliminate unnecessary external wiring and connections which helps build more reliability into the unit.

As mentioned before, input signals may be analog or digital. Analog signals are produced by small AC electrical generators called **permanent magnet (PM) generators**. The most common ones on a vehicle are the WSS and vehicle speed sensor (VSS).

Wheel Sensors

The PM wheel speed sensor (also called a magnetic reluctance sensor) is composed of an ALNICO permanent magnet and a coil of copper wire. ALNICO is an abbreviation for the aluminum, nickel, and cobalt elements used in the metal alloy (**Figure 10-10**). The ALNICO magnet is very stable; that is, it loses only a very small percentage of its strength each year. This characteristic makes the ALNICO magnet ideally suited for its use in the wheel sensor because of its longevity.

Flux lines are the magnetic force that surrounds a magnet.

The **reluctor** is the metal-toothed ring used to influence the magnetic flux lines of the PM generator.

A metal-toothed ring (tone ring or tone wheel) is placed in proximity to the permanent magnet and coil. One of the metal teeth on this ring attracts the magnetic lines of force of the magnet. These magnetic lines of force are called **flux lines**. The flux lines must pass across the copper wires in the coil as they are attracted to the metal tooth of the **reluctor**. The principle of magnetic induction states that whenever a magnetic flux line crosses a conductive wire, it induces voltage in that wire. When that wire is in an electrical circuit, it generates an AC electrical current, which is termed an analog signal.

The tooth rotates and aligns the valley between the teeth with the magnet. The distance between the metal of the reluctor and the permanent magnet is great enough to allow the flux lines to return to their original location. The flux lines reverse themselves as the magnetic pull weakens; as they travel in the reverse direction across the copper wire, they induce voltage once more. However, this time the voltage induced by the flux lines is the opposite of the initial induced voltage. Movement of the flux lines between the permanent magnet and the reluctor produces an AC voltage for use by the ABS (**Figure 10-11**).

The PM generator circuit was designed to self-test at power on self-test (POST). The technician needs to understand how this circuit works in case a wheel sensor code is stored in the diagnostic program of the microprocessor. Service technicians who do not understand the circuit are likely to start yanking expensive components, such as the microprocessor, the wiring, and the sensor, or, in a worst-case scenario, all three, and

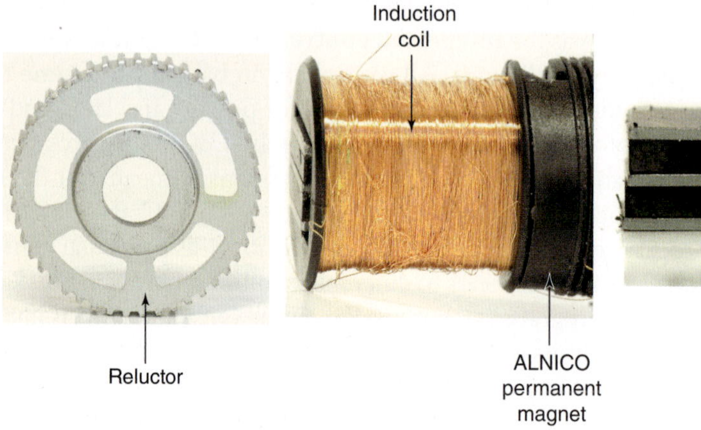

Figure 10-10 PM generator.

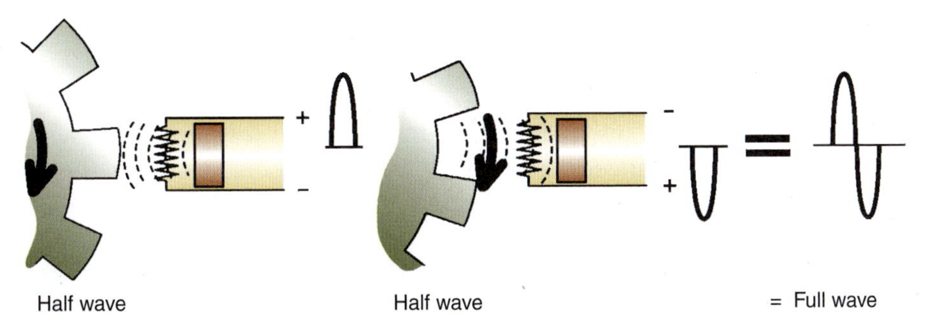

Figure 10-11 Wheel sensor output.

replacing these perfectly good components with identical parts. This is both expensive for the customer and labor intensive for the service technician.

The microprocessor has a regulated power supply that produces a stable direct current (DC) voltage. The voltage signal is usually 1.5 volts. That voltage begins its journey at the power supply inside the microprocessor and travels through an electronic circuit board until it arrives at an R_1 fixed-value resistor. The R_1 resistor value is usually 10,000 ohms. The current flow continues its forward movement from the microprocessor until it arrives at the PM generator. The resistance of the coil in the PM generator is usually 1,000 ohms. The current returns to the microprocessor power supply through the return wiring and electronic circuit board. This completes the series circuit. A typical ABS wheel speed sensor circuit is illustrated in **Figure 10-12**.

Anything that consumes electricity (voltage) is called a load. Voltage drops occur across each resistance in a series circuit. The resistance may be a coil of wire or it may be a carbon fixed-value resistor. The total of the voltage drops in a circuit must equal the power supply. A voltage-monitoring detection circuit sees the voltage drop. The monitor expects to see a voltage of 136.3 millivolts. The technician can detect that voltage by test probing the body wiring with a voltmeter. If the monitor sees no voltage, it assumes there is a problem. That problem could be with the power supply or with the electronic circuit board at the resistor. The problem also could be with the body wiring to the PM generator. The generator coil could be shorted to ground or open.

The PM generator (magnetic reluctance sensor) is a variable reluctance sensor in that the amount of voltage produced is directly related to the speed of the wheel. As the wheel slows, the strength of the voltage may not be detectable by the EBCM. So if the wheel begins to skid at a low speed, the ABS may not function correctly. Although this may be acceptable with standard ABSs, it is not acceptable with the brake systems that are discussed later. Continental Teves uses a magnetoresistive sensor as its WSS in the

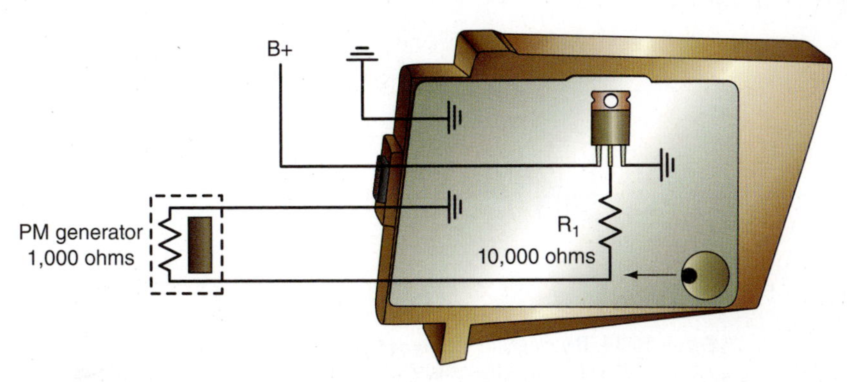

Figure 10-12 PM generator circuit.

The magnetoresistive sensor produces a digital signal and has its own power supply. The amplitude of the signal is constant, which means the voltage level does not change with wheel speed, only the frequency. A magnetic reluctance sensor can produce a readable signal even with the wheel stopped.

Caution

Servicing a magnetoresistive sensor is different from servicing a PM generator sensor. Follow the manufacturer's instructions closely or damage to the sensor or electronic circuits could occur. Details about testing and diagnosis are discussed in the *Shop Manual*.

Continental Teves Mk60/Mk70, brake system (**Figure 10-13**). As a matter of fact, most all systems that have stability control use magnetoresistive sensors. The magnetoresistive sensor cannot produce a voltage, so it must be supplied with 12-volt power provided by the EBCM (**Figure 10-14**). This creates an electromagnet at the head of the sensor. The return voltage or signal is changed based on the resistance or disturbance of the magnetic field around the sensor and the relationship of a tone ring tooth to the sensor. This sensor has two integrated circuits (ICs) that increase or amplify the resistance change into a DC signal (digital) that is returned to the EBCM. The output signal of one IC is a constant 7 mA at 0.9 volt when no tooth on the tone ring is near the sensor. When a tooth aligns or enters the sensor's magnetic field, the second IC produces the same amount of current and voltage. As a result, the return signal voltage is increased to 14 mA at 1.65 volts (**Figure 10-15**). Because the voltage output signal strength is not dependent on how fast the wheel turns, this sensor is more accurate and allows for the precise braking control demanded by active brake and active suspension systems. The speed of the wheel is calculated by how often or how fast the voltage changes from low voltage, 0.9 volt, to high voltage, 1.65 volts, or the cycle frequency. A fast wheel produces very quick repetitive cycles (high frequency), whereas slow cycles represent a slow wheel (low frequency). It should be noted that some sensors may have different voltage levels, so always consult service information for the specific vehicle being diagnosed.

At times, like when turning a corner, all four wheels may be traveling at different speeds. A four-wheel ABS has to take this speed difference into account on braking actions on curves and when cornering. Even simple RWAL/RABS had to allow for this condition. It is probable that most roads have more curves than straight sections. On the other hand, most panic stops are done when the vehicle is traveling straight ahead even if only for short distances.

Brake Switch. Most switch inputs to computer logic circuits are grounding devices. They complete the circuit to ground. The computer monitors a circuit with a voltage-dropping resistor. The ABS switch is different from these circuits because it sends a B1 voltage signal

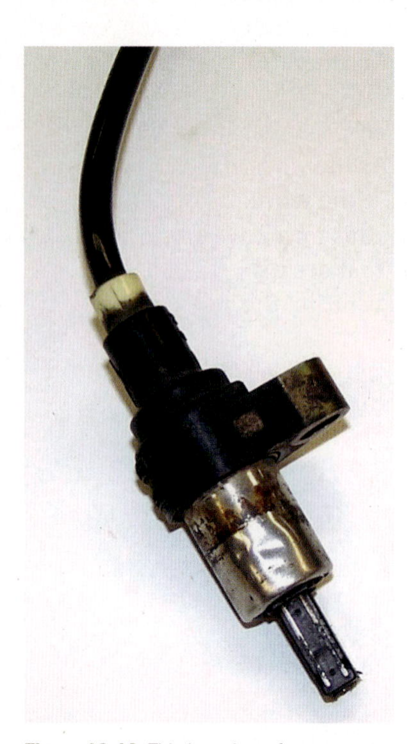

Figure 10-13 This is a view of a magnetoresistive wheel speed sensor.

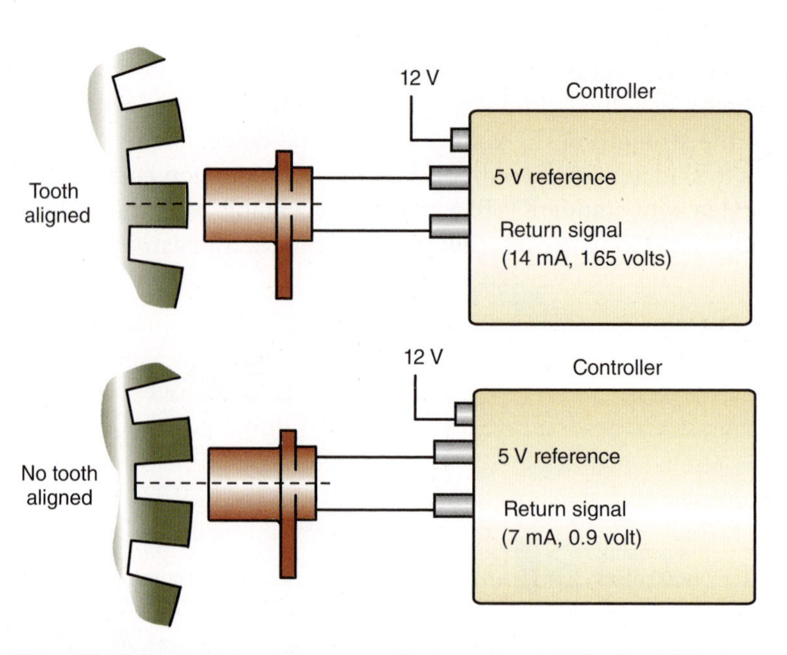

Figure 10-14 The magnetoresistive sensor will return a signal of 14 mA at 1.65 volts with a tooth aligned with the sensor head. With no tooth aligned, the return signal is 7 mA at 0.9 volt.

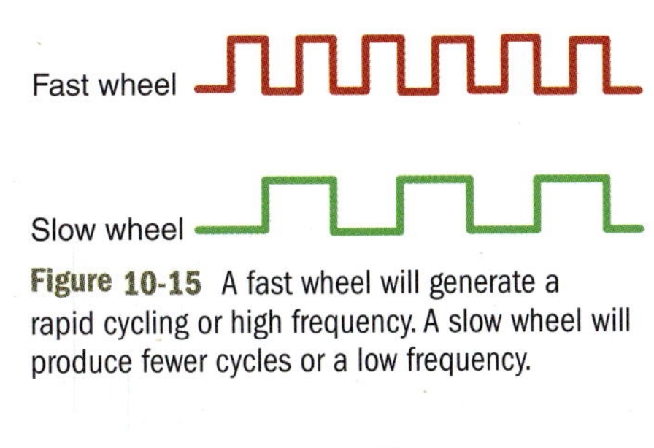

Figure 10-15 A fast wheel will generate a rapid cycling or high frequency. A slow wheel will produce fewer cycles or a low frequency.

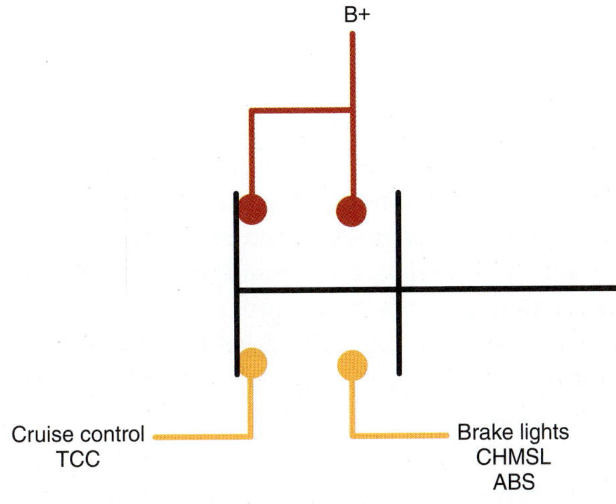

Figure 10-16 Brake switch.

to the microprocessor when the brake pedal is depressed. This is a complex switch that contains two sets of contacts: one open circuit and one closed circuit (**Figure 10-16**).

The normally open contacts in the brake switch close when the brake pedal is depressed. This sends B+ power to the stop lamp circuit, the center high-mounted stop lamp (CHMSL), and the ABS microprocessor. When the normally closed contacts open, they cancel the vehicle cruise control and torque converter lock-up features. The latest ABS units use a brake pedal position switch to determine how fast the brake pedal has been depressed. If a panic stop is anticipated, some ABS units can increase the amount of brake pressure applied to the brakes through the hydraulic modulator. The brake pedal position switch is also a critical component of vehicle stability control.

> **Caution**
>
> Many magnetoresistive sensors have a mercury acceleration sensor internal to the sensor. Any sensor with mercury must be considered a hazardous material when stored or in use. It must be treated as hazardous waste when determined to be unserviceable.

Pumps and Accumulators

The last major components of an ABS are internal or external hydraulic pumps and accumulators. Many ABSs have pumps to supply their own brake boost pressure instead of vacuum boosters.

High-Pressure Pumps and Accumulators. Some ABS units boost pressure comes from a high-pressure electric pump. It charges the accumulator, with brake fluid from the reservoir. Anytime the accumulator pressure drops below a minimum point, a pressure switch grounds a power relay coil or sends a signal to the EBCM to command the pump on. This connects the high-pressure pump to the ignition B+ power. The ABS boost pressure operates the high-pressure pump when the key is turned on if the vehicle has not been started for a time and the pressure has bled below the minimum point.

Low-Pressure Return Pumps. An ABS without an accumulator or pressure switch but with an electrohydraulic unit that contains an electric motor is called a low-pressure system. The motor is connected to a low-pressure pump that quickly returns fluid to the master cylinder and the reservoir. On some systems, this pump is used to apply light pressure to an isolated hydraulic circuit of an individual wheel. This feature is used to correct traction problems during acceleration. If the vehicle is equipped with a lateral accelerometer, the ABS is programmed to eliminate fishtail during acceleration. It applies light brake pressure to the spinning drive wheel.

Dynamic Rear Proportioning. The DRP replaces the standard hydraulic proportioning valve found on many older vehicles. It is an electronic operation within the hydraulic control unit and is used on most late-model vehicles.The DRP uses active control and the existing ABS to regulate brake pressure to the base rear wheels. As with the hydraulic valve, this keeps most of the braking force on the front wheels during routine stops. The red brake warning light will illuminate if there is a hydraulic problem within the base brake system.

AUTHOR'S NOTE The DRP is active for every situation that may cause a wheel locking condition. The engineers who designed the proportioning valve system had to assumptions based on vehicle loading and road conditions when they designed the proportioning valve. With dynamic rear proportioning the actual conditions can be read by the computer and the best outcome can be had based on these conditions.

Active brakes can warn the driver of an impending crash, or actually applying the brakes if necessary to avoid a crash.

Lamps and Communications

The lamps are not essential to ABS operation; however, they are really the only contact between the vehicle's electronic package and the operator. Communications, on the other hand, is essential to the electronic package because without communications between the various components there would be no operation.

Warning Lamps. Most people understand that the red brake warning lamp is used by the foundation brake system. This lamp indicates problems to the driver. It comes on when brake fluid levels are low and when the parking brakes are applied. This lamp also can be used as a brake pad wear indicator lamp. Some ABSs use the red warning lamp to indicate ABS problems when the amber ABS warning lamp or its circuit is nonoperational.

The amber warning lamp is used in several situations: to indicate that the microprocessor is testing the system; to indicate ABS operation; and to alert the driver about a malfunction in the ABS. Some systems use a remote lamp driver to control the amber warning lamp. If the microprocessor is functional, it is able to command the remote lamp driver to turn the amber warning lamp off after the POST. If the microprocessor or its related remote lamp driver wiring is not functional, it is unable to turn off the amber warning light, so it remains on.

COMMUNICATIONS

The majority of ABSs share data with other computers on the vehicle through the vehicles communication network. The ABS system communicates with other computers such as the ECM, BCM, and stability control system. This reduces the number of sensors needed and helps the vehicle to react as a system. An example of this sharing is the vehicle speed information. The ECM can share this information with the EBCM, eliminating the need for two separate sensors. The technician can communicate with all the computers in the network through the use of a scan tool. This allows the technician to retrieve codes and records but also to command functions, read data lists, and take data snapshots. Some ABSs can hold as many as six failure codes for up to 100 ignition cycles.

Different carmakers use many variations in computer communication. The vehicle service information is still the best source for specific vehicle repair information. It is the best source of information for code retrieval, data list interpretation, and testing.

TRACTION CONTROL SYSTEM

The TCS is an outgrowth of four-wheel four-channel ABSs. Most of the same components are used in both systems with the major changes being in the controller and its programming. The TCS also interfaces with the engine (PCM) and transmission (TCM) control modules to request that specific actions be taken.

Shop Manual
page 462

The TCS is designed to control wheel spin or positive wheel slip. If the vehicle is accelerated hard from a dead stop, the wheels tend to spin on the road surface assuming there is enough power available from the engine. This has led to some broken axles and other damage if the spinning wheel was on dirt and suddenly moved onto dry pavement. Although TCS can reduce wheel spin during hard acceleration, that is not its underlying purpose. It is designed to control wheel spin during snow, ice, wet pavement, and other road conditions.

Before the days of traction control a driver with a manual transmission could start out in a gear higher than first on slippery pavement. This reduced the amount of torque at the wheels and may limit wheel spin. The TCS controller can limit torque applied by retarding timing and reducing injector pulse width, and with an automatic transmission shift to a higher gear ratio as needed.

Going back to the discussion on isolation/dump valves, it was noted that the accumulator in that valve could apply some pressure to the wheel brakes to reduce wheel speed. The TCS has taken that a step further by using a hydraulic modulator modified from the one used strictly for the ABS. If a wheel sensor indicates that a drive wheel's speed is greater than that of the other wheels, the TCS controller can request that the ECM reduce the ignition timing and or injector pulse width. If the wheel is still exhibiting a positive wheel spin, then the TCS controller will command the hydraulic modulator to apply hydraulic fluid to that wheel. In effect, the brakes are applied. Because of the gear setup in the vehicle's differential, the power is transferred to the other drive wheel(s). When both or all four of the drive wheels approach the same speed, the TCS reduces or eliminates the pressure to the spinning wheel brakes. It can reapply the brakes if necessary or leave them off, depending on the wheel speed.

AUTHOR'S NOTE The TCS is an electronic version of the old, and still used, limited slip differential that was common on many production performance vehicles. This type of differential has some problems because of its mechanical nature, the locking clutches, and, in many cases, the type of lubricant. A differential without any provision for limited slip is called an "open differential." With an open differential, if one wheel has no traction, then it spins and the other wheel simply sets still. TCS can apply the brakes to the spinning wheel and allow some torque to be transferred to the wheel with traction, allowing the vehicle to escape the slippery situation.

In the case of automatic transmissions, the TCS controller will signal the TCM to shift up to a higher gear, reducing the torque being supplied to the drive wheels. In either case, power is reduced to the drive wheels, and it is hoped the wheels can regain traction. It is possible that all three actions—engine reduction, higher gear ratio, and brake application—may be required to control positive wheel slip. This is especially helpful in snow and mud conditions.

The major differences between the ABS and the TCS are different hydraulic modulator and the controller. Most TCSs and ABSs use the same controller with more sophisticated programming (**Figure 10-17**).

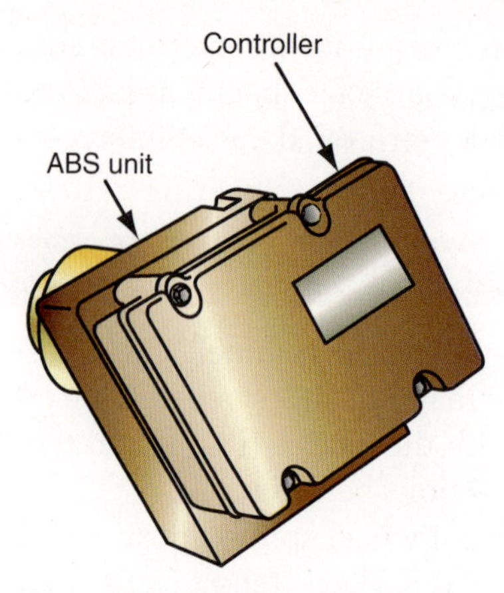

Figure 10-17 The ABS controller on this unit is mated to the hydraulic modulator.

AUTHOR'S NOTE Educating the owner of the vehicle often becomes an important aspect of doing business. Sometimes senior citizens with a new vehicle may complain of a serious braking noise and pedal movement during hard braking or when braking on a gravel road surface. They may not be aware of exactly how an ABS works. After questioning the owner on exactly what is happening and a complete inspection of the brake system reveals no faults, it may be best to explain to the customer exactly how an ABS works. It is also wise to inform them that if they are in a dangerous situation when panic braking is needed, they should not release the brake pedal until the danger is gone or the vehicle is stopped.

The **BPMV** is the unit that houses the ABS and TCS control valves.

DELPHI DBC-7 ABS

The Delphi DBC-7 is the oldest ABS/TCS system discussed in this chapter. It should be interesting to compare it to the Teves Mk60/70, and the Bosch ABS 9.0.

The Delphi DBC-7 is used almost exclusively on General Motors vehicles. It is a nonintegral system that houses the valve body and controller into a single housing. The DBC-7 can be used on three- and four-channel ABS and can be adapted to TCS. The DBC-7 is lighter and less costly than its predecessors. The DBC-7 is a major change from the Delco-VI it replaced.

The most obvious engineering change is the change from Delco VI is a motor pack hydraulic modulator back to the more standard modulator using solenoids The valves and pumps are in the valve block known as the **brake pressure modulator valve (BPMV)**. The valve solenoids and controller are mounted to the valve body and are called the EBCM. If the system is fitted with TCS, this control unit is listed as the **electronic brake traction control module (EBTCM)**.

Shop Manual
page 242

Even though EBTCM is the name given to the electrohydraulic modulator with TCS, most in the field will continue to call the unit electrohydraulic modulator because TCS has been standard equipment for some time.

The ABS relay, which is mounted separately in the Delco-VI, is now mounted within the EBCM or EBTCM. Each of these two units can be replaced as separate assemblies if necessary. In cavity 11 of the EBCM (EBTCM) wiring harness, a small vent tube has been installed to stop buildup of pressure or vacuum between the controlling unit and the valve housing.

In the BPMV are an inlet and an outlet valve for each brake hydraulic channel. Two accumulators are placed in the BPMV, one for each front/rear or diagonal hydraulic circuit. This is a total of eight valves and two accumulators for a four-channel system with six valves and two accumulators for a three-channel. Outlet valves are normally closed and inlets are normally open. The EBCM/EBTCM provides a ground to the solenoids to close the solenoid circuits.

Four-channel systems are used on General Motors passenger cars equipped with the DBC-7 ABS. Each BPMV is connected to a separate brake line and the lines are color coded: L/R is purple, R/R is yellow, L/F is red, and the R/F is green. A brake pedal switch is used to signal that the brakes are applied. This signal is not needed for ABS functioning, but the signal will deactivate the TCS. Additional sensors are installed on some vehicles. One is an accelerometer sensor that measures the forward and reverse motions of the vehicle so the EBCM/ EBTCM can adjust brake pressure according to how fast the vehicle is decelerating. The sensor is provided with a 5-volt reference signal and returns a portion of the 5 volts to the EBCM/EBTCM. If the voltage is high, then the pressure to the front brake can be reduced. This helps prevent nose diving at the front of the vehicle and provides better vehicle stability and control. Like all ABSs wheel speed is sensed by speed sensors. Wheel sensors are used and may be integrated into the wheel bearing assembly (hub) or of the typical plug-in type. The integrated sensors provide a more concise signal that is read directly by the EBCM/ EBTCM instead of being converted from AC to DC like the ABS VI systems. Integrated WSSs are used on higher-end GM FWD vehicles (**Figure 10-18**). Neither type of sensor is adjustable and all have a fixed air gap when installed.

The DBC-7 works like most ABSs in that it controls brake fluid pressure to a wheel(s) that is slowing faster than the other wheels. In the pressure hold phase, the inlet valve is closed to prevent further pressure buildup from the master cylinder and to hold the current brake line pressure. The pressure decrease phase opens the outlet valve to allow some fluid to flow from the affected brake line into the accumulator to relieve some of the pressure. This releases the brake and allows the wheel to turn faster. During pressure increase, the inlet valve is opened and the outlet valve closed, allowing master cylinder pressure to the affected brake line. The pump is switched on during the three phases to pump fluid from the accumulator back into the brake's hydraulic system. This cycle is repeated as long as the pedal is depressed and the ABS is required.

The DBC-7 uses **dynamic rear proportioning (DRP)** to control rear brake pressure during normal braking. This eliminates the mechanical/hydraulic proportioning valve and is much better at precise pressure control. The proportioning process is done by cycling the inlet and outlet valve(s) to the rear wheel(s) much the same as the ABS actions used to control wheel lockup.

Dynamic rear proportioning replaces the old mechanical proportioning valves with ABS action to prevent rear wheel lock up during heavy braking..

The DBC-7 has two possible traction control systems: **enhanced traction system (ETS)** and TCS. The TCS can control wheel slippage by signaling the PCM to reduce power and then applying brake pressure to the spinning wheel. This is a typical TCS operation using a program in the EBCM. The ETS can signal the PCM to reduce engine power output only by retarding spark timing, shutting down selected cylinders, leaning air-fuel mixture, reducing the throttle angle or upshifting the transmission. One or any combination of these five actions may be commanded by the PCM. Of note is the fact that no braking action is taken in the ETS to control wheel spin.

Enhanced traction system can control wheel slippage in some cases by simply reducing the output torque of the engine by reducing ignition timing, cutting cylinders, reducing the throttle opening, or reducing fuel delivery...

If the EBCM senses wheel slip when the brakes are not applied, its first step is to request the PCM to reduce engine torque. The PCM can alter spark timing and open selected fuel injector circuits. The reduction in engine torque is signaled to the EBCM by the PCM. This signal is called the delivered torque signal.

ETS can also be called torque management.

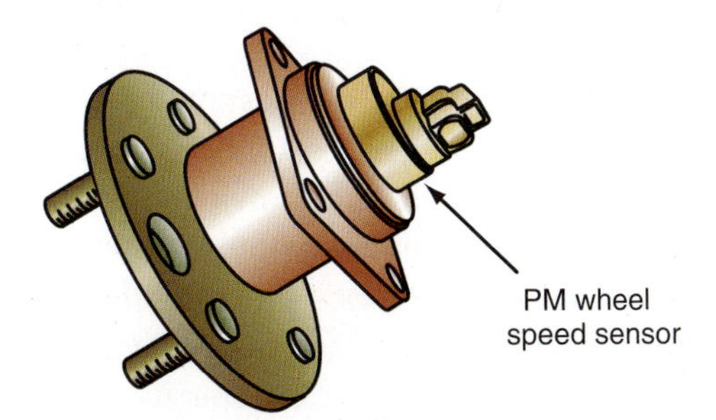

Figure 10-18 Shown is a GM hub-and-bearing assembly with an integrated PM wheel speed sensor.

Figure 10-19 The traction control lights shown are typical of General Motors vehicles and are very similar in other vehicle brands.

If the wheel(s) continues to slip, the EBCM can apply the brakes to the affected wheel(s). This is done by closing the ABS inlet valve. The prime valve is opened and the pump is switched on to charge the accumulator. The EBCM then opens and closes the inlet and outlet valves to accomplish three phases: pressure hold, pressure increase, and pressure decrease. The different phases are cycled until the wheel slippage stops.

The TCS has two dash-mounted lights and an audible warning (**Figure 10-19**). One light is the TRAC OFF. When this lamp is lit, it means the TCS programming has been deactivated because of a sensed malfunction. At the same time the EBCM requests the BCM to sound an alarm by using the radio. The TRACTION ACTIVE light is lit when the EBCM module senses wheel slippage and begins to apply the brakes in an attempt to regain traction.

A **protocol** is a computer language that modules use to communicate with each other.

Another major change between the DBC-7 and previous GM ABSs is the use of **Class 2 protocol**. Class 2 protocol is a computer network language. Each wired-in module can share information by different pulse signals used to identify each module on the net. Class 2 is faster than the older electronic communication and operates at a higher voltage. The communication speed is 10,400 bits per second (bps) versus the old 8,192 bps. The DBC-7 operates on 7 volts instead of 5 volts. Although these changes do not seem significant on the surface, when the speed of electrons is approximately 186,000 miles per second it is a huge change in the amount of information that can be transferred. The system does require a scan tool capable of reading Class 2 data. Now the prevailing network communication type is CAN, although several networks still use Class 2 in their systems, but they have largely been replaced by CAN as far as real-time networking needs like VSC. Although the DBC-7 ABS used class 2 networking, Bosch 9.0 and Teves mk60/70 use CAN networking.

CAN Protocol

The control area network (CAN) is the most common network communication type at this point, and it is incorporated into most all vehicles. CAN helps to save wiring and sensors by sharing information over the vehicle network and allowing multiple modules to communicate and work together. The CAN system uses either a single wire of a pair of wires to send messages to other computers (modules). CAN connects most all the modules in the vehicle. Since CAN is a recognized communication protocol (computer language), it can allow the use of generic modules that can be programmed to perform specific functions. There are two types of CAN. Low-speed CAN operates at 125 kb/s (kilobits per second) and is also known as ISO 11898-3. High-speed CAN, which is needed for "real time" computer interaction, is rated at

1Mb/second. High-speed CAN is used for critically fast computation speeds like ABS and stability control operations. Compare this to the Class 2 discussion later in the chapter, which has a speed of 10.4kb/s, low-speed CAN is much faster than normal Class 2 systems. Some vehicles still use Class 2 for some functions, but CAN has been fairly dominant since 2008, especially when speed is a factor. With the increasing use of ABS/TCS, ARC, and electrical steering, that train of thinking is becoming to consider the vehicle as a system rather than a series of components with specific jobs. The entire vehicle is working and communicating together for the good of the driver.

Continental Teves Mk60/Mk70

The Teves Mk60/Mk70 is used on some compact cars for the ABS, TCS, and stability control systems. The wheel speed sensors are of the magnetoresistive type, and produce a square wave signal. The magnetoresistive wheel speed sensors are much more accurate at very low speeds. The hydraulic modulator controls brake pressures at all four wheels, including dynamic rear proportioning, which eliminates the need for separate proportioning valves. The system uses Electronically Variable Rear Proportioning (EVRP). During EVRP, the EBCM controls rear brake pressure through the solenoid valves during hard rear braking to prevent rear wheel lock up. The EBCM controls the system and supplies power to the control solenoids and hydraulic pump. The brake hydraulic modulator contains isolation and dump valves for all four wheels, and two traction control isolation/supply valves for the front wheels. (Stability control is discussed further in the next chapter.) The hydraulic modulator controls pressure to each wheel to prevent wheel lockup by reducing, holding, or increasing pressure. In this mode, the modulator cannot increase the pressure beyond that supplied from the master cylinder. The Continental Teves Mk60/Mk70 system has hydraulic brake assist that increases pressure to the brakes whenever a panic stop is detected through the brake pedal sensor. (The hydraulic assist system was explained in the chapter on power brakes.)

Traction control is accomplished by monitoring the wheel that is spinning under acceleration. The ABS computer will first request that engine torque is reduced through the vehicle network. If the wheel continues to slip, the EBCM commands the pump motor to run and applies braking pressure through the appropriate valve to stop the wheel from spinning.

AUTHOR'S NOTE Up until about 1993–1994, the engine was engineered with fuel economy and emissions as the core with little regard to the transmission. Electronic engine controls and their maintenance were designed and performed with the same idea. A transmission was engineered based on how best to deliver engine power to the wheels and further the emission and economy goals of the engine. The engine and transmission were treated as separate stand-alone components. It became obvious during the 1990s that the vehicle could not meet government and consumer requirements without engineering these two major components into one system. By the early 2000s, the term *powertrain* became more commonplace because the engine and transmission had to function as one organized unit. For the repair technician, this meant that a drivability problem had to involve diagnosis of the system as opposed to one component. Today an incorrect shift point in the transmission may be directly linked to or caused by a malfunction of an engine control.

Bosch

The Bosch ABS 9.0 is a four-channel ABS system. This system is capable of handling ABS, TCS, VSC, Hill start assist, DRP and brake assist. Bosch 9.0 also houses pressure sensors and some of the needed components to perform VSC duties. The Bosch 9.0 unit looks similar to earlier hydraulic modulators, but because of the introduction of rare earth magnets into the unit, it is approximately 30 percent smaller, and at the same time it is lighter. The hydraulic pump for the Bosch 9.0 system has nine pistons instead of two that were used on earlier systems, allowing it to move more fluid quicker.

Bosch ABS 9.0 Components

A diagram of the Bosch 9.0 system is shown in **Figure 10-20**. Electronic Brake Control Module or EBCM is used to interpret inputs and command outputs from the ABS/TCS/VSC system. It is usually mounted on the hydraulic modulator (GM calls this the brake pressure modulator valve or brake modulator). The hydraulic modulator contains four inlet and outlet valves (one for each wheel). The hydraulic modulator also has two traction/stability control isolation valves and two traction/stability control supply valves. The hydraulic modulator maintains a diagonal split LF and RR circuits and RF and LR circuits that are separated hydraulically, just like a conventional braking system. The BCM monitors the brake pedal position sensor (BPP) and sends the information over the vehicle network to the EBCM. An internal brake pressure sensor tells the EBCM how hard the driver is pushing on the brake pedal. The ECM sends vehicle speed information to the brake control module, the BCM and the instrument cluster. The steering angle sensor sends information to the EBCM that will be used for VSC calculations. The TCM is also on the vehicle network and relays gear position information to the EBCM. Vehicle accelerometers are part of the inflatable restraint system, and are also used as part of the VSC system. The VSC system will be discussed more in depth in the next chapter. The wheel speed sensors are magnetoresistive sensors that use a 12-volt power supply from the EBCM to the sensors. As you recall, the magnetoresistive sensors are more accurate at low speeds, and can, in fact, produce a waveform from a stopped wheel. Bosch ABS 9.0 uses the same methodology as other ABS systems to control wheel skid: isolation (holding) of the wheel brake pressure; then, if the wheel continues to slow down, the pressure is reduced; and finally, when the wheel is turning at a speed near the other wheels, the pressure is allowed to increase. With traction control, there are two isolation/apply valves for the drive wheels, which allow the pump motor to apply one brake and isolate the other from pressure as often as necessary to control positive wheel spin.

Bosch POST

Bosch ABS 9.0 does a power-on self-test POST when the key is first turned, and then an initialization test of all the electrical solenoids and motors as soon as one of the wheel speeds exceeds 10 mph.

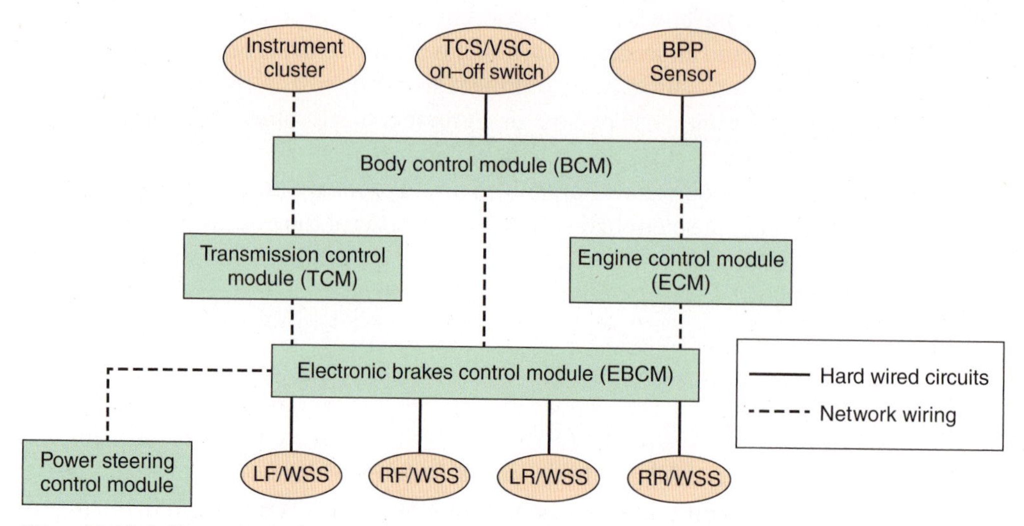

Figure 10-20 A diagram for the Bosch 9.0 system.

SUMMARY

- A sensor may measure electrical energy directly or convert a mechanical action into electrical energy.
- Signals may be analog or digital; digital signals have the advantage of having a consistent amplitude regardless of wheel speed.
- Frequency is the number of times a cycle occurs within one second.
- Duty cycle indicates the amount of time on versus time off.
- Actuators are normally used to change electrical energy into mechanical action.
- A controller is the computer module for a particular device, system, or group of systems.
- *Multiplexing* is the term applied to the communications on the vehicle computer system. More than one message can be sent at the same time.
- Control area network (CAN) is a popular communication protocol for vehicles.
- The hydraulic modulator is the actuator that directly controls brake fluid pressure during electronically controlled braking.
- Commands are electrical signals generated by the controller and sent to an actuator to perform an action.
- An ABS offers drivers a braking system that can help control and stabilize their vehicles.
- Integrated and nonintegrated ABSs usually indicate the placement of the hydraulic modulator in relation to the master cylinder and power booster.
- Four-wheel ABSs come in two configurations: three-channel and four-channel.
- Controllers may be known as electronic brake control module, controller antilock brake, or similar designations desired by the ABS manufacturers.
- An electrohydraulic unit is the combined assembly of a controller and hydraulic modulator.
- Permanent magnet sensors generate an ac voltage based on the speed of the wheel and are the most common wheel speed sensors.
- Magnetoresistive sensors have a low and a high voltage and current signal regardless of wheel speed.
- Wheel speed is calculated based on the frequency of change between waveform cycles generated by the magnetoresistive sensor. ABSs may be equipped with high-pressure pumps and accumulators. The mechanical proportioning valve has been replaced on many vehicles with electrically controlled proportioning programming known as dynamic rear proportioning. Common TCS/ABS components are WSS, ICU, and the mechanical/hydraulic service brake assemblies.
- The DBC-7 hydraulic modulator with TCS has two additional valves and an electronic brake traction control module.
- The DBC-7 uses integrated PM sensors for wheel speed.
- The DBC-7 ABS provides programming for dynamic rear proportioning, tire inflation monitoring system, and two TCS programs named enhanced traction system and traction control system.
- Class 2 wiring is used within the DBC-7 system for shared communications between networked controllers.

REVIEW QUESTIONS

Short-Answer Essays

1. Describe the general operation of a magnetic reluctance (PM) wheel speed sensor.
2. Describe the general operation of a magnetoresistive wheel speed sensor.
3. Explain the job of wheel speed sensors in the ABS system.
4. Describe the construction differences between the integrated ABS and the nonintegrated ABS.
5. Describe the difference in the DBC-7 without TCS and the DBC-7 with TCS.
6. What can the Bosch ABS 9.0 unit control as far as braking/stability control enhancements?
7. Explain why some ABS systems no longer use proportioning valves.
8. What is the advantage of vehicle networking?
9. What is the first thing the traction control can do to help eliminate positive wheel spin?

Fill in the Blanks

1. If the EBCM senses wheel slip when the brakes are not applied, its first step is to request the PCM to reduce_____________ _____________.
2. The DBC-7 operates on _____________ volts instead of _____________ volts.
3. High-speed CAN, which is needed for "real time" computer interaction is rated at _____________
4. A slow wheel as monitored by the wheel speed sensor will produce a _____________frequency signal, whereas a faster wheel will cause a comparatively _____________ frequency.
5. The _____________ replaces the standard hydraulic proportioning valve found on many older vehicles.
6. The magnetoresistive sensor produces a _____________ signal and has its own power supply. The amplitude of the signal is constant.
7. Magnetic lines of force are called _____________ _____________.
8. The Continental Teves Mk60/Mk70 system has hydraulic brake assist that increases pressure to the brakes whenever a _____________ _____________ is detected through the brake pedal sensor.
9. The hydraulic pump for the Bosch 9.0 system has _____________ pistons instead of two that were used on earlier systems.
10. The TCS is designed to control wheel spin or _____________ wheel slip.

Multiple Choice

1. Technician A says with digital signals only the frequency changes. Technician B says an analog voltage signal is infinitely variable; in other words, it flows smoothly as it changes voltage levels. Who is correct?

   A. A only
   B. B only
   C. Both A and B
   D. Neither A nor B

2. Sensors are being discussed. Technician A says that sensors are used to measure some sort of physical action and change it to an electrical signal. Technician B says that the simplest sensors are often on–off switches. Who is correct?

   A. A only
   B. B only
   C. Both A and B
   D. Neither A nor B

3. Technician A says that an actuator converts mechanical action to electric energy. Technician B says that an actuator uses electric current to perform a mechanical action. Who is correct?

   A. A only
   B. B only
   C. Both A and B
   D. Neither A nor B

4. Actuators and sensors are being discussed. Technician A says that a too low current may mean that an actuator or sensor is not functioning properly. Technician B says that the controller cannot monitor actuator and sensor operation electrically. Who is correct?

   A. A only
   B. B only
   C. Both A and B
   D. Neither A nor B

5. Command signals are being discussed. Technician A says that automotive controllers (computers) are simple compared to those used at work or school. Technician B says that most automotive computers work without sharing information with other computers on the network. Who is correct?

   A. A only
   B. B only
   C. Both A and B
   D. Neither A nor B

6. While discussing TCS as an outgrowth of four-wheel four-channel ABS, Technician A says most of the same components are used in both systems. Technician B says the TCS also interfaces with the engine (PCM) and transmission (TCM) control modules to request that specific actions be taken. Who is correct?

   A. A only
   B. B only
   C. Both A and B
   D. Neither A nor B

7. Technician A says that an integrated ABS unit has many serviceable parts. Technician B says that the nonintegrated ABS has many serviceable parts. Who is correct?

   A. A only
   B. B only
   C. Both A and B
   D. Neither A nor B

8. There are several different names in use for the ABS controller. All of the following names have been used *except:*

   A. BACU
   B. ICU
   C. EBCM
   D. CAB

9. While discussing ABS brakes, Technician A says ABS is excellent in directional stability. Technician B says ABS is excellent in directional control. Who is correct?

   A. A only
   B. B only
   C. Both A and B
   D. Neither A nor B

10. Technician A says that PM generator (variable reluctance) sensor produces a digital signal. Technician B says that a PM generator used an electromagnet as part of the sensor. Who is correct?

    A. A only
    B. B only
    C. Both A and B
    D. Neither A nor B

CHAPTER 11
ADVANCED BRAKING SYSTEMS

Upon completion and review of this chapter, you should be able to:

- Compare how fatality rates for miles traveled have been reduced.
- Describe the history of stability control through ABS/TCS systems.
- Describe the need for stability control system.
- Explain the purpose of each of the stability control sensors.
- Describe active braking.
- Describe cruise control systems that utilize active braking.
- Explain the operation of a regenerative braking system on a hybrid vehicle.

Terms To Know

Distronic

Pre-brake

Rolling diameter

INTRODUCTION

The public and the federal governments have demanded more from vehicles and their manufacturers since the development of the automobile. First, we saw the demand for cleaner vehicles and the development of emission controls in the early 1960s. In the mid-1960s, safety belts became mandatory equipment in new vehicles. More fuel-efficient vehicles were mandated after the fuel shortages in the 1970s. In the 1980s we started to see fuel efficiency and performance go hand in hand as new fuel injection and ignition systems were developed. Next, the need for safer vehicles came to the forefront. We have seen the development of air bags, ABS and traction control system (TCS), and lately active braking systems and stability control. The automotive industry continues to develop new and innovative technology that few other industries can match. Of course, the automotive industry did have help from the development of faster computers that could meet the challenge of "real time" reaction and integrated networks inside the vehicle that allowed the ECM, BCM, ABS, and stability control (VSC) modules to react as one to prevent crashes. According to the National Highway Traffic Safety Administration (NHTSA), the fatality rate per 100 million miles traveled was 5.5 in 1966, 1.10 in 2011, 1.25 in 2016. The fatality rate would have been less, except for the fact that some people still do not use their safety belts or are doing something very unwise like texting or drinking while driving. The biggest cause of fatalities is driver inattention. By some estimates as many as 95 percent of all accidents are caused by driver inattention.

STABILITY CONTROL SYSTEMS

History

Shop Manual
page 515

The first development on the road to stability control, as you have seen in this text, was antilock braking. As you probably recall from earlier chapters, the development of ABS came at a time when the prime concern was the loss of vehicle control because of the possibility of the rear wheels locking and causing a spin from which the driver could not correct, particularly on wet pavement. Many vehicles were transitioning to front-wheel drive, meaning that the front wheels did nearly all the braking. Compounding that problem was the fact that pickup trucks were gaining in popularity. Many of these trucks never hauled anything heavier than groceries, whereas many trucks were still in use by the construction, farming, and hauling industry. Work trucks were likely to be heavily loaded. This posed a huge problem for the designers, as the same truck might do both jobs. The original solution was a height-sensing proportioning valve intended to help compensate for heavy loads by measuring the sag on the rear springs because of loading. Rear-wheel antilock brakes were eventually established for full-size pickup trucks. These were relatively simple by today's standards. The rear wheels were controlled as a pair; no matter which wheel was skidding, both rear wheels were treated in the same way. The way in which the rear brake pressure was modulated has not changed that much. This is the familiar isolation (when the brake pressure from the master cylinder was isolated) and not allowed to increase regardless of how much pressure was applied. If the wheel continued to slow down too fast (as reported by the wheel speed sensor), then the pressure was reduced to the rear wheels. This cycle could take place several times per second and greatly increased the safety of the vehicle. These early ABS systems were single-channel ABS. They controlled only the rear brakes, and then only as a pair.

Three- and Four-Channel ABS

The next major development of the ABS system was the advent of three-channel and four-channel ABSs. Three-channel ABSs were the most common when ABS were first introduced in passenger cars. Both front wheels were controlled independently of each other, but the rear wheels were controlled as a pair. If one rear wheel started to lock up, then both rear wheels were modulated in the same manner. Finally, for stability control, four-channel ABS is the rule. All four wheels are controlled independently. ABS, as you recall, does not necessarily stop a vehicle faster but allows for control of the vehicle in a hard-braking situation. If the front wheels lock, the vehicle cannot be steered; if the rear wheels lock, then the vehicle tends to spin out of control (**Figure 11-1**).

ABS and Stability Control

You could say that ABSs in themselves were a start of stability control, because if the vehicle remained drivable during a panic stop it is far less likely to crash. Development of stability control is the story of integration of several individual systems that have the ability to work together.

Figure 11-1 ABS brakes and stability control have saved countless accidents over the years.

Traction Control

Traction control was born out of the ABS system and appeared just a few years after ABS became common on most vehicles. Traction control did not require a large amount of hardware to accomplish; it required only a pump (if it was not already present) and a couple of valves. The rest was software enhancements. Traction control is basically the reverse of ABS; it looked for a wheel spinning too *fast* and then applied the brakes to slow it down (**Figure 11-2**). Wheels spinning because of a loss of traction is called positive wheel slip. Traction enhancements have been around for many years on rear-wheel drive vehicles as limited-slip differentials, although these new electronic systems are faster and do not have clutches to wear out or lubricant to change. Traction control also has one more development that is important; it was one of the first systems to interface with the ECM. The first response to a wheel slip is a torque reduction request from the EBCM whenever a wheel is spinning. This torque reduction usually comes about by a reduction in ignition timing and throttle control. Electronic throttle control allows the vehicle network to control the throttle position. If this does not correct the wheel slip, then the brake pressure modulator valve (BPMV) is commanded to apply brake pressure to the slipping wheel. The ECM can also tell the transmission to shift to a higher gear (also reducing torque) at the request of the ABS computer. It is important to note that on most vehicles, traction control is only effective at relatively low speeds, but some sports cars can have traction control at much high speeds.

AUTHOR'S NOTE Old-timers had their own way to get some traction whenever they were stuck in the mud or snow. On a vehicle with a conventional differential, if one wheel has no traction it will spin by itself and the vehicle will not move. Knowing this, they would apply the emergency brake slightly. Of course, this would apply the brakes on both wheels, but sometimes it was enough to slow the wheel with no traction just enough that the wheel with traction can pull the vehicle out of the slippery situation. Another technique was to start the vehicle out in a gear higher than first (with a manual transmission). This would reduce the torque applied to the spinning wheels. Today, the modern TCS works in the same ways as illustrated, just higher tech.

Figure 11-2 Positive wheel spin can be stopped with traction control.

Figure 11-3 A typical steering angle sensor measures the driver's intended direction.

Figure 11-4 The actuator is solenoid used to control power steering fluid flow from the pump to the steering gear.

Computer-Controlled Hydraulic Steering/Electric Steering

Shop Manual page 521

Electronic or electrically controlled steering began in the 1990s. The electric steering computer regulated the flow of pressurized power fluid to the steering gear. During low-speed operation, full power steering flow was allowed for parking maneuvers. At highway speeds the power steering flow is cut back for better road feel. Full power steering pressure can be delivered during an emergency maneuver by utilizing the steering wheel angle sensor. The steering angle sensor for one of these hydraulic systems is located at the bottom of the steering column and measures steering rotation in both degree of turn and speed of steering wheel rotation (**Figure 11-3**). Using these data and other information such as VSS, the electric steering computer regulates the flow of pressurized power fluid to the steering gear using a solenoid mounted on the power steering pump (**Figure 11-4**). Other electrical steering assist can be done by attaching an electric motor to the steering shaft or gear that can apply force to the steering mechanism.

Electric Power Steering

Electric power steering serves two purposes for automakers. The electric power steering uses no energy when not in use, such as when driving straight down the road, and, for systems that depend on computer control such as lane departure and steering a vehicle back into the correct lane. The electric steering effort is appropriate to the road speed; the same is true for the speed sensitive type hydraulic system, which also uses the VSS to determine road speed. There are three types of electrical power steering: the column mount, (**Figure 11-5**), the rack mount (**Figure 11-6**), and the pinion-mounted electric motor (**Figure 11-7**). The electrical power steering control module (PSCM) is connected with the vehicle's network to share information with the other stability control components, such as driver input steering torque and steering wheel position and road speed from the ECM (**Figure 11-8**). The torque sensor is a dual sensor that is mounted on a torsion bar inside the steering column, which tells the VSC module how hard the driver is turning the steering wheel. The steering torque sensor produces two output signals from

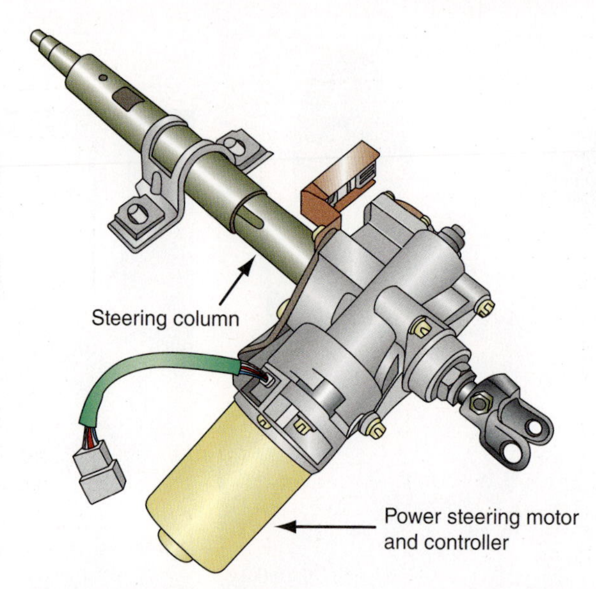

Figure 11-5 A column style electric power steering system used on some GM vehicles. Similar systems were used on several different vehicles.

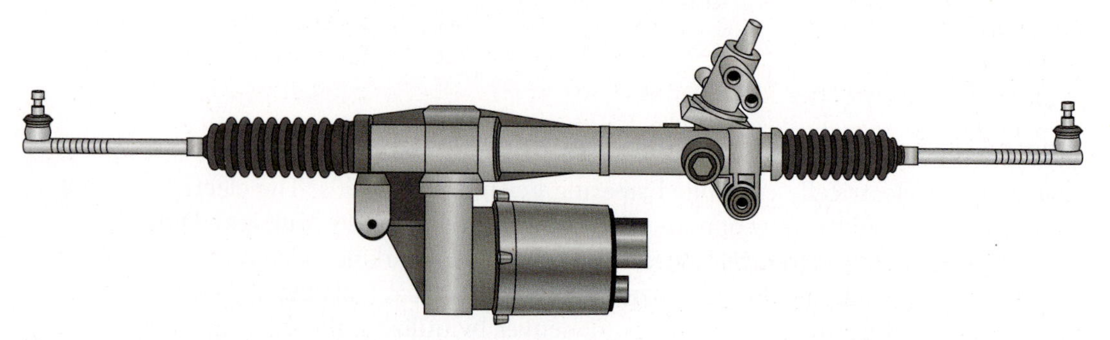

Figure 11-6 A rack-mounted motor style electric rack and pinion.

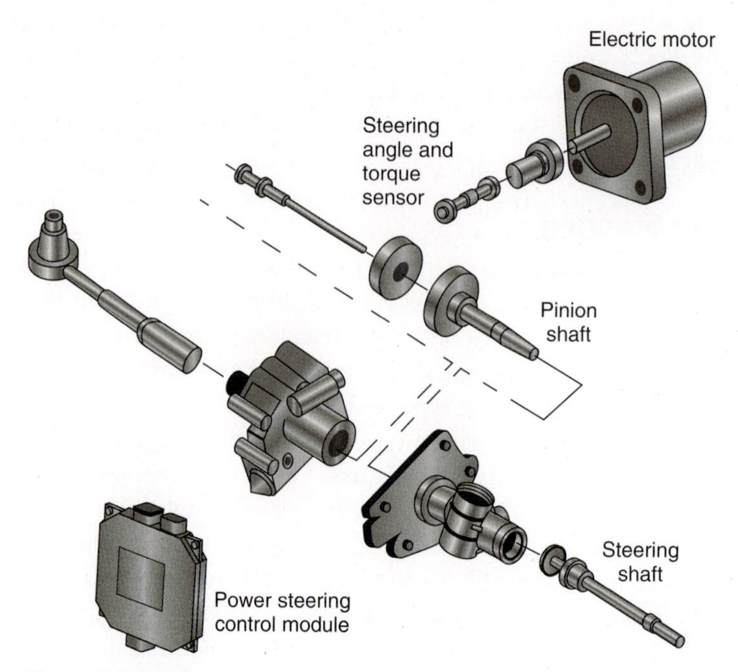

Figure 11-7 A pinion-mounted electric power steering motor.

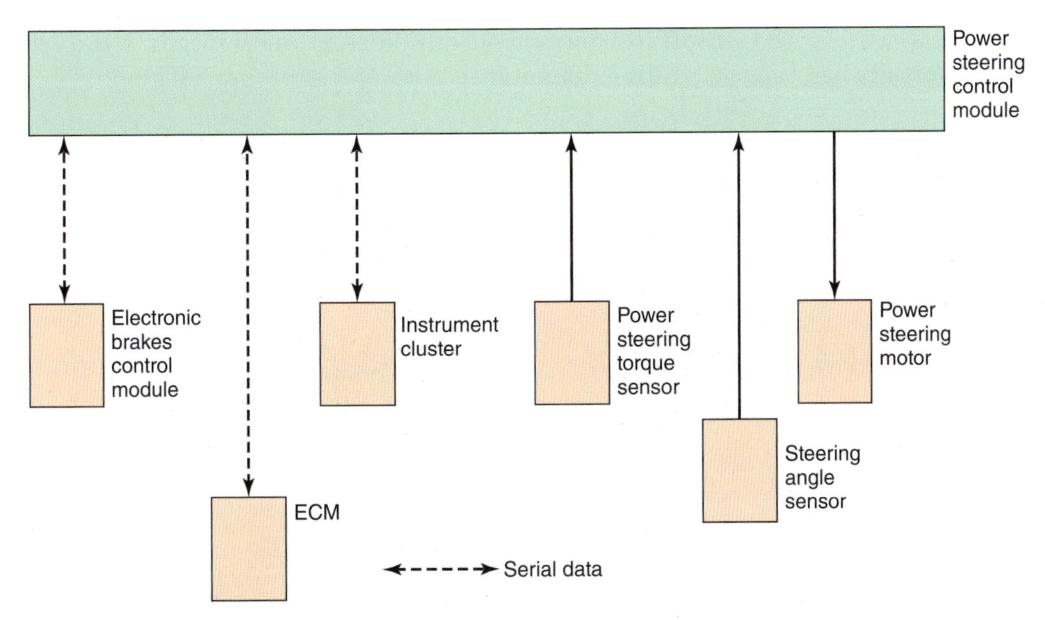

Figure 11-8 Electric power steering network of modules, inputs, and outputs.

about 0.25 volts to almost 5 volts. Output signal 1 voltage goes up and output signal 2 goes down during a right turn; during a left turn, output signal 1 goes down and output signal 2 goes up. In this way, the computer has two signals to check for validity of this important signal. The steering wheel position sensor is also used as an input to the stability control. **Figure 11-9** shows a scan tool graph of a steering angle sensor as the wheel is turned to the left.

Tire Pressure Monitoring Systems

Tire pressure monitoring systems (TPMS) have been around for the past several years on high-end vehicles and have been introduced into passenger vehicles. The system monitors and detects low tire pressure in the vehicle's tires. The system has been designed to prevent a low tire from overheating and blowing out, which has been implicated in causing several roll-over crashes, particularly involving SUVs. Additionally, having tires with unequal pressures also affects vehicle handling. In order to maximize the effect of stability control and ABS braking, external factors such as tire pressures need to be at the correct pressures. Stability control is difficult enough while not adding the complication of underinflated tires. The **direct TPMS** operates by using transmitters in each wheel, which are generally located in the tire valve stems (**Figure 11-10**). A radio signal that contains tire pressure information is transmitted to the TPMS module. On some systems, the tire position must be relearned, if the tires are rotated. Some systems called **indirect TPMS** can

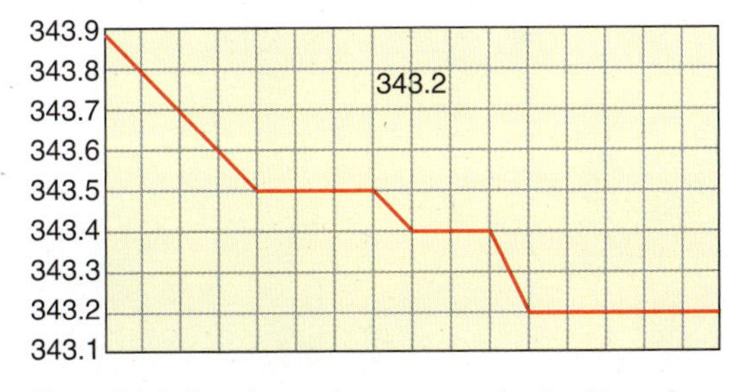

Figure 11-9 Steering angle sensor graph, wheel turned fully to the left.

Figure 11-10 A typical tire pressure sensor.

use the ABS wheel speed sensors to determine slightly different wheel speeds. A tire with low air pressure has a smaller **rolling diameter** than a wheel at proper inflation.

Stability control is not a substitute for careful driving. Remind your customers that stability control is a driving aid; much like a rear-view mirror, just much higher tech.

Stability Control

Stability control is the controlled application and release of brakes at the correct time, independently of each other, to prevent an out of control vehicle during hard maneuvering. There are two common terms for loss of steering control through a corner: understeer and oversteer.

AUTHOR'S NOTE It is important to remember that no matter what kind of sophisticated braking or traction control system is present on a vehicle, worn out tires will defeat the ABS, TCS, and stability control systems every time.

You can think of stability control action like the way a bulldozer turns; one tread stops and pulls the bulldozer to the side with the stopped tread.

Understeer

Understeer, simply stated, means that the vehicle did not make it through a turn and went off the road on the outside of the turn. It appears almost as though the driver did not turn hard enough, but more than likely was driving too fast for weather conditions or the condition of the tires or vehicle and the front wheels lost traction (**Figure 11-11**). In an understeer situation, the inside rear-wheel brake can be applied by the stability control system to help pull the vehicle around the curve.

Oversteer

Oversteer is the loss of traction at the rear wheels leading to a vehicle spin (**Figure 11-12**). Once again, the driver might have steered into the corner too hard, the rear tires might

Figure 11-11 Understeer means the vehicle is going to go off the road on the outside edge—usually because the front wheels have lost traction.

Figure 11-12 Oversteer means the vehicle is going into a spin because the rear wheels have lost traction.

be worn dangerously thin, or the driver might have been driving too fast for conditions. The outer-front-wheel brake is applied by the stability control system to help counter oversteer.

STABILITY CONTROL HARDWARE

Steering Angle Sensor

As was illustrated earlier, the steering angle sensor measures the actual direction the steering wheel turned, as well as how fast the wheel is turned. This information can be used by the electronic stability control (VSC) module and compared to the yaw sensor. (**Figure 11-13**).

Yaw Sensor

Shop Manual
page 519

The yaw sensor can measure the difference between the actual direction that the vehicle is traveling and the direction the driver is trying to steer the vehicle by way of the steering angle sensor. This difference is called the **slip angle**. The VSC computer then calculates the amount of steering correction needed. Some yaw sensors also contain lateral and longitudinal acceleration or "G"-sensors (**Figure 11-14**) that can measure side force the vehicle is under during a turn. The longitudinal acceleration sensor can determine the rate of acceleration and deceleration. **Figure 11-15** is a screen shot from a scan tool showing the steering angle, yaw rate, and lateral and longitudinal acceleration. To understand yaw, imagine a pin through the roof of the car and into the pavement below; the yaw rate is the angle and speed that the vehicle is rotating around the pin through the center of the vehicle. Think of a vehicle sliding sideways down the road. The front of the vehicle is facing off the road and the vehicle is sliding down the roadway. The rate that the vehicle is spinning around its vertical axis is the yaw rate. The slip angle is the difference between the intended direction of travel and the desired direction of travel (**Figure 11-16**). Think of a vehicle that is drifting around a corner; the slip angle is desired during this kind of driving (**Figure 11-17**). The rear of the car sliding around in the corner is yaw.

Beam-Type Hall-Effect Yaw Sensors

The newest yaw sensors are beam-type Hall-effect sensors that receive a 5-volt signal from the EBCM and return some portion of that voltage as a signal. Usually the portion returned is 2.5 volts, representing a zero gravity force in a lateral direction. In other words, the vehicle is

Figure 11-13 A yaw rate sensor measures the actual turning rate of a vehicle.

Figure 11-14 A G-sensor, also known as a lateral acceleration sensor, that works with the VSC. The G-Sensor measures the rate of acceleration and the direction of movement.

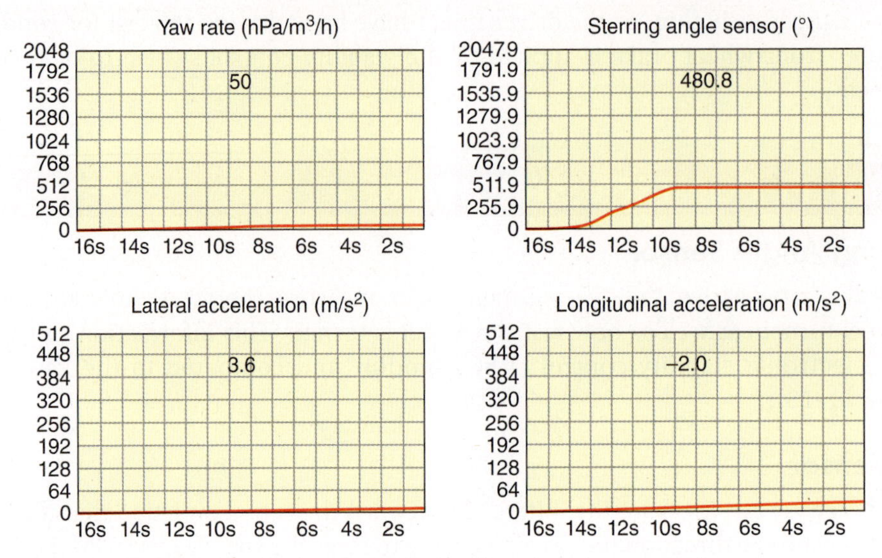

Figure 11-15 A typical reading for a longitudinal and lateral accelerometer and yaw sensor.

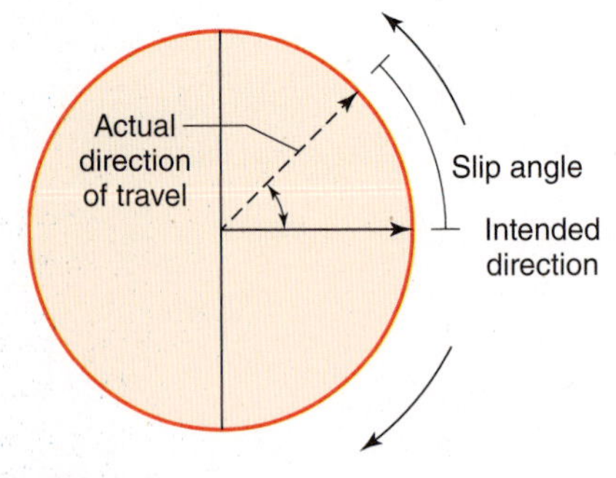

Figure 11-16 Illustration of yaw and slip angle.

Figure 11-17 A vehicle drifting around a corner is illustrating yaw rate.

completely level and moving straight ahead or sitting still. The variation from 2.5 volts is interpreted by the EBCM as the actual motion of the vehicle. The beam-type Hall-effect sensor produces an analog output signal that is interpreted by the VSC computer.

Hydraulic Modulator (BPMV)

The hydraulic modulator, also called a brake pressure modulator valve (BPMV, **Figure 11-18**), is generally composed of a valve body that contains the control valves used to isolate, decrease, and increase pressure at the individual wheels that can control the application of brake pressure according to the demands placed on the vehicle by the driver, and the optimum corrections as calculated by the VSC computer. The hydraulic modulator can also contain a pump motor and, in many cases, the VSC module (EBCM) is part of the assembly.

Throttle Actuator Control

The ECM has control of the throttle position on most late-model vehicles using the throttle actuator control (TAC) (**Figure 11-19**). The VSC module can request the ECM to find

Figure 11-18 The hydraulic modulator, also known as a BPMV, can apply or release pressure to any of the four brakes.

Figure 11-19 A throttle actuator control unit.

Figure 11-20 The accelerator pedal position sensor is usually located on the accelerator pedal.

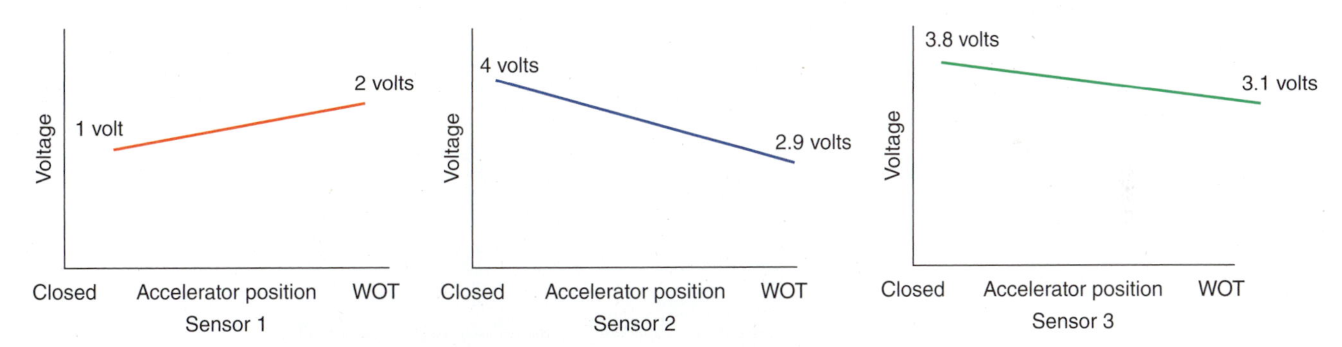

Figure 11-21 The accelerator pedal position is made up of three sensors in one housing.

the most appropriate throttle angle for vehicle control. The accelerator pedal position (APP) sensor (**Figure 11-20**) relays the driver's demands for throttle to the ECM. The APP sensor has at least two and sometimes three separate sensors to make sure the signal is reliable (**Figure 11-21**). If the driver is requesting full throttle in a situation that is resulting in a wheel spin, the ECM can request that the throttle be backed down to an acceptable level regardless of the driver's input. The VSC module and the ECM can keep track of the throttle position by using the throttle position sensor (TP sensor). The TP sensor has two sensors, each using at least two separate sensors (**Figure 11-22**). The ignition timing can also be retarded to remove some torque from the wheels when required.

Roll Control/Ride Control

It would be simple to prevent a vehicle from rolling in a turn by making the suspension very stiff and lowering the vehicle, like road-racing vehicles. This is not acceptable to most drivers in their daily vehicle. Some systems incorporate a feature called automatic ride control (ARC). Automatic ride control works with the vehicles struts/shock absorbers to prevent roll in hard turns but can still provide a comfortable ride.

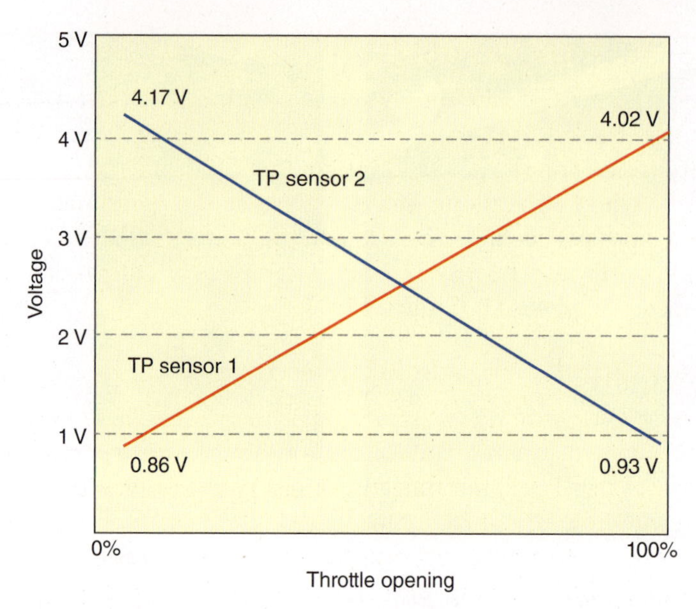

Figure 11-22 Dual TP sensor voltage rates. (Note: There are several types of sensors; always use the chart for your specific vehicle.)

A separate ride control module controls the shock absorbers' compression rate. These systems use hydraulic, pneumatic, or magnetic controls that can rapidly change the rate of the shocks. An air spring suspension (**Figure 11-23**) can control body roll and suspension rate, as well as control ride height. Some of the suspension control systems also have a control on the dash to change the overall feel of the suspension, such as sport and normal settings.

Figure 11-23 The air spring can be inflated or deflated to control body lean and ride height for better control and a better ride.

Cadillac Magnetic Ride Control

Cadillac and several other GM divisions have a magnetic suspension system that can alter the stiffness of the shock absorbers electrically and, therefore, the amount of body roll during hard cornering. The shock absorbers contain fluid called **magneto-rheological fluid** that stiffens when a magnetic field is exposed to it. The magnetic field can be changed at 1,000 times per second. The strength of the magnetic field can be changed by varying current flow (**Figure 11-24**). The ride height is constantly monitored by suspension position sensors, as shown in **Figure 11-25**.

VSC Computer and the Vehicle Network

Shop Manual
page 517

The VSC computer works in cooperation with the other computers in the vehicle network to provide stability control by commanding the appropriate brake pressures for ABS, TCS, and VSC purposes, as well as communicating with the instrument panel control module (IPC) and the ECM. The communication with the IPC tells the VSC module of the driver's demands, such as turning the VSC off, and communicating messages to the IPC such as illumination of the "ABS" warning lamps, "TCS" active lamp and "VSC active" lamp depending on the application. The reason the warning lights are activated is to tell the driver that the situation is being corrected for them. The idea is that the driver may get too confident in their driving ability without the notifications. Through communication with the ECM, the VSC can request lower engine torque and/or reduced throttle angle to

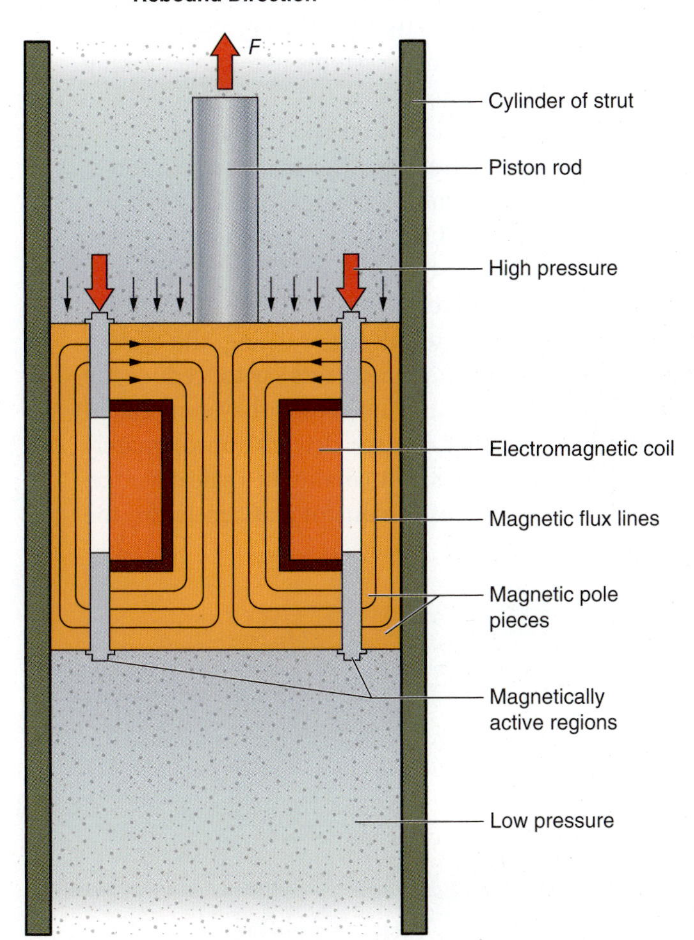

Figure 11-24 The flow control valve can vary the magneto-rheological fluid's viscosity by applying an electric field.

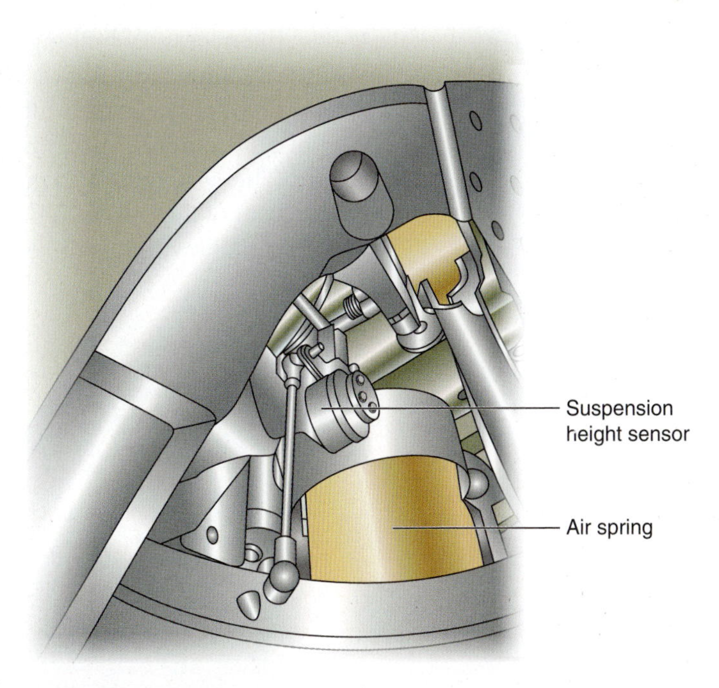

Figure 11-25 Height sensor shown on rear wheel. Air shock on this vehicle is for load leveling the vehicle when trunk is loaded.

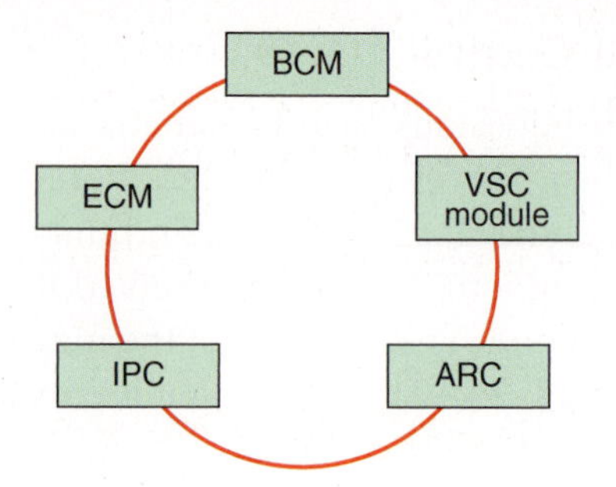

Figure 11-26 Vehicle stability control depends on the interaction of many control modules on the vehicle's network.

help keep the vehicle under control. The ECM can also share information such as brake pedal position, throttle position sensor (TPS) and vehicle speed sensor (VSS) signals with the VSC and ARC computers (**Figure 11-26**).

ACTIVE BRAKING SYSTEMS

Several new systems have been developed in recent years to add to the active braking systems that have been developed over the past several years. Many vehicles use a system with high-definition cameras that can scan the road ahead and warn the driver of a possible collision. Manufacturers have developed radar systems to warn drivers and can even apply the brakes to avoid a collision, if necessary. Radar cruise systems use active braking systems that utilize the cruise control not only to keep a set speed, but to maintain a safe distance to vehicles ahead. Active braking systems were also discussed in Chapter 6.

There are several types of active braking systems available such as Subaru's **"Eyesight"** or Mercedes **"Pre-brake** that prepare the braking system for a sudden stops.

Active braking systems will build increased pressure from the master cylinder in anticipation of a panic stop, when the danger of an impending crash is possible as measured by the vehicle's short-range radar system (**Figure 11-27**). Some of these systems, like Mercedes Benz's **Pre-brake** or Subaru's **Eyesight** system, will prepare the brakes to apply only the necessary braking pressure when the driver responds, thus preventing a panic stop situation. If the driver does not apply the brakes after a warning, the system will do it for him or her to prevent an accident, or as the manufacturers say to "reduce the severity of an accident if it occurs." Mercedes also has a cruise control system called **Distronic**. The Distronic is a cruise control system that can see drivers approximately 600 feet in front of the vehicle and apply the brakes if necessary to maintain a safe distance. Once the safe following distance is reestablished, the cruise control can accelerate the

Distronic is the name that Mercedes gave their cruise control that has the ability to maintain a safe distance from other vehicles on the road ahead with no input from the driver.

Figure 11-27 An active brake warning system.

vehicle back to its programmed speed. If the gap between vehicles is closing too fast, the Distronic cruise control can calculate the force needed to avoid the vehicle and apply just the force needed to stop the vehicle. Distronic cruise control can apply the brakes if the driver fails to respond to visual and audible warnings. The calculated brake force can stop the vehicle, but not too fast for drivers behind to be able to stop on their own.

REGENERATIVE BRAKING SYSTEMS

Introduction

The idea behind regenerative braking is to recover some of the energy that is lost as heat when the brakes are applied and the vehicle is stopped. Most hybrid and electric vehicles have a way to recover this lost energy as electrical power stored in the battery. The general idea is that part of the braking is actually accomplished by using generators attached to the wheels or engine that engage to slow the vehicle down during moderate stops. A hydraulic system is still needed for panic stops and to bring the vehicle to a complete stop.

The amount of regenerative braking is tailored by the EBCM and depends on how much braking force is being demanded from the driver. The braking is done in a blended manner; some of the braking is done regeneratively, and some is done hydraulically. If braking is very heavy, then most of the braking will be done by the hydraulic brakes. If braking requested is very light, then most of the braking is done by the regeneration braking.

Prius

The Toyota Prius uses two motor generators: MG1 and MG2. MG1 is driven by the gasoline engine and generates the power to charge the main storage battery. MG2 is generally used as an electric motor drive, but will act as a generator when the drive wheels are turning the generator. The Toyota Prius uses regenerative braking that begins as soon as the accelerator pedal is released by activating the motor/generator MG2, which is connected through the vehicle's driveline. This allows the vehicle to recover about 30 percent of the energy previously lost as heat on braking. The motor generator produces alternating current, which is then transferred to the hybrid vehicle battery through an inverter. The inverter converts the AC voltage to DC. Direct current has to be used to recharge a battery. The power conserved depends on the state of charge of the battery and is controlled by the power management ECU. Speed downhill can be controlled by shifting the transmission into the "B" range when going downhill and can convert more energy to the battery as well. The hybrid Prius works with VSC in the same manner as nonhybrid vehicles do, through the hydraulic modulator.

Caution

Make certain that all safety precautions are followed whenever working on or around a hybrid vehicle. Always disable the high-voltage system by removing the service plug or switching the service disconnect to "off." Wear high-voltage gloves and shoes as prescribed by the manufacturer.

KIA Optima

The KIA Optima uses active hydraulic brakes. The system uses a brake pedal position sensor to determine the amount of braking force needed. The braking control module blends the hydraulic braking with the regenerative braking to produce the total braking force required. The KIA Optima Hybrid uses a belt-driven starter generator (**Figure 11-28**) to help slow the vehicle during deceleration; hydraulic brakes are used to bring the vehicle to a complete stop and during hard braking. The starter generator has coolant circulating through it to keep it cool.

Chevrolet Volt

The Chevrolet Volt uses one electric motor to drive the wheels. The gasoline engine is used only to charge the battery. Any time the accelerator pedal is released, the drive motor becomes a generator.

Figure 11-28 A belt-driven starter generator can be used for regenerative braking by loading the engine on deceleration.

SUMMARY

- Highway fatalities have been dropping due to vehicle safety improvements like ABS brakes and stability control.
- ABS was the starting point for traction control and stability control systems.
- The development of faster computers that could meet the challenge of "real time" reaction and computer networks inside the vehicle allowed the ECM, BCM, ABS, and stability control modules to react as one to prevent crashes.
- Stability control helps prevent oversteer and understeer.
- Understeer means that the vehicle did not make it through a turn and went off the road on the outside of the turn.
- Oversteer is the loss of traction at the rear wheels leading to a vehicle spin.
- Electronic/hydraulic and electric steering systems have been incorporated into the stability control systems. The steering angle, torque, and speed are valuable input to the stability control system.
- Tire pressure monitoring systems help owners maintain proper tire pressures, which increases safety, as well as the effectiveness of the stability control system.
- Active braking systems like Mercedes Benz's Pre-brake system will actually prepare the brakes to apply only the necessary braking pressure when the driver responds, thus preventing a panic stop.
- Smart cruise control systems have the ability to maintain distance behind the vehicle in front of them by applying the brakes as necessary to slow the vehicle.
- Hybrid vehicles utilize regenerative braking to help recharge the batteries.

REVIEW QUESTIONS

Short-Answer Essays

1. Compare how fatality rates for miles traveled have been reduced.
2. Explain what brought on the development of ABS braking and why.
3. Explain how the rear wheels were controlled on the original rear-wheel antilock brakes such as RWAL or RABS.
4. Explain the difference between three- and four-channel ABS brakes.
5. Explain what is meant by *understeer*.
6. Explain what is meant by the term *oversteer*.
7. Describe the first response to wheel slip by the traction control.
8. Explain how computer-controlled hydraulic steering works.

9. Briefly explain how active braking systems operate.
10. Explain the use of a yaw sensor.

Fill in the Blanks

1. By some estimates, as many as ______________ percent of all accidents are caused by driver inattention.
2. Computer ______________ inside the vehicle allowed the ECM, BCM, ABS, and stability control (VSC) modules to react as one to prevent crashes.
3. Traction control is basically the ______________ of ABS.
4. The ______________ sensor is a dual sensor that is mounted on a torsion bar inside the steering column.
5. The Toyota Prius uses ______________ braking that begins as soon as the accelerator pedal is released by activating the motor/generator.
6. ______________ braking systems will build increased pressure from the master cylinder in the event of a panic stop.
7. The Toyota Prius uses two motor generators: ______________ and ______________.
8. The KIA Optima Hybrid uses a ______________ driven starter generator to help slow the vehicle during deceleration.
9. Electronic or electrically controlled steering began in the ______________.
10. Active ride control works with the vehicles struts/shock absorbers to prevent ______________ roll ______________ in hard turns.

Multiple Choice

1. Technician A says that the ABS was the first step on the road to stability control systems. Technician B says that ABS brakes allow the driver to control the vehicle when braking. Who is correct?

   A. A only
   B. B only
   C. Both A and B
   D. Neither A nor B

2. Technician A says that a direct TPMS sensor transmits a signal to the TPMS module concerning tire pressure. Technician B says that some TPMSs use temperature sensors to determine an overheating tire. Who is correct?

   A. A only
   B. B only
   C. Both A and B
   D. Neither A nor B

3. Technician A says the steering torque sensor tells the VSC module how hard the driver is turning the steering wheel. Technician B says the steering wheel position sensor is also used as an input to the stability control. Who is correct?

   A. A only
   B. B only
   C. Both A and B
   D. Neither A nor B

4. Technician A says an air spring suspension can control body roll. Technician B says an air spring suspension cannot be used to control ride height. Who is correct?

   A. A only
   B. B only
   C. Both A and B
   D. Neither A nor B

5. While discussing the VSC module, Technician A says the VSC computer works in cooperation with the other computers in the vehicle network to provide stability control. Technician B says that the VSC module stands alone and does not need to communicate with the vehicle network. Who is correct?

   A. A only
   B. B only
   C. Both A and B
   D. Neither A nor B

6. Technician A says that reducing torque to the wheels can help prevent loss of traction. Technician B says that starting out in a lower gear can reduce torque to the wheels. Who is correct?

   A. A only
   B. B only
   C. Both A and B
   D. Neither A nor B

7. The Toyota Prius uses two electric motor generators: MG1 and MG2. Technician A says that MG1 is used as a generator when braking and on deceleration. Technician B says that MG2 is used as the main drive electric motor. Who is correct?

   A. A only
   B. B only
   C. Both A and B
   D. Neither A nor B

8. Technician A says the TAC module has direct control of the throttle position. Technician B says that the ECM can command the TAC to find the best throttle position for the conditions. Who is correct?

   A. A only C. Both A and B
   B. B only D. Neither A nor B

9. Technician A says the Distronic system is a cruise control system. Technician B says the Distronic system can see drivers approximately 6,000 feet in front of the vehicle.

   A. A only C. Both A and B
   B. B only D. Neither A nor B

10. Technician A says that the yaw sensor measures the actual turning angle that the vehicle is traveling. Technician B says that the steering angle sensor measures the direction that the driver is trying to steer. Who is correct?

    A. A only C. Both A and B
    B. B only D. Neither A nor B

GLOSSARY
GLOSARIO

Note: **Terms are highlighted in bold,** followed by Spanish translation in color.

ABS An ABS is a service brake system that modulates hydraulic pressure to one or more wheels as needed to keep those wheels from locking during braking. If the wheel locks during braking, steering control becomes very difficult. ABS allows the driver to maintain control of the vehicle during a panic stop.

ABS Un ABS es un sistema de servicio de frenos que modula la presión hidráulica a una o más ruedas según sea necesario para evitar que esas ruedas se enllaven durante el frenado. Si la rueda se enllava durante el frenado, el control de la dirección se vuelve muy difícil. El ABS premite al conductor mantener el control del vehículo durante una parada de emergencia.

accumulator A container that stores hydraulic fluid under pressure. It can be used as a fluid shock absorber or as an alternate pressure source. A spring or compressed gas behind a sealed diaphragm provides the accumulator pressure. In an ABS, the accumulator is a gas-filled chamber that acts as both a storage container for system fluid and a reserve pressure chamber to provide smooth antilock operation and dampen pressure pulses from the system pump.

acumulador Recipiente que contiene líquido hidráulico bajo presión. Se puede utilizar como líquido amortiguador o como fuente de presión alternativa. Un resorte o gas comprimido detrás de un diafragma sellado proporciona presión al acumulador. En un sistema ABS, el acumulador es una cámara llena de gas que actúa a la vez como recipiente de almacenamiento de líquido para el sistema y como cámara de reserva de presión para proporcionar una operación suave de antibloqueo y amortiguar los pulsos de presión que provienen de la bomba del sistema.

active brakes A brake system that warns the driver of an impending collision with a vehicle ahead. The system can also apply the brakes to slow the vehicle if the driver does not respond.

frenos activos Sistema de frenos que advierte al conductor de una colisión inminente con un vehículo que se encuentra al frente. El sistema también puede activar los frenos para que se disminuya la velocidad del vehículo si el conductor no responde.

active braking systems Active braking systems can apply the brakes without the driver touching the brake pedal. Meant to be used during an emergency stop.

sistemas de frenado activo Los sistemas de frenado activo pueden accionar los frenos sin que el conductor toque los pedales de freno. Destinado para utilizarse durante una parada de emergencia.

actuator(s) Any device that receives an output signal or command from a computer and does something in response to the signal.

impulsor Cualquier dispositivo que reciba un comando o una señal de salida de un computador y haga algo como respuesta a dicha señal.

adjustable pedal system (APS) Mechanical devices capable of moving the brake, accelerator, and clutch pedal forward (up) and backward (down) to accommodate different drivers. Usually computer controlled based on the driver's input.

sistema de pedal adaptable (APS) Dispositivo mecánico capaz de mover los pedales del freno, el acelerador y el clutch o embrague hacia adelante (arriba) y hacia atrás (abajo) para acomodar a los diferentes conductores. Usualmente está controlado por una computadora en el mando de entrada del conductor.

adsorption The condensation of a gas on the surface of a solid.

adsorción La condensación de un gas en la superficie de un sólido.

air brakes Braking system used on large trucks that utilize air pressure instead of hydraulics to operate.

frenos de aire Sistema de frenado utilizado en grandes camiones que usa presión de aire en lugar de presión hidráulica para funcionar.

ampere (A) The unit for measuring electric current. One ampere equals a current flow of 6.241×10^{18} electrons per second.

amperio (A) Unidad de medida de la corriente eléctrica. Un amperio equivale a un flujo de corriente de $6{,}28 \times 10^{18}$ electrones por segundo.

analog A signal that varies proportionally with the information that it measures. In a computer, an analog signal is voltage that fluctuates over a range from high to low.

analógica Señal que varía proporcionalmente con la información que mide. En un computador, una señal analógica es la tensión que fluctúa en un margen de alto a bajo.

antilock brake control module (ABCM) The computer that controls the ABS operation. May be used on some systems for traction control.

módulo de control antibloqueo de frenos (AC) Computador que controla el funcionamiento de los frenos ABS. En algunos sistemas se puede utilizar para controlar la tracción.

antilock brake system (ABS) A service brake system that modulates hydraulic pressure to one or more wheels as needed to keep those wheels from locking during braking. An antilock brake system can improve vehicle control during hard braking and eliminate or reduce the tendency for the vehicle to skid.

sistema antibloqueo de frenos (ABS) Sistema de frenos de servicio que modula la presión hidráulica a una o más ruedas cuando es necesario para evitar que éstas se bloqueen al frenar. Un sistema antibloqueo de frenos puede mejorar el control del vehículo durante un frenado brusco y eliminar o reducir la tendencia del vehículo a patinar.

aramid fibers A family of synthetic materials that are stronger than steel but weigh little more than half what an equal volume of fiberglass weighs.

fibras de aramida Familia de materiales sintéticos más fuertes que el acero pero que pesan poco más de la mitad que un mismo volumen de fibra de vidrio.

arcing The process of grinding or forming drum brake linings to conform to the drum diameter and provide clearance where needed.

formación de arco Proceso de limar o formar los forros de frenos de tambor para que se adapten al diámetro del tambor y den tolerancia donde sea necesario.

asbestos The generic name for a silicate compound that is very resistant to heat and corrosion. Its excellent heat dissipation abilities and coefficient of friction make it ideal for automotive friction materials such as clutch and brake linings. Asbestos fibers are a serious health hazard if inhaled.

amianto Nombre genérico de un compuesto de silicatos muy resistente al calor y a la corrosión. Su excelente capacidad de disipación de calor y su coeficiente de fricción lo hacen ideal para materiales de fricción para automoción, como el embrague y los forros de frenos. Las fibras de amianto constituyen un serio peligro para la salud si se inhalan.

asbestosis A progressive and disabling lung disease caused by inhaling asbestos fibers over a long period of time.

asbestosis Enfermedad pulmonar progresiva que provoca la incapacidad de la persona a causa de la inhalación de fibras de amianto.

aspect ratio The ratio of the cross-sectional height to the cross-sectional width of a tire expressed as a percentage.

relación entre dimensiones Relación entre la altura y la anchura de la sección transversal de un neumático, expresada en un porcentaje.

atmospheric pressure The weight of the air that makes up the Earth's atmosphere.

presión atmosférica Peso del aire que constituye la atmósfera de la Tierra.

atmospheric suspended A term that describes a power brake vacuum booster in which atmospheric pressure is present on both sides of the diaphragm when the brakes are released. An obsolete type of vacuum booster.

de suspensión atmosférica Término que describe un reforzador de vacío para frenos de potencia en el cual hay presión atmosférica a ambos lados del diafragma cuando se sueltan los frenos. Tipo obsoleto de reforzador de vacío.

automatic ride control (ARC) An electronically controlled suspension system designed to improve vehicle ride and increase vehicle stability.

control automático del movimiento (ARC) Sistema de suspensión controlado electrónicamente diseñado para mejorar el manejo y la estabilidad del vehículo.

Automotive Friction Material Edge Code A series of codes on the side of a brake lining (disc or drum) that identifies the manufacturer, the lining material, and coefficient of friction. These codes are for lining identification and comparison; they do not indicate quality.

Código de Borde para Material de Fricción en Automóviles Serie de códigos situados en el lado de un forro de frenos (de disco o de tambor) que identifica el fabricante, el material del forro y el coeficiente de fricción. Estos códigos identifican y comparan los forros; no indican calidad.

backing plate The mounting surface for all other parts of a drum brake assembly except the drum.

placa de refuerzo Superficie en que se montan todas las partes de un freno de tambor, excepto el tambor.

ball and ramp A common kind of caliper-actuated parking brake.

bola y rampa Tipo común de freno de estacionamiento accionado por calibre.

banjo fitting A round, banjo-shaped tubing connector with a hollow bolt through its center that enables a brake line to be connected to a hydraulic component at a right angle.

ajuste de banjo Conector redondo, en forma de banjo, atravesado en el centro por un perno hueco, que permite conectar en ángulo recto una línea de frenos a un componente hidráulico.

bearing cage The steel component that holds the rollers together in a tapered roller bearing.

jaula de cojinetes Componente de acero que mantiene juntas las bolillas en un cojinete de bolillas cónicas.

bearing cone The inner race of a tapered roller bearing; usually an integral assembly with the rollers and the cage.

cono del cojinete Canaleta interior de un cojinete de bolillas cónicas; por lo general, conjunto integrado de bolillas y jaula.

bearing cup The outer race of a tapered roller bearing; usually pressed into the wheel hub.

taza del cojinete Canaleta exterior de un cojinete de bolillas cónicas; por lo general está comprimida en el buje de la rueda.

belted bias ply tire Tire construction that incorporates the belts used in radial ply tires with the older bias ply construction.

neumático encintado de capas contrapuestas Construcción de neumáticos que incorpora las cintas usadas en los neumáticos de capas radiales a la construcción más antigua de capas contrapuestas.

bias ply tire Tire construction in which the cords in the body plies of the carcass run from bead to bead at an angle from 26 to 38 degrees instead of 90 degrees as in a radial ply tire.

neumático de capas contrapuestas Construcción de neumáticos en la cual los cables de las capas del cuerpo de la carcasa van de una a otra nervadura en un ángulo de 26 a 38 grados, en vez de uno de 90 grados como en los neumáticos radiales.

bimetallic drum A composite brake drum made of cast iron and aluminum.

tambor bimetálico Tambor de frenos hecho de hierro fundido y aluminio.

binary system The mathematical system that uses only the digits 0 and 1 to present information.

sistema binario Sistema matemático que utiliza sólo los dígitos 0 y 1 para presentar información.

binder Adhesive or glue used in brake linings to bond all the other materials together.

aglomerante Adhesivo o pegamento que se utiliza en los forros de frenos para unir todos los demás materiales.

bit A binary digit (0 or 1). Bit combinations are used to represent letters and numbers in digital computers. Eight bits equal one byte.

bit Dígito binario (0 o 1). Las combinaciones de bits se emplean para representar letras y números en los computadores digitales. Ocho bits equivalen a un byte.

bleeding A service procedure that removes air from the hydraulic system.

extracción de vapor Procedimiento de mantenimiento que extrae aire del sistema hidráulico.

bonded brake linings Bonded brake linings are attached to the steel backing with adhesive.

revestimientos de freno unidos Los revestimientos de freno unidos están pegados a la placa de resfuerzo de acero con remaches.

bonded lining Brake lining attached to the pad or shoe by high-strength, high-temperature adhesive.

forro adherido Forro de freno unido a la pastilla o zapata por un adhesivo muy fuerte, de alta temperatura.

brake assisted (BA) A sensor mounted on the vacuum brake booster to detect brake pedal motion and speed of motion for faster brake application.

freno, asistido (BA) Sensor instalado en el sistema booster del freno para detectar el movimiento del pedal y la velocidad del movimiento para una más rápida aplicación del freno.

brake caliper The part of a disc brake system that converts hydraulic pressure back to mechanical force that applies the pads to the rotor. The caliper is mounted on the suspension or axle housing and contains a hydraulic piston and the brake pads.

calibre del freno Parte de un sistema de frenos de disco que vuelve a convertir la presión hidráulica en fuerza mecánica que aplica las pastillas al rotor. El calibre va montado en el alojamiento del eje o la suspensión, y contiene un pistón hidráulico y las pastillas de freno.

brake fade The partial or total loss of braking power caused by excessive heat, which reduces friction between the brake linings and the rotors or drums.

pérdida de freno Pérdida parcial o total de la potencia de freno debido a un calor excesivo, el cual reduce la fricción entre los forros de los frenos y los rotores o tambores.

brake pad The part of a disc brake assembly that holds the lining friction material that is forced against the rotor to create friction to stop the vehicle.

pastilla de freno Parte de un conjunto de frenos de disco que aloja el material de fricción del forro que se fuerza contra el rotor para crear la fricción que detendrá el vehículo.

brake pedal position (BPP) sensor Sensor used to determine how fast and how far the brake pedal has been depressed.

sensor de posición del pedal de freno (BPP) Sensor utilizado para determinar cuán rápido y qué tanto se presiona el pedal de freno.

brake pressure modulator valve (BPMV) A valve assembly used in the Teves Mark 20 ABS to control braking of the wheels. Commonly referred to as the hydraulic modulator.

válvula moduladora de la presión de los frenos (BPMV) Válvula que se usa en el sistema de ABS del Teves Mark 20 para controlar el frenado de las ruedas. Comúnmente se le llama modulador hidráulico.

brake shoes The curved metal parts of a drum brake assembly that carry the friction material lining.

zapatas de freno Partes metálicas curvas de un conjunto de frenos de tambor que llevan el forro de material de fricción.

buffer An isolating circuit used to avoid interference between a driven circuit and its driver circuit. Also a storage device, or circuit, that compensates for a difference in the rate of data transmission. A buffer can absorb data transmitted faster than a receiving circuit or device can respond.

separador de interferencias Circuito aislante que se emplea para evitar posibles interferencias entre un circuito controlado y su circuito controlador. También un dispositivo, o circuito, de almacenamiento que compensa una diferencia en la velocidad de transmisión de datos. Un separador de interferencias puede absorber datos transmitidos más rápidamente de lo que puede responder un circuito o dispositivo receptor.

bulkhead Steel panel that separates the engine compartment from the passenger compartment.

mampara Panel de acero que separa el compartimiento del motor del compartimiento del pasajero.

caliper The major component of a disc brake system. It houses the piston(s) and supports the brake pads.

calibre Componente principal de un sistema de frenos de disco. Contiene el pistón (o pistones) y soporta las pastillas de freno.

caliper support The bracket or anchor that holds the brake caliper.

soporte del calibre Mordaza o ancla que aloja el calibre de freno.

camber The inward or outward tilt of the wheel measured from top to bottom and viewed from the front of the car.

inclinación Inclinación de la rueda hacia dentro o hacia fuera, medida de arriba abajo y vista desde la parte frontal del vehículo.

cam-ground lining A brake shoe lining that has been arced or formed so that it is thinner at the ends than at the center, and the lining surface is not a portion of a circle with a constant radius.

forro elíptico Forro de zapata de freno que se ha arqueado o formado de tal manera que es más delgado en los extremos que en el centro, y cuya superficie de recubrimiento no forma parte de un círculo de radio constante.

carcass The steel beads around the rim and layers of cords or plies that are bonded together to give a tire its shape and strength.

carcasa Nervaduras de acero alrededor de la llanta y las capas de cordones, o capas que están unidas para dar forma y resistencia a un neumático.

casing Layers of sidewall and undertread rubber added to a tire carcass.

cubierta Capas de caucho añadidas a la carcasa del neumático.

caster The backward or forward angle of the steering axis viewed from the side of the car.

inclinación del eje Inclinación hacia delante o hacia atrás del eje de dirección visto desde el costado del auto.

center high-mounted stoplamp (CHMSL) Stoplamp mounted above the taillamps or tailgate to aid in visibility.

luz de freno central elevada (CHMSL) Luz de freno colocada sobre las luces o la cajuela trasera para mejorar la visibilidad.

central processing unit (CPU) The calculating part of any computer that makes logical decisions by comparing conditioned input with data in memory.

unidad central de proceso (UCP) Parte de cálculo de un computador que llega a decisiones lógicas comparando los datos de entrada condicionados con los datos que tiene en la memoria.

ceramic Ceramic brake pads are made from ceramic and copper which make less noise and cause less damage to rotors than semimetallic pads.

Cerámica Los tacos cerámicos de freno están hechos de cerámica y cobre, los cuales hacen menos ruido y causan menos daño a los rotores en comparación a los tacos semi-metálicos.

ceramic brake pads Pads consisting of a combination of ceramic material and copper or some other metal fibers.

cojines de freno de cerámica Cojines que consisten en una combinación de materiales y del cobre de cerámica, o algunas otras fibras del metal.

channel Individual legs of the hydraulic system that relay pressure from the master cylinder to the wheel cylinder.

canal Terminales del sistema hidráulico que transmiten presión desde el cilindro maestro al cilindro de la rueda.

check valve Valve that allows fluid or air to flow in one direction but not in the opposite direction.

válvula de control Válvula que permite el paso de líquidos o aire en un sólo sentido contrario.

chlorinated hydrocarbon solvents A class of chemical compounds that contain various combinations of hydrogen, carbon, and chlorine atoms; best known as a class of cleaning solvents.

solventes clorohidrocarbonados Clase de compuestos químicos que contienen diferentes combinaciones de átomos de hidrógeno, carbono y cloro; más conocidos como un tipo de solventes limpiadores.

class 2 wiring A multiplexing network in which each wired-in module can share information through different pulse signals that are used to identify each module on the net.

cableado clase 2 Red múltiple en la cual cada módulo conectado pueden compartir información mediante diferentes señales de pulsación que se usan para identificar cada módulo en la red.

coefficient of friction A numerical value that expresses the amount of friction between two objects, obtained by dividing tensile force (motion) by weight force. A coefficient of friction can be either static or kinetic.

coeficiente de fricción Valor numérico que expresa la cantidad de fricción que hay entre dos objetos, obtenido mediante la división de la fuerza de tracción (movimiento) por la del peso. Los coeficientes de fricción pueden ser estáticos o dinámicos.

cold inflation pressure The tire inflation pressure after a tire has been standing for 3 hours or driven less than 1 mile after standing for 3 hours.

presión de inflado en frío Presión de inflado del neumático después de haber estado en reposo durante tres horas o haber recorrido menos de una milla después del mismo tiempo de reposo.

combination valve A hydraulic control valve with two or three valve functions in one valve body.

válvula de combinación Válvula de control hidráulico con dos o tres funciones de paso en el mismo cuerpo de válvula.

command An electrical signal or output signal from a computer (controller) to an actuator.

comando Señal eléctrica o señal de salida de una computadora (servo-regulador) a un servo-motor.

dynamic rear proportioning and electronic brake distribution Dynamic rear proportioning and electronic brake distribution are names for the electronic function of the proportioning valve by the ABS braking system.

dosificación dinámica trasera y distribución electrónica de frenos La dosificación dinámica trasera y la distribución electrónica del freno son nombres de la función electrónica de la válvula dosificadora por el sistema de frenos ABS.

eccentric Not round, or concentric.

excéntrico Que no es redondo ni concéntrico.

electronic brake control module (EBCM) A term used to designate an ABS control module or the ABS computer.

módulo de control electrónico de freno (EBCM) Término que se usa para designar un módulo de control ABS o una computadora ABS.

electronic brake system (EBS) A hydraulic modulator used in conjunction with the brake assist sensor.

sistema de frenado electrónico (EBS) Modulador hidráulico que se usa en conjunto con el sensor de asistencia de frenado.

electro-hydraulic brake (EHB) An electrically controlled hydraulic braking system that is being used to combine ABS and TCS into one system.

freno electrohidráulico Sistema de control hidráulico controlado eléctricamente que se usa para combinar el ABS y el TCS en un sistema.

electrohydraulic unit The microprocessor and hydraulic unit are combined in one unit.

unidad electrohidráulica Microprocesador y unidad hidráulica combinados en un solo elemento.

electromagnetic induction The generation of voltage in a conductor by relative motion between the conductor and a magnetic field.

inducción electromagnética Generación de tensión en un conductor mediante un movimiento relativo entre el conductor y un campo magnético.

electromagnetic interference (EMI) A magnetic force field that influences a signal being sent to the microprocessor.

interferencia electromagnética (EMI) Campo de fuerza magnética que influye en una señal enviada al microprocesador.

electronic brake traction control module (EBTCM) Computerized module that controls the traction and antilock brakes.

módulo de control electrónico de tracción de frenos (EBTCM) Módulo computarizado que controla la tracción y los frenos antibloqueo.

electronic steering Power steering that uses an electric motor for assist.

dirección electrónica Dirección asistida que utiliza un motor eléctrico.

enhanced traction system (ETS) A traction control system used with the DBC-7 ABS.

sistema de tracción mejorada (ETS) Sistema de control de tracción que se usa con el ABS del DBC-7.

equalizer Part of the parking brake linkage that balances application force and applies it equally to each wheel. The equalizer often contains the linkage adjustment point.

compensador Parte del acoplamiento del freno de estacionamiento que equilibra la fuerza ejercida y la aplica por igual en cada rueda. Con frecuencia, el compensador incluye el punto de ajuste del acoplamiento.

erasable programmable read-only memory (EPROM) Computer memory program circuits that can be erased and reprogrammed. Erasure is done by exposing the integrated circuit chip to ultraviolet light.

memoria de sólo lectura borrable y programable (EPROM) Circuitos de programa de memoria del computador que se pueden eliminar y reprogramar. La eliminación es posible exponiendo a la luz ultravioleta el chip de circuito integrado.

federal motor vehicle safety standards (FMVSS) U.S. government regulations that prescribe safety requirements for various vehicles, including passenger cars and light trucks. The FMVSS regulations are administered by the U.S. Department of Transportation (DOT).

normas federales de seguridad de vehículos a motor (FMVSS) Normas gubernamentales de EE.UU. que dictan los requisitos de seguridad para diversos vehículos, incluyendo los automóviles de pasajeros y los camiones ligeros. El Departamento de Transportes de los EE.UU. (DOT) es el que administra las normas FMVSS.

filler A class of materials used in brake linings to reduce noise and improve heat transfer.

relleno Tipo de material usado en los forros de frenos para reducir el ruido y mejorar la transferencia de calor.

fittings The term applied to all plumbing connections on the car or in the house.

ajustes Término aplicado a las conexiones de tuberías del auto o de la casa.

fixed caliper brake A brake caliper that is bolted to its support and does not move when the brakes are applied. A fixed caliper must have pistons on both the inboard and the outboard sides.

freno de calibre fijo Calibre de freno que se fija al soporte con un perno y no se mueve al aplicar los frenos. Un calibre fijo debe tener pistones tanto en el lado exterior como en el interior.

fixed rotor A rotor that has the hub and the rotor cast as a single part.

rotor fijo Rotor en el que el buje y el rotor forman una única pieza.

fixed seal A seal for a caliper piston that is installed in a groove in the caliper bore and that does not move with the piston.

junta fija Junta para un pistón de calibre que se instala en una hendidura del hueco del calibre y que no se mueve con el pistón.

floating caliper A caliper that is mounted to its support on two locating pins or guide pins. The caliper slides on the pin in a sleeve or bushing. Because of its flexibility, this kind of caliper is said to float on its guide pins.

calibre flotante Calibre que se monta al soporte mediante dos pasadores de posición o pasadores guía. El calibre se desliza dentro del pasador a través de un manguito o casquillo. Debido a su flexibilidad, se dice que este tipo de calibre flota en los pasadores guía.

floating rotor A rotor and hub assembly made of two separate parts.

rotor flotante Conjunto de rotor y buje en dos piezas distintas.

flux lines Lines of magnetism.

líneas de flujo Líneas de magnetismo.

force Power working against resistance to cause motion.

fuerza Energía que trabaja contra la resistencia para producir movimiento.

free play The distance the brake pedal moves before the master cylinder primary piston moves.

holgura Distancia recorre el pedal del freno antes de que se mueva el pistón primario del cilindro maestro.

frequency The number of times, or speed, at which an action occurs within a specified time interval. In electronics, frequency indicates the number of times that a signal occurs or repeats in cycles per second. Cycles per second are indicated by the symbol *hertz* (Hz).

frecuencia Número de veces, o velocidad, a la que ocurre una acción dentro de un intervalo de tiempo especificado. En electrónica, la frecuencia indica el número de veces que una señal se da o se repite en ciclos por segundo. Los ciclos por segundo se indican con el símbolo "herzios" (Hz).

friction The force that resists motion between the surfaces of two objects or forms of matter.

fricción Fuerza que se opone al movimiento entre las superficies de dos objetos o formas de materia.

friction materials Friction materials help provide the frictional stopping power, such as graphite, powdered metal, and nut shells.

materiales de fricción Los materiales de fricción ayudan a proporcionar la energía de frenado friccional, como el grafito, metal en polvo y cáscaras de nuez.

friction modifier A class of materials used in brake linings to modify the final coefficient of friction of the linings.

modificador de rozamiento Tipo de material usado en los forros de frenos para modificar el coeficiente final de fricción de las envolturas.

fulcrum The pivot point of a lever.

punto de apoyo Punto de apoyo de una palanca.

gas fade It is caused by hot gas and dust, which reduce between the brake shoe and drum or rotor and brake pad during prolonged hard braking.

pérdida de gas es causada debido al gas caliente y al polvo que se reducen entre la zapata del freno y el tambor o rotor, y la pastilla del freno durante un frenado fuerte prolongado.

geometric centerline A static dimension represented by a line through the center of the vehicle from front to rear.

línea media geométrica Dimensión estática representada por una línea en el centro del vehículo desde el frente hasta la parte de atrás.

gross vehicle weight rating (GVWR) Total weight of a vehicle plus its maximum rated payload.

peso bruto del vehículo (GVWR) Peso total de un vehículo más la carga de régimen máxima.

Hall-effect switch A device that produces a voltage pulse dependent on the presence of a magnetic field. Hall-effect voltage varies as magnetic reluctance varies around a current-carrying semiconductor.

conmutador de efecto Hall Aparato que produce una variación rápida de tensión dependiente de la presencia de un campo magnético. La tensión de efecto Hall varía al cambiar la resistencia magnética de alrededor de un semiconductor que lleva corriente.

height-sensing proportioning valve A proportioning valve in which hydraulic pressure is adjusted automatically according to the vertical movement of the chassis in relation to the rear axle during braking; sometimes also called a weight-sensing proportioning valve.

válvula dosificadora de detección de altura Válvula dosificadora en la que la presión hidráulica se ajusta automáticamente según el movimiento vertical del chasis en relación con el eje trasero durante el frenado; a veces también se la denomina válvula dosificadora de detección de peso.

hold-down springs Small springs that hold drum brake shoes in position against the backing plate while providing flexibility for shoe application and release.

resortes de sujeción Pequeños resortes que mantienen las zapatas de frenos de tambor en posición contra la placa de refuerzo, a la vez que dan flexibilidad para aplicar y soltar la zapata.

hydraulic modulator The common term for an ABS computer-controlled set of valves used to control brake pressure to various wheels.

modulador hidráulico Término común para el conjunto de válvulas controladas por computadora del ABS que se usan para controlar la presión del frenado a las distintas ruedas.

hydraulic system mineral oil (HSMO) A brake fluid made from a mineral oil base, used by a few European carmakers. DOT specifications do not apply to HSMO fluids, and HSMO fluids cannot be mixed with DOT fluids.

aceite mineral del sistema hidráulico (HSMO) Líquido de frenos a base de aceite mineral que utilizan algunos fabricantes de automóviles europeos. Las especificaciones de los líquidos DOT no se aplican a los líquidos HSMO, y éstos últimos no se pueden mezclar con los primeros.

hydro-boost A hydraulic power brake system that uses the power steering hydraulic system to provide boost for the brake system.

reforzador hidráulico Un sistema hidráulico de frenos de potencia que utiliza el sistema hidráulico de dirección de potencia para alimentar el sistema de frenado.

hydroplane The action of a tire rolling on a layer of water on the road surface instead of staying in contact with the pavement. Hydroplaning occurs when water cannot be displaced from between the tread and the road.

aquaplaning Acción de un neumático que rueda sobre una capa de agua sobre la superficie vial en lugar de mantenerse en contacto con el pavimento. El "aquaplaning" se produce cuando no se puede desplazar el agua entre los dibujos del neumático y la calle.

hydroplaning Hydroplaning is the sliding or skidding caused by the vehicle riding on a thin film of water, risking a serious loss of traction. Hydroplaning is aggravated by having worn tires and excessive speed.

Hidroplaneo El Hidroplaneo es el deslizamiento o resbalo causado por una fina capa de agua, arriesgándose a una grave pérdida de tracción. El hidroplaneo es agravado por el uso de llantas desgastadas y la excesiva velocidad.

hygroscopic The chemical property or characteristic of attracting and absorbing water, particularly out of the air. Polyglycol brake fluids are hygroscopic.

higroscópico Que posee la propiedad química o característica de atraer y absorber agua, en especial del aire. Los líquidos de frenos con poliglicol son higroscópicos.

inertia The tendency of an object in motion to keep moving and the tendency of an object at rest to remain at rest.

inercia Tendencia de un objeto en movimiento a seguir moviéndose y la de un objeto en reposo a permanecer en reposo.

integrated ABS An antilock brake system in which the ABS hydraulic components, the standard brake hydraulic components, and a hydraulic power booster are joined in a single, integrated hydraulic system.

ABS integrados Sistema antibloqueo de frenos en el que los componentes hidráulicos de los frenos ABS, los componentes hidráulicos de los frenos estándar y un reforzador hidráulico de potencia se unen en un único sistema hidráulico integrado.

integrated circuit (IC) A complete electronic circuit of many transistors and other devices, all formed on a single silicon semiconductor chip.

circuito integrado (IC) Completo circuito electrónico de muchos transistores y otros dispositivos, todos formados sobre un único chip semiconductor de siliconas.

integrated control unit (ICU) A controller for an ABS and TCS.

integrado control unidad (ICU) UN controlador por un ABS y TCS.

intermediate lever Part of the parking brake linkage under the vehicle that increases application force and works with the equalizer to apply it equally to each wheel.

palanca intermedia Parte del acoplamiento del freno de estacionamiento, situada debajo del vehículo, que aumenta la fuerza de aplicación y que trabaja con el compensador para aplicarla por igual a cada rueda.

ISO fitting A type of tubing flare connection in which a bubble-shaped end is formed on the tubing; also called a bubble flare.

ensanche ISO Tipo de conexión con ensanche del tubo en el cual un extremo toma forma de burbuja; también se le llama ensanche de burbuja.

kinetic energy The energy of mechanical work or motion.

energía cinética Energía del trabajo mecánico o movimiento.

kinetic friction Friction between two moving objects or between one moving object and a stationary surface.

fricción cinética Fricción entre dos objetos en movimiento, o entre uno en movimiento y una superficie estacionaria.

lands The raised surfaces on a valve spool.

partes planas Superficies elevadas en un carrete de válvula.

lateral accelerometer Vehicle sensor used to measure the speed of the vehicle's lateral movement during operation.

acelerómetro lateral Un acelerómetro lateral es un sensor del vehículo usado para medir la velocidad del movimiento lateral del vehículo durante la operación.

lathe-cut seal A fixed seal for a caliper piston that has a square or irregular cross section; not round like an O-ring.

junta torneada Junta fija para un pistón de calibre que tiene una sección transversal cuadrada o irregular, no redonda como en un toroide.

leading shoe The first shoe in the direction of drum rotation in a leading-trailing brake. When the vehicle is going forward, the forward shoe is the leading shoe, but the leading shoe can be the front or the rear shoe depending on whether the drum is rotating forward or in reverse and whether the wheel cylinder is at the top or the bottom of the backing plate. The leading shoe is self-energizing.

zapata tractora Primera zapata en la dirección de giro del tambor en un freno de tracción-remolque. Cuando el vehículo avanza, la zapata delantera es la tractora, pero la zapata tractora puede ser la delantera o la trasera dependiendo de si el tambor está girando hacia delante o hacia atrás y de si el cilindro de la rueda está en la parte superior de la placa de refuerzo o en la inferior. La zapata tractora es autónoma en cuanto a energía.

leading-trailing brake A drum brake that develops self-energizing action only on the leading shoe. Brake application force is separate for the leading and the trailing shoes. Also called a partial-servo or a nonservo brake.

freno tracción-remolque Freno de tambor que desarrolla una acción autónoma sólo sobre la zapata tractora. La fuerza de aplicación de freno es independiente para las zapatas tractoras y para las de remolque. También se conoce como freno parcialmente asistido o freno no asistido.

leverage The use of a lever and fulcrum to create a mechanical advantage, usually to increase force applied to an object. The brake pedal is the first point of leverage in a vehicle brake system.

transmisión por palancas Utilización de una palanca y un punto de apoyo para crear una ventaja mecánica, generalmente aumentar la fuerza que se aplica a un objeto. El pedal del freno es el primer punto de transmisión por palancas en un sistema de frenos de vehículos.

lining fade Brake fade because of a loss of brake lining coefficient of friction caused by excessive heat.

pérdida por envolturas Pérdida de frenado debido a una disminución del coeficiente de fricción de los forros de freno a causa de un calor excesivo.

lockup A condition in which a wheel stops rotating and skids on the road surface.

bloqueo Condición en la cual una rueda deja de girar y patina sobre la superficie vial.

lockup or negative wheel slip Lockup or negative wheel slip is a condition in which a wheel stops rotating and skids on the road surface.

enllave o deslizamiento negativo de la rueda El enllave o deslizamiento negativo de la rueda es una condición en la cual una rueda deja de girar y se desliza sobre la superficie de la carretera.

mass The measure of the inertia of an object or form of matter or its resistance to acceleration. Also the molecular density of an object.

masa Medida de la inercia de un objeto o forma de materia, o su resistencia a la aceleración. También la densidad molecular de un objeto.

master cylinder The liquid-filled reservoir in the hydraulic brake system or clutch where hydraulic pressure is developed when the driver depresses a foot pedal.

cilindro principal El cilindro principal es el depósito llenado del líquido en el sistema de frenos hidráulico donde se desarrolla la presión hydráulica cuando el conductor presiona un pedal con su pie.

mechanical advantage Increase in force through the use of leverage, such as at the brake pedal.

Ventaja Mecánica Aumento de la fuerza a través del uso de apalancamiento, como en el pedal de freno.

mechanical fade Brake fade because of heat expansion of a brake drum away from the shoes and linings. Mechanical fade does not occur with disc brakes.

pérdida mecánica de frenado Pérdida de frenado debido a la dilatación térmica de un tambor de freno, que lo separa de las zapatas y los forros. La pérdida mecánica no se produce con los frenos de disco.

metallic lining Brake friction material made from powdered metal that is formed into blocks by heat and pressure.

forro metálico Material de fricción para frenos hecho de metal en polvo que se transforma en bloques por medio de calor y presión.

metering valve A hydraulic control valve used primarily with front disc brakes on RWD vehicles. The metering valve delays pressure application to the front brakes until the rear drum brakes have started to operate.

válvula de dosificación Válvula de control hidráulica que se usa principalmente con frenos delanteros de disco en vehículos RWD. La válvula de dosificación demora la aplicación de presión a los frenos delanteros hasta que hayan comenzado a funcionar los frenos traseros de tambor.

microprocessor A digital computer or processor built on a single integrated circuit chip. A microprocessor can perform functions of arithmetic logic and control logic. It is the basic building block of a microcomputer system.

microprocesador Computador digital o procesador construido sobre un único chip de circuitos integrados. Un microprocesador puede realizar funciones de lógica aritmética y lógica de control. Es el elemento básico de construcción de un sistema de microprocesador.

mold-bonded lining A pad assembly made by applying adhesive to the pad and then pouring the uncured lining material onto the pad in a mold. The assembly is cured at high temperature to fuse the lining and adhesive to the pad.

forro adherido al molde Conjunto de pastillas formado por la aplicación de adhesivo a las pastillas y vertiendo luego el material de envoltura no vulcanizado sobre éstas en un molde. El conjunto se vulcaniza a una temperatura elevada para fundir el forro y el adhesivo a la pastilla.

momentum The force of continuing motion. The momentum of a moving object equals its mass times its speed.

impulsor Fuerza de un movimiento continuo. El impulso de un objeto en movimiento es igual a la masa por la velocidad.

multiplexing The network used by multiple computers so that only one component is transmitting at a time.

multiplexación La red usada por las computadoras múltiples de modo que solamente una componente esté transmitiendo a la vez.

National Highway Transportation and Safety Agency (NHTSA) A federal agency assigned to develop regulations for highway safety including vehicle safety features.

Agencia Nacional de Transporte y Seguridad en Carreteras (NHTSA) Agencia federal asignada a desarrollar regulaciones para la seguridad en las carreteras incluyendo detalles de seguridad vehicular.

negative wheel slip Wheel lockup that occurs when too much braking force is applied to a wheel and the tire skids on the pavement. ABS controls negative wheel slip by modulating (decreasing and increasing) the hydraulic pressure to the wheel or wheels that is/are skidding.

deslizamiento negativo de la rueda Bloqueo de la rueda que tiene lugar cuando se aplica un frenado demasiado fuerte a la rueda y la llanta patina sobre el pavimento. Los frenos ABS controlan el deslizamiento negativo de la rueda regulando (aumentando y disminuyendo) la presión hidráulica de la rueda o las ruedas que patina/n.

network The channel through which several computers share information.

red Canal a través del cual diversos computadores comparten la información.

nonintegrated ABS An antilock brake system in which the ABS hydraulic components are attached to, but separate from, the normal brake hydraulic system and power booster.

ABS no integrado Sistema antibloqueo de frenos en el que los componentes hidráulicos del ABS se conectan, pero estando separados, al sistema hidráulico normal de frenos y al reforzador de potencia.

O-ring A circular rubber seal shaped like the letter "O."

toroide Junta de goma circular con la forma de la letra "O."

Occupational Safety and Health Administration (OSHA) A division of the U.S. Department of Labor that establishes and enforces workplace safety regulations.

Administración de Seguridad y Salud en el Trabajo (OSHA) División del Departamento de trabajo de los EE.UU. que establece y hace cumplir las normas de seguridad en el lugar de trabajo.

ohm The unit used to measure the amount of electrical resistance in a circuit or an electrical device. One ohm is the amount of resistance present when one volt forces one ampere of current through a circuit or a device. Ohm is abbreviated with the Greek letter omega (Ω).

ohmio Unidad usada para medir la cantidad de resistencia eléctrica de un circuito o de un aparato eléctrico. Un ohmio es la cantidad de resistencia presente cuando un voltio hace pasar un amperio de corriente a través de un circuito o de un aparato. El ohmio se abrevia con la letra griega omega (Ω).

Ohm's law The mathematical formula for the relationships between voltage, current, and resistance; often stated simply as E (voltage) $= I$ (current) $\times$ R (resistance).

ley de Ohm Fórmula matemática que expresa las relaciones entre tensión, corriente y resistencia; frecuentemente se enuncia simplemente por E (tensión) = I (corriente) × R (resistencia).

organic lining Brake friction material made from nonmetallic fibers bonded together in a composite material.

forro orgánico Material de fricción del freno hecho de fibras no metálicas adheridas en un material compuesto.

overload spring Spring in the end of the cable in cable-operated adjusters that lets the cable move without breaking if the pawl or star wheel is jammed.

resorte de sobrecarga Resorte al final del cable de los reguladores operados por cables que permite que éste se mueva sin romperse si el trinquete o la rueda de estrella están agarrotados.

P-metric system The most common modern system to specify passenger car tire sizes.

sistema métrico P Sistema métrico más moderno y común para especificar el tamaño de los neumáticos de los automóviles.

pad hardware Miscellaneous small parts such as antirattle clips and support clips that hold brake pads in place and keep them from rattling.

hardware de cojines Piezas pequeñas surtidas, como grapas antiresonancia y grapas de apoyo que mantienen los cojines de los frenos en su lugar y evitan que suenen.

pad wear indicators Devices that warn the driver when disc brake linings have worn to the point where they need replacement. Wear indicators may be mechanical (audible) or electrical.

indicadores del desgaste de los cojines Dispositivos que advierten al conductor de que los forros de los frenos de disco se han gastado tanto que es necesario cambiarlos. Los indicadores de desgaste pueden ser mecánicos (acústicos) o eléctricos.

parking brake control The pedal or lever used to apply the parking brakes.

control del freno de estacionamiento Pedal o palanca que se usa para aplicar los frenos de estacionamiento.

parking brakes The braking system that is used to hold the vehicle stationary while parked.

pawl A hinged or pivoted component that engages a toothed wheel or rod to provide rotation or movement in one direction while preventing it in the opposite direction.

trinquete Componente articulado o embisagrado que se engrana a una rueda o varilla dentada para ofrecer rotación o movimiento en un sentido mientras que lo impide en el sentido contrario.

permanent magnet (PM) generator A reluctance sensor. A sensor that generates a voltage signal by moving a conductor through a permanent magnetic field.

generador de magneto permanente (PM) Sensor de reluctancia. Sensor que genera una señal de tensión al mover un conductor a través de un campo magnético permanente.

perpetual energy Energy that can be produced or converted without using additional energy forever.

energía perpetua Energía que puede producirse/convertise sin usar energía adicional.

phenolic plastic Plastic made primarily from phenol, a compound derived from benzene and also called carbolic acid.

plástico fenólico Plástico hecho principalmente de fenol, compuesto derivado del benceno, llamado también ácido carbólico.

piston stop A metal part on a brake backing plate that keeps the wheel cylinder pistons from moving completely out of the cylinder bore.

tope de pistón Parte metálica de la placa de refuerzo del freno que impide que los pistones del cilindro de la rueda se salgan del hueco del cilindro.

polyglycol A mixture of several alcohols. Polyalkylene-glycol-ether brake fluids that meet specifications for DOT 3 and DOT 4 brake fluids.

poliglicol Mezcla de varios alcoholes. Líquidos de frenos de polialquilenglicoléter que están dentro de las especificaciones DOT 3 y DOT 4 para líquidos de frenos.

positive wheel spin The excessive wheel spin that occurs during acceleration as a wheel loses traction and spins on the pavement.

deslizamiento positivo de la rueda Giro excesivo de la rueda que se produce durante una aceleración cuando una rueda pierde tracción y gira sobre el pavimiento.

pre-brake A Mercedes-Benz system that will actually prepare the brakes to apply only the necessary braking pressure when the driver responds, thus preventing a panic stop.

prefreno Sistema de Mercedes-Benz que prepara los frenos para aplicar solo la presión de frenado necesaria una vez que responde el conductor, lo que previene frenadas intempestivas.

pressure Force exerted on a given unit of surface area. Pressure equals force divided by area and is measured in pounds per square inch (psi) or kilopascals (kPa).

presión Fuerza ejercida sobre una unidad de superficie determinada. La presión es igual a la fuerza dividida por el área y se mide en libras por pulgadas cuadradas (psi) o kilopascales (kPa).

pressure differential The difference between two pressures on two surfaces or in two separate areas. The pressures can be either pneumatic (air) or hydraulic.

diferencial de presión Diferencia entre dos presiones en dos superficies o áreas distintas. La presión puede ser neumática (aire) o hidráulica.

pressure differential valve A hydraulic valve that reacts to a difference in pressure between the halves of a split brake system. When a pressure differential exists, the valve moves a plunger to close the brake warning lamp switch.

válvula de diferencial de presión Válvula hidráulica que reacciona ante una diferencia de presión entre las dos partes de un sistema de frenos dividido. Cuando hay una diferencia de presión, la válvula mueve un pistón que cierra el conmutador de la luz indicadora de freno.

pressure-sensor base (PSB) Has a pressure sensor located on each wheel and reads tire pressure directly. It is a software that measures tire pressure with on-wheel sensors and relays the pressure to an onboard computer.

base de presión sensitiva Software que mide la presión de las llantas con sensores en la rueda y transmite la presión a una computadora del vehículo.

primary shoe The leading shoe in a duo-servo brake. The primary shoe is self-energizing and applies servo action to the secondary shoe to increase its application force. Primary shoes have shorter linings than secondary shoes.

zapata primaria Zapata guía en un servofreno dual. La zapata primaria es autónoma y aplica una servoacción a la zapata secundaria para aumentar su fuerza de aplicación. Las zapatas primarias tienen forros más pequeños que las secundarias.

proportioning valve A hydraulic control valve that controls the pressure applied to rear drum brakes. A proportioning valve decreases the rate of pressure application above its split point as the brake pedal is applied harder.

válvula de dosificación Válvula de control hidráulico que controla la presión aplicada a los frenos de tambor traseros. Una válvula de dosificación disminuye la proporción de aplicación de presión por encima de su punto de separación cuando se presiona más fuerte el pedal del freno.

protocol A protocol is a computer language that modules use to communicate with each other.

protocolo Un protocolo es un lenguage de computadora que modula el uso para comunicarse mutuamente.

quick take-up master cylinder A dual master cylinder that supplies a large volume of fluid to the front disc brakes on initial brake application, which takes up the clearance of low-drag calipers.

cilindro maestro de tensor rápido Cilindro maestro doble que proporciona una gran cantidad de líquido a los frenos de disco delanteros en la primera frenada, compensando la holgura de los calibres de baja resistencia.

quick take-up valve The part of the quick take-up master cylinder that controls fluid flow between the reservoir and the primary low-pressure chamber.

válvula de tensor rápido Parte del cilindro maestro del tensor rápido que controla el flujo de líquido entre el depósito y la cámara de baja presión primaria.

radial ply tire Tire construction in which the cords in the body plies of the carcass run at an angle of 90 degrees to the steel beads in the inner rim of the carcass. Each cord is parallel to the radius of the tire circle.

neumático de capas radiales Fabricación de neumáticos en los que los cordones de las capas del cuerpo de la carcasa se fijan con un ángulo de 90 grados a las nervaduras de acero en borde interior de la carcasa. Cada cordón es paralelo a los radios del círculo del neumático.

reaction disc (or reaction plate and levers) The components in a vacuum power booster that provide pedal feel or feedback to the driver.

disco de reacción (o placa y palancas) Componentes de un reforzador de vacío que proporcionan sensaciones en el pedal o respuesta al conductor.

rear wheel antilock (RWAL/RABS) A two-wheel ABS used on the rear wheels of light-duty pickup trucks and some SUVs. One of the best known Kelsey–Hayes systems.

antibloqueo de ruedas traseras (RWAL) Frenos ABS para dos ruedas utilizado en las ruedas traseras de camionetas ligeras y en algunos SUV. Uno de los sistemas de Kelsey–Hayes más conocidos.

regenerative braking Converting the energy (heat and mass) of a moving vehicle to stop that vehicle.

frenar regenerador El frenar regenerador está convirtiendo la energía (calor y masa) de un vehículo móvil para parar ese vehículo.

relay An electromagnetic switch that uses a small amount of current in one circuit to open or close a circuit with greater current flow. Relays are used as remotely controlled switches for circuits.

relé Un conmutador electromagnético que toma una pequeña cantidad de corriente de un circuito para abrir o cerrar otro con mayor flujo de corriente. Los relés se utilizan como conmutadores de control remoto para los circuitos.

reluctor A metal tooth ring used to influence the magnetic flux lines of the PM generator.

reluctor Anillo metálico dentado que se usa para influir en las líneas de flujo magnético del generador de PM.

replenishing port The rearward port in the master cylinder bore; also called other names.

puerto de abastecimiento Puerto situado más atrás en el hueco del cilindro maestro; también recibe otros nombres.

reservoir Storage tank for the master cylinder.

depósito Tanque de almacenamiento para el cilindro maestro.

residual pressure check valves Residual pressure check valves were once used to hold slight pressure on the drum brake pistons to maintain seal contact with the walls of the wheel cylinder. Cup expanders are used now.

válvula de retención de presión residual Las válvulas de retención de presión residual alguna vez fueron utilizadas para mantener una ligera presión sobre los pistones del tambor de frenos para mantener el contacto del sello con las paredes del cilindro de la rueda. Ahora su utilizan expansores de taza.

return spring A strong spring that retracts a drum brake shoe when hydraulic pressure is released.

resorte de vuelta Resorte fuerte que retrae una zapata de freno de tambor cuando se libera presión hidráulica.

riveted lining Brake lining attached to the pad or shoe by copper or aluminum rivets.

forro remachado Forro de freno unido a la pastilla o zapata por remaches de cobre o de aluminio.

rolling diameter The effective diameter of a tire when placed on the vehicle. It can vary with tire pressure.

diámetro del elemento rodante El diámetro efectivo de un neumático cuando se coloca en el vehículo. Puede variar con la presión.

rolling resistance The resistance of the vehicle's weight and mass and the friction of all moving components to the movement of the vehicle.

coeficiente de resistencia a la rodadura Resistencia del peso y masa de un vehículo y de la fricción de todos los componentes movibles al movimiento del vehículo.

rotor The rotating part of a disc brake that is mounted on the wheel hub and contacted by the pads to develop friction to stop the car. Also called a disc.

rotor Parte giratoria de un freno de disco que va montada en el buje de la rueda y entra en contacto con las pastillas para causar el rozamiento que detiene el vehículo. También llamado disco.

run-flat A tire that can be run at reduced speed, even when flat.

neumático runflat o antipinchazo Neumático que se puede utilizar a velocidad reducida, aun después de haber sufrido un pinchazo.

screw-and-nut A common kind of caliper-actuated parking brake.

tornillo y tuerca Tipo común de freno de estacionamiento accionado por calibre.

scrub radius The distance from the tire contact patch centerline to the point where the steering axis intersects the road.

radio de fricción Distancia desde la línea central de la banda de rodadura del neumático hasta el punto en que el eje de dirección corta la vía.

secondary shoe The trailing shoe in a duo-servo brake. The secondary shoe receives servo action from the primary shoe to increase its application force. Secondary shoes provide the greater braking force in a duo-servo brake and have longer linings than primary shoes.

zapata secundaria Zapata de remolque en un servofreno dual. La zapata secundaria recibe acción asistida de la primaria para aumentar su fuerza de aplicación. Las zapatas secundarias proporcionan mayor potencia de frenado en un servofreno dual y tienen forros más grandes que las primarias.

section width The width of a tire across the widest point of its cross section, usually measured in millimeters.

anchura de sección Anchura de un neumático en el punto más amplio de su sección transversal; se suele medir en milímetros.

self-adjusters A cable, lever, screw, strut, or other linkage part that provides automatic shoe adjustment and proper lining-to-drum clearance as a drum brake lining wears.

autoreguladores Cable, palanca, tornillo, puntal u otro mecanismo de unión que proporciona un ajuste automático a la zapata y la holgura adecuada entre forro y tambor cuando se desgasta el forro de un freno de tambor.

self-energizing operation The action of a drum brake shoe when drum rotation increases the application force of the shoe by wedging it tightly against the drum.

operación autoactivada Acción de una zapata de freno de tambor cuando la rotación del tambor aumenta la fuerza de aplicación de la zapata al ajustarse como una cuña contra el tambor.

semimetallic lining Brake friction material made from a mixture of organic or synthetic fibers and certain metals; these linings do not contain asbestos.

forro semimetálico Material de fricción del freno hecho con una mezcla de fibras orgánicas o sintéticas y ciertos metales; estos forros no contienen amianto.

service brakes The disc or drum brakes operated by the driver to stop the vehicle.

frenos de servicio Frenos de disco o de tambor sobre los que actúa el conductor para detener el vehículo.

servo action The operation of a drum brake that uses the self-energizing operation of one shoe to apply mechanical force to the other shoe to assist its application. Broadly, servo action is any mechanical multiplication of force.

servoacción Acción de un freno de tambor en el que una zapata actúa de forma autónoma al aplicar una fuerza mecánica a la otra zapata y ayudarla en su funcionamiento. Más ampliamente, una ser-voacción es cualquier multiplicación mecánica de una fuerza.

setback A difference in wheelbase from one side of a vehicle to the other.

retroceso Diferencia en la distancia entre ejes en un lado y otro del vehículo.

shoe anchor The large pin, or post, or block against which a drum brake shoe pivots or develops leverage.

anclaje de zapata Pasador grande, o poste, o bloque contra el que la zapata de un freno de tambor pivota o desarrolla transmisión por palancas.

sliding caliper A caliper that is mounted to its support on two fixed sliding surfaces or ways. The caliper slides on the rigid ways and does not have the flexibility of a floating caliper.

calibre de desplazamiento Calibre que se monta en el soporte sobre dos superficies o vías fijas de deslizamiento. El calibre se desliza por vías rígidas y no tiene la flexibilidad de un calibre flotante.

slope The numerical ratio or proportion of rear drum brake pressure to full system pressure that is applied through a proportioning valve. If half of the system pressure is applied to the rear brakes, the slope is 1:2 or 50 percent.

atenuación diferencial Razón numérica o proporción entre la presión del freno de tambor trasero y la presión total del sistema que se aplica a través de una válvula de dosificación. Si la mitad de la presión del sistema se aplica a los frenos traseros, la atenuación diferencial es 1:2 ó del 50 por ciento.

solid rotor A rotor that is a solid piece of metal with a friction surface on each side.

rotor sólido Rotor formado por una pieza sólida de metal con una superficie de fricción a cada lado.

split point The pressure at which a proportioning valve closes during brake application and reduces the rate at which further pressure is applied to rear drum brakes.

punto de separación Presión a la que la válvula de dosificación se cierra durante la aplicación de los frenos y se reduce la relación con la que se aplica más presión a los frenos de tambor traseros.

spool valve A cylindrical sliding valve that uses lands and valleys around its circumference to control the flow of hydraulic fluid through the valve body.

válvula de carrete Válvula cilíndrica de deslizamiento que usa partes planas y hundimientos en su circunferencia para controlar el flujo del líquido hidráulico a través del cuerpo de la válvula.

square-cut piston seal A fixed seal for a caliper piston that has a square cross section.

junta cuadrada de pistón Junta fija para un pistón de calibre con sección transversal cuadrada.

star wheel A small wheel that is part of a drum brake adjusting link. Turning the star wheel lengthens or shortens the adjuster link to position the shoes for proper lining-to-drum clearance.

rueda en estrella Rueda pequeña que forma parte de un acoplamiento de ajuste de un freno de tambor. Al hacer girar la rueda en estrella se alarga o acorta el acoplamiento para poner en posición las zapatas y conseguir una tolerancia conveniente entre forro y tambor.

static friction Friction between two stationary objects or surfaces.

fricción estática Fricción entre dos objetos o superficies estáticas.

steering axis inclination (SAI) The angle formed by the steering axis of a front wheel and a vertical line through the wheel when viewed from the front with the wheels straight ahead.

inclinación del eje de dirección (SAI) Ángulo formado por el eje de dirección de una rueda delantera y una línea vertical que pasa por la rueda cuando se mira desde la parte delantera con las ruedas directamente hacia delante.

steering knuckle The outboard part of the front suspension that pivots on the ball joints and lets the wheels turn for steering control.

Mangueta de Dirección La parte externa de la suspensión delantera que pivota en las articulaciones esféricas y permite que las ruedas giren para controlar la dirección.

steering wheel position sensor Steering wheel position sensor determines which direction the driver is trying to steer.

sensor de posición del volante El sensor de posición del volante determina en qué dirección el conductor está intentando dirigirse.

stroke sensor Stroke sensor informs the brake controller how fast and how much pressure the driver applied to the brakes.

sensor de golpes El sensor de golpes informa al controlador de freno qué tan rápido y cuánta presión el conductor aplicó a los frenos.

stroke simulator Stroke simulator provides brake pedal feedback to the driver.

estimulador de golpes El estimulador de golpes proporciona retroalimentación del pedal de frenos al conductor.

Subaru's "Eyesight" or Mercedes "Pre-brake" There are several types of active braking systems available such as Subaru's "Eyesight" or Mercedes "Pre-brake that prepare the braking system for a sudden stop. These systems can actually apply the brakes to stop the vehicle or lessen the severity of a crash.

"Acance de vista" del Subaru o "Pre-freno" del Mercedes Hay varios tipos de sistemas de frenado activo disponibles, tales como el "Alcance de vista del subaru" o el "pre-freno" del Mercedes los cuales preparan el sistema de frenos para una parada repentina. Esos sistemas de hecho pueden accionar los frenos para detener el vehiculo o disminuir la severidad de un choque.

swept area The total area of the brake drum or rotor that contacts the friction surface of the brake lining.

zona barrida Área total del freno de tambor o del rotor que toca la superficie de fricción del forro del freno.

synthetic lining Brake friction materials made from nonorganic, nonmetallic, and nonasbestos materials; typically fiberglass and aramid fibers.

forro sintético Materiales de fricción de frenos hechos de materiales no orgánicos, no metálicos y sin amianto; por lo común, fibra de vidrio y fibras de aramida.

table The outer surface of a brake shoe to which the lining is attached.

tabla Superficie exterior de una zapata de freno a la que se une el forro.

tandem Two or more devices placed one behind the other in line.

en serie Dos o más dispositivos colocados en línea uno detrás de otro.

tandem booster A power brake vacuum booster with two small diaphragms in tandem to provide additive vacuum force.

reforzador en serie Reforzador de vacío para frenos con dos pequeños diafragmas en serie que proporcionan mas fuerza al vacío.

tensile force The moving force that slides or pulls an object over a surface.

fuerza de tracción Fuerza de movimiento que desliza o arrastra un objeto sobre una superficie.

thermal energy The energy of heat.

energía térmica Energía del calor.

thrust angle The angle between the geometric centerline and the thrust line of a vehicle.

ángulo de empuje El ángulo formado por la línea geométrica central y la línea de empuje de un vehículo.

thrust line The bisector of total toe on the rear wheels, or the direction in which the rear wheels are pointing.

línea de empuje Bisectriz de la separación total de las ruedas traseras, o la dirección en que señalan dichas ruedas.

tire load range The load-carrying capacity of a tire, expressed by the letters A through L. The load range letters replace the older method of rating tire strength by the number of plies used in its construction.

límites de carga de neumáticos Capacidad de carga de un neumático, expresada por letras de la A a la L. Las letras de los límites de carga reemplazan el método anterior de clasificar la resistencia de los neumáticos mediante el número de capas usadas en su construcción.

tire pressure monitoring system (TPMS) A system of devices and software used to alert the driver of underinflated tire(s).

sistema de supervisión de la presión del neumático (TPMS) Un sistema de los dispositivos y del software que alerta el conductor de neumáticos inflados inferiores.

toe angle The difference in the distance between the centerlines of the tires on either axle (front or rear) measured at the front and rear of the tires and at spindle height.

ángulo de separación Diferencia de la distancia entre las líneas centrales de los neumáticos en cada eje (delantero o trasero) medida en las partes delantera y trasera de los neumáticos y a la altura del husillo.

toe-out on turns (turning radius) The difference between the angles of the front wheels in a turn.

divergencia en los giros (radio de giro) Diferencia entre los ángulos de las ruedas delanteras al girar.

traction control systems Traction control systems are designed to prevent wheel slip in low-traction situations by applying the brakes to the free spinning wheel.

sistemas de control de tracción Los sistemas de control de tracción están diseñados para prevenir el deslizamiento de las ruedas en situaciones de baja tracción al aplicar los frenos a la rueda giratoria libre.

trailing shoe The second shoe in the direction of drum rotation in a leading-trailing brake. When the vehicle is going forward, the rear shoe is the trailing shoe, but the trailing shoe can be the front or the rear shoe depending on whether the drum is rotating forward or in reverse and whether the wheel cylinder is at the top or the bottom of the backing plate. The trailing shoe is non-self-energizing, and drum rotation works against shoe application.

zapata de remolque Segunda zapata en la dirección de giro del tambor en un freno de tracción-remolque. Cuando el vehículo avanza, la zapata trasera es la remolcada, pero la zapata de remolque puede ser la delantera o la trasera dependiendo de si el tambor está girando hacia delante o hacia atrás y de si el cilindro de la rueda está en la parte superior de la placa de refuerzo o en la inferior. La zapata de remolque no es autónoma, y el giro del tambor funciona contra la aplicación de la zapata.

tread The layer of rubber on a tire that contacts the road and contains a distinctive pattern to provide traction.

rodadura Capa de goma de un neumático que está en contacto con la carretera y contiene un patrón distintivo que favorece la tracción.

tread contact patch The area of the tire tread that contacts the road; determined by tire section width, diameter, and inflation pressure.

banda de rodadura Superficie de la rodadura del neumático que se pone en contacto con la carretera; viene determinada por la anchura de la sección del neumático, el diámetro y la presión de inflado.

tread wear indicator A continuous bar that appears across a tire tread when the tread wears down to the last $\frac{2}{32}$ ($\frac{1}{16}$) inch. When a tread wear indicator appears across two or more adjacent grooves, the tire should be replaced.

indicador de desgaste de la rodadura Barra continua que aparece transversalmente en la rodadura del neumático cuando ésta se desgasta más de $\frac{2}{32}$ ($\frac{1}{16}$) pulgadas. Cuando aparece el indicador de desgaste de rodadura cruzando dos o más hendiduras adyacentes, se debe cambiar el neumático.

unidirectional rotor A rotor with cooling fins that are curved or formed at an angle to the hub center to increase cooling airflow. Because the fins work properly only when the rotor turns in one direction, unidirectional rotors cannot be interchanged from right to left on the car.

rotor unidireccional Rotor con álabes refrigerantes curvadas o formando un ángulo con el centro del buje para aumentar el flujo de aire de refrigeración. Como los álabes sólo funcionan adecuadamente cuando el rotor gira en una sola dirección, los rotores unidireccionales no se pueden intercambiar entre la derecha y la izquierda del auto.

unidirectional tread pattern A tire tread pattern that can be rotated only in one direction. Thus, left- and right-side tires with unidirectional tread cannot be interchanged.

patrón de rodadura unidireccional Patrón de rodadura de un neumático que se puede girar en una única dirección. Es decir, los neumáticos de la izquierda y de la derecha con rodadura unidireccional no se pueden intercambiar.

uniform tire quality grading (UTQG) indicators Letters and numbers molded into the sidewall of a tire to indicate relative tread life, wet weather traction, and heat resistance.

indicadores del grado de calidad de neumáticos uniformes (UTQG) Letras y números moldeados en los lados de un neumático para indicar la duración relativa de un neumático, la tracción en tiempo lluvioso y la resistencia al calor.

vacuum In automotive service, vacuum is generally considered to be air pressure lower than atmospheric pressure. A true vacuum is a complete absence of air.

vacío En relación con los automóviles, el vacío se suele considerar como una presión de aire menor que la atmosférica. Un vacío verdadero es la completa ausencia de aire.

vacuum suspended A term that describes a power brake vacuum booster in which vacuum is present on both sides of the diaphragm when the brakes are released. The most common kind of vacuum booster.

de suspensión de vacío Término que describe un reforzador de vacío para frenos en el que existe vacío a ambos lados del diafragma cuando se sueltan los frenos. El tipo más común de reforzador de vacío.

valleys The annular grooves or recessed areas between the lands of a valve spool.

hundimientos Hendiduras anulares o áreas ahuecadas entre las partes de una válvula de carrete.

vehicle stability control Vehicle stability control is an electronically assisted braking system that can selectively apply and release brakes during critical maneuvers to help the driver maintain control.

control de estabilidad del vehiculo El control de estabilidad del vehiculo es un sistema eléctronico de frenado asistido que puede accionar o liberar los frenos durante maniobras críticas para ayudar al conductor a mantener el control.

vehicle stability system The integration of electronically controlled steering, braking, and suspension systems that provide for enhanced vehicle control and safety.

sistema de estabilidad vehicular Integración de la dirección controlada electrónicamente, los frenos y el sistema de suspensión que proporciona un control mejorado de los vehículos y de la seguridad.

vent port The forward port in the master cylinder bore; also called by other names.

válvula de ventilación Válvula delantera de la parte interior del cilindro maestro; también recibe otros nombres.

ventilated rotor A rotor that has cooling fins cast between the braking surfaces to increase the cooling area of the rotor.

rotor ventilado Rotor que tiene álabes refrigerantes entre las superficies de frenado para aumentar la superficie refrigerante del rotor.

voltage The electromotive force that causes current to flow. The potential force that exists between two points when one is positively charged and the other is negatively charged.

voltio Unidad usada para medir la cantidad de fuerza o energía eléctrica.

water fade Brake fade that occurs when water is trapped between the brake linings and the drum or rotor and the coefficient of friction is reduced.

pérdida de agua Pérdida de frenado debido a que el agua queda atrapada entre los forros de los frenos y el tambor o el rotor, reduciéndose el coeficiente de rozamiento.

web The inner part of a brake shoe that is perpendicular to the table and to which all of the springs and other linkage parts attach.

membrana Parte interior de una zapata de freno perpendicular a la tabla y en la que se fijan todos los resortes y otras partes de acoplamiento.

weight The measure of the Earth's gravitational force or pull on an object.

peso Medida de la fuerza de gravedad de la Tierra o de la atracción sobre un objeto.

wheel cylinder The hydraulic slave cylinder mounted on the backing plate of a drum brake assembly. The wheel cylinders convert hydraulic pressure from the master cylinder to mechanical force that applies the brake shoes.

cilindro de la rueda Cilindro hidráulico auxiliar montado en la placa de refuerzo de un conjunto de frenos de tambor. Los cilindros de la rueda convierten la presión hidráulica del cilindro maestro en la fuerza mecánica que se aplica a las zapatas de freno.

wheel offset The distance between the centerline of the rim and the mounting plane of the wheel.

desajuste de la rueda Distancia entre la línea central de la llanta y el plano de montaje de la rueda.

wheel speed base A software program in an ABS used to determine the inflation of a tire.

base de velocidad de la rueda Programa de software en un ABS que se usa para determinar la cantidad de aire de una llanta.

wheel speed sensor A sensor used to determine the rotating speed of a wheel.

sensor de velocidad de las ruedas Sensor utilizado para determinar la velocidad de rotación de una rueda.

yaw Swinging motion to the left or to the right of the vertical centerline or rotation around the vertical centerline.

derrape Movimiento de balanceo de izquierda a derecha de la línea central vertical o giro alrededor de dicha línea.

INDEX

Note: Page numbers in bold print reference non-text material.

C

T

U

V

W

Y